International and European Environmental Policy Series

Edited by
Alexander Carius
R. Andreas Kraemer

Springer
Berlin
Heidelberg
New York
Barcelona
Hong Kong
London
Milan
Paris
Singapore
Tokyo

Alexander Carius Kurt M. Lietzmann (Eds.)

Environmental Change and Security

A European Perspective

With Support by Kerstin Imbusch and Eileen Petzold-Bradley

Foreword by the German Federal Minister for the Environment, Nature Conservation and Nuclear Safety Jürgen Trittin

With 23 Figures and 7 Tables

Springer

EDITORS

Alexander Carius
Ecologic
Pfalzburger Str. 43/44
10717 Berlin
Germany

Dr. Kurt M. Lietzmann
Federal Ministry for the Environment,
Nature Conservation and Nuclear Safety
Department RS I 5
PO Box 12 06 29
53048 Bonn
Germany

Translated from the German by Christopher Hay, Übersetzungsbüro für Umweltwissenschaften, Darmstadt

ISBN-13: 978-3-642-64314-9 e-ISBN-13: 978-3-642-60229-0
DOI: 10.1007/978-3-642-60229-0
The book was originally published under the title „Umwelt und Sicherheit. Herausforderungen für die internationale Politik". © Springer-Verlag 1998

Library of Congress Cataloging-in-Publication Data Applied For
Environmental Change and Security: A European Perspective / Eds.: Alexander Carius; Kurt M. Lietzmann, with support by Kerstin Imbusch and Eileen Petzold-Bradley. Foreword by Jürgen Trittin. - Berlin; Heidelberg; New York; Barcelona; Hong Kong; London; Milan; Paris; Singapore; Tokyo: Springer, 1999
ISBN-13: 978-3-642-64314-9

Softcover reprint of the hardcover 1st edition 1999

Cover Design: Erich Kirchner
Typesetting: Camera-ready by Thomas Leppert, Ecologic, Berlin
SPIN: 10698334 30/3136 – 5 4 3 2 1 0

About Ecologic

Ecologic is a not-for-profit institution for applied environmental research and policy consultancy, seated in Berlin. Ecologic conducts implementation-focused research projects and prepares expert opinions on a range of issues relating to nature conservation and environmental protection. These include new approaches in environmental policy, ecologically sustainable resource policy, international environmental agreements and institutions, environmental planning and the integration of environmental concerns into other policy realms. Ecologic's work focuses in particular on analyzing the environmental policy of the European Union and its member states and enhancing the effectiveness of international environmental regimes.

Ecologic is part of the network of Institutes for European Environmental Policy with offices in a number of principal European cities, and is part of numerous other issue-focused and project-related international networks. Ecologic works primarily for international and supranational organizations, parliaments, governments, national agencies and local authorities, industry federations, trade unions, and environmental or conservation NGOs. Ecologic pursues exclusively academic and non-profit aims. It is an economically and academically independent, non-partisan institute.

Ecologic publishes current research reports and studies in its book series "Contributions to international and European environmental policy".

Foreword

Jürgen Trittin

At the threshold of the 21st Century, global environmental change and its socio-economic effects such as rapid population growth, poverty, famine, poor health conditions, and migration are likely to intensify particularly in developing countries and transition economies in the future. Global environmental degradation (including climate change and ozone depletion, biodiversity loss, deforestation and desertification, transboundary pollution, and natural disasters) as well as depletion of renewable resources (including water, fuelwood, cropland and fish) and non-renewable resources (including ores and minerals) are increasingly seen as major contributing factors to conflicts.

The intensity as well as the interdependence of these problems will have effects on an international scale and the capacity of individual nations to deal with them in a comprehensive manner is limited. Therefore, these environmental challenges call for new partnerships and enhanced co-operative action between international institutions and regional organisations in both industrialised and developing countries. Efforts for policy integration and co-ordination among international and regional organisations to respond to environmental change needs to be further developed. In particular, enhancing the relevant frameworks available for communication among foreign and security policy actors and institutions with relevant development and environmental organisations and stakeholders within civil society is of paramount importance.

Reducing global inequalities and taking preventive action at all levels, environmental and development policy plays a key role in the prevention of conflict and contributes to stabilising the socio-economic and political context of actors experiencing environmental stress. In particular, implementing environmental and development policy at an early stage when environmental stress is still a matter of dispute without the use of violence, is more effective and the range of measures that can be applied is much broader. However, environmental and development policy at the same time, can be positively employed in post-conflict phases by supporting political, economic and administrative reforms to change structures which have contributed to conflict in the past (i.e. in the case of post conflict rehabilitation and nation building required in the Balkan States). To prevent environmental conflicts, there are a number of possible sustainable policy measures that should be implemented, ranging from sustainable economic growth and poverty reduction programs to strengthening equity and role of women in society, and to strengthening democratic processes allowing for the creation of a climate and the capacity for constructive interaction between civil society and government.

It is also essential to intensify efforts for the sustainable management of natural resources within the context of international environmental regimes, particularly in pursuit of solving global environmental problems with local and regional impacts. At the global and regional level, agreements on the management of natural resources such as the UN Convention on Highly Migratory Fish Stocks, the UN Framework Convention on Climate Change, the UN Convention on Biological Diversity (CBD) and the Convention to Combat Desertification (CCD) should be strengthened and further developed. Environmental agreements in the framework of the UN Economic Commission for Europe (UNECE) and various co-operative institutions for the transboundary management and protection of water resources (i.e. the adoption of the Petersberg Declaration in March 1999) are other examples of confidence building mechanisms that have been instrumental in avoiding conflict escalation between countries.

The issue of the relationship between environment and security has been discussed academically on the international political agenda since the publication of the Brundtland-Report in 1987. At the United Nations Conference on Environment and Development in Rio de Janeiro in 1992, there was a general acceptance of the idea that environmental, economic and social issues are interdependent and cannot be pursued separately. As a result, numerous international organizations such as the United Nations Development Programme (UNDP) and the United Nations Environment Programme (UNEP), UN Economic Commission for Europe (UNECE), Organisation for Economic Co-operation and Development (OECD), the European Commission (EU) and the World Bank and national governments have addressed this issue, with a diverse array of approaches. In particular, the need to manage environmental stress and its consequences for security has also been acknowledged by the principal European security organisations, including the North Atlantic Treaty Organization (NATO) and the Organisation for Security and Co-operation in Europe (OSCE). For example, a study process commenced in 1995 with a Pilot Study commissioned by NATO's Committee on the Challenges of Modern Society (CCMS), which is examining the complex of "Environment and Security in an International Context".

However, this build-up of political pronouncements on the security policy relevance of environmental change has yet to be matched by a similar intensity of concrete political action. Despite regular exhortations of the dangers of global environmental conflict, governmental expenditure for preventive environmental protection has risen just as little as has the commitment to international environmental agreements, such as on climate protection. Therefore, urgently required is the need to intensify international cooperation through the various institutional frameworks in order to operationalise comprehensive and preventative policy approaches. To solve or prevent environmental conflicts, it is necessary to more consistently integrate environmental concerns in other relevant policy sectors, in particular in development, foreign and security policy, but also in agricultural, energy and social policy. These measures need to be equally applied and implemented over the long-term at the relevant international and national levels. Therefore, I welcome and encourage on-going discussions for forging new partnerships and enforcing institutions with greater shared but differentiated responsibilities and also generating new approaches for strengthening environment mediation and resolution as an effective approach to conflict prevention and sustainable development.

Jürgen Trittin

Federal Minister for the Environment, Nature Conservation
and Nuclear Safety, Germany

Editors' Preface

Towards the end of this decade, the role of global, regional and local environmental change in triggering violent conflict has emerged as a major topic of the social and political sciences. This debate centers on the question of the socio-economic conditions under which environmental changes in different regions of the world might lead to violent conflict. Is there a link between environmental degradation, resource scarcity and violent conflict? Which regions of the world are particularly susceptible to such conflict? Which demands do these new security impacts place upon polity and society? How can environmentally induced conflicts be prevented?

Environmental changes generally do not lead directly to violent conflict; they are rather one element within a complex web of causality that includes an array of socio-economic problems such as overpopulation, poverty, mass migration, refugee movements, famine, political instability and ethno-political tensions. In this web, environmental degradation and resource scarcities are not only a cause but also an outcome of these socio-economic problems. Environmental disasters such as the reactor accident at Chernobyl are not the prime driving force, although that disaster certainly also gave impetus to this debate. Of greater significance are the environmental changes and their consequences that take place gradually, are potentially global and can only be reversed over long time frames. These changes call for a vigorous political and societal response.

The link between environmental degradation and security was initially studied above all in Canada and the USA in the late 1970s. In the middle of the 1980s, this debate became an important issue on the international policy agenda under the headings of "environmental security" or "ecological security". The debate revolved mainly around new, non-military security impacts and the consequent changes in foreign and security policy tasks of the institutions and actors concerned. In the Brundtland Report, released in 1987, the World Commission on Environment and Development was the first international institution to explicitly refer to the link between environmental degradation and conflict. Since then, numerous international organizations and national governments have tackled this topic from many different aspects. These organizations have included NATO, OECD, OSCE and the United Nations Development Programme (UNDP).

In German-speaking countries, too, both the political and natural sciences have turned their attention to selected elements of this research complex. While in these countries both scientific studies and the broader debate in society concentrated initially, starting in the 1970s, on the environmental consequences of crisis and war, greater attention has also been given in recent years to the broader interrelations between environmental policy on the one hand and foreign and security policy on the other. This enhanced interest in the topic gave rise to an international conference on "Environment and Security" in July 1997 at the Science Centre Berlin, convened jointly by the German Federal Ministry for the Environment, Nature Conservation and Nuclear Safety and Ecologic, the Centre for International & European Environmental Research. More than 50 eminent scientists of various disciplines came together with policy advisors and diplomats from Germany, Austria, Switzerland, Denmark and Sweden to debate the complex causes of environmentally induced conflicts and ways to address them in politics and society. This was the first event in Germany to bring together representatives of environmental policy research, development cooperation and peace and conflict research, each of whom approaches the topic from different perspectives. A further stimulating feature of the conference was that it provided a forum at which to juxtapose and debate scientific analyses on the one hand and the need for concrete political advice of decision-makers on the other.

In his opening address, *Walter Hirche*, Parliamentary State Secretary at the German Environment Ministry, underscored the political importance accorded to the issue in Germany. The event has spawned a series of further in-depth round-table debates, workshops and conferences.

The present volume documents the results of this event. The contributions outlined briefly in the following reflect the professional background of their respective authors, and the diversity of perspectives from which the debate on environment and security is conducted.

In the first part of this volume, the authors address fundamental issues relating to the nexus between environment and security (Part A). In their introductory chapter, *Alexander Carius* and *Kerstin Imbusch* initially set out the various dimensions under which the issue of "Environment and Security" is debated by academics of various disciplines and by policy-makers. The subsequent analysis of the contextual conditions and characteristics of environmentally induced conflicts, viewed as a specific expression of the interrelations between environment and security, illustrates the complexity of the issue – an issue so complex that it can only be addressed adequately by means of an integrated approach of actors in various policy sectors. *Kurt Lietzmann*, who, as the director of the NATO/CCMS pilot study "Environment and Security in an International Context", is well acquainted with the debate in both the academic and political realms, explains the contribution that this pilot study made to the political and academic debates and to the further development of the issue in general. He describes the role of the CCMS (Committee on the Challenges of Modern Society), which unites decision-makers from the foreign, security and environmental policy arenas and is thus particularly suited as a forum in which to address the issue. The pilot study examined the links between environmental stress and conflict and developped appropriate preventive policy measures for action in both the environment/development policy and foreign/security policy.

Lothar Brock outlines the prospects of environmental conflict research as a branch of study in its own right. He argues that it is necessary to study conflicts from the perspective of the amenability of environmental problems to resolution, instead of a discourse on 'environmental conflicts'. He posits that in the course of the modernization process it is not so much resource degradation that leads to violent disputes in societies (so-called environmental conflicts). Conflicts rather emerge over the way in which environmental problems are handled and thus it is on this that environmental and conflict research should focus.

Michael Windfuhr's article gives a comparative overview of the approaches to the issue of 'environment and security' taken by peace and conflict research on the one side and environmental policy research on the other. Windfuhr classifies these approaches according to the understanding they reflect of the linkage between the two spheres: a) environmental stress as a causal determinant of violent conflict (short-term dimension), b) environmental stress as an indirect, catalyzing factor (medium-term dimension) and c) potentially conflict-engendering impacts of non-sustainable patterns of production and consumption (long-term dimension). After discussing various approaches to the typological classification of environmental conflicts and their determinants, he drafts a wide-ranging program for future research.

Manfred Wöhlcke addresses the environment and security issue from the specific perspective of population growth. This is an aspect that has tended to be neglected in the debate, and is politically volatile. The perspective that he outlines is accordingly pessimistic. Growth in world population will have considerable negative impacts upon the availability and quality of natural resources, and upon food security. In his view, it is essential to strengthen international environmental policies with an equitable distribution of burdens, but also to enforce population policy measures in order to reduce the potential for conflict generated by these issues.

The second part of the volume draws together scientific approaches for characterization and typological classification of environmental conflicts (Part B). In his contribution on environmental degradation in the South as a cause of violent conflict, *Günther Bächler* summarizes the findings of his major empirical research project on this issue. His findings show that environmentally induced conflicts only escalate into violence in specific unfavorable socio-economic contexts. Particular contextual circumstances that can contribute to resource degradation leading to violent conflict include: inescapable socio-economic conditions; lack of conflict resolution mechanisms in society; instrumentalization of a scarce resource by parties to a conflict; capacity of actors to organize and engage in conflict; and existing conflictual constellations. Thus, environmentally induced conflicts can be described as a confluence of several of these factors.

Frank Biermann also pursues an approach of integrating the ecological and socio-economic factors that influence the emergence of environmentally induced conflicts. He discusses the possibility of utilizing in conflict research for the study of environmentally induced conflicts the interdisciplinary syndrome approach developed by the German Advisory Council on Global Change (WBGU) as an analysis matrix for the study of global change. The syndromes characterized by the Council can be used to place the above-mentioned multiple influencing factors in relation to each other and to examine their interactions. Biermann presents syndrome analysis as a suitable research tool by which to develop early warning systems for security-relevant situations of environmental stress.

In the third and final essay in this part of the volume, *Christoph Rohloff* discusses the general issue of whether environmental conflicts can be integrated in the typological classifications of peace and conflict research. Rohloff suggests that it is difficult to integrate environmentally induced conflicts within existing typologies of conflict, as these types of conflict have complex causes and dynamics and are frequently resolved peacefully. To analyze environmentally induced conflicts in empirical peace and conflict research, it is thus necessary to use a graduated concept of conflict that also embraces the possibility of non-violent conflict resolution.

Part C of the volume is devoted to the methodological concretization of the nexus between environment and security, and to the question of whether and how environmentally induced conflicts can be modeled. In his methodological notes, *Wolf-Dieter Eberwein* discusses the complex interrelations among environment, conflict and security by asking specifically when an environmentally induced conflict becomes a security problem. In view of the complex interdependence among the ecological, human and political systems, Eberwein sees, from a methodological perspective, a need for "integrated modeling" of these conflicts. This can use simulation models to analyze the possible interactions among all three spheres, thus providing a basis for timely, preventive political intervention.

Detlef Sprinz concentrates on ways to identify the specific environmental component of environmentally induced conflicts, thus isolating them from other factors influencing conflictual processes, classifying environmentally induced conflicts as a category of conflict in its own right. Sprinz presents the concept of environmental threshold values as an integral part of an empirical-quantitative research design. Such threshold values can be used to identify correlations between environmental stress and violent conflict. They can then be analyzed more systematically than is possible by means of empirical-qualitative methods, real propensity for conflict can then be tested and appropriate recommendations for action developed.

Jürgen Scheffran, by contrast, includes in his approach to modeling environmentally induced conflict the diverse, complex cause-effect relationships which environmental and conflict research have been able to identify through comparative case studies. The conflict model developed, serves to structure these relationships and to reduce them to core hypotheses. However, it also illustrates the limits of modeling such conflicts in this way. It becomes apparent that essential variables, such as the goals of actors and the interests that guide their actions, can only be quantified in a rudimentary fashion. Thus even empirical-quantitative, model-driven environmental and conflict research must ultimately content itself with plausibility appraisals.

The foregoing chapters concerned with a more theoretical discussion of the nexus between environment and security, with causes and effects and with the identification and modeling of environmentally induced conflicts, are followed by papers whose common theme is that of potential political strategies by which to prevent such conflicts. Foreign and security policy approaches are discussed first (Part D).

Bernd Wulffen sets out in his contribution that in the foreign policy arena, but also in international environmental diplomacy, the debate has been slow to address aspects of environmentally induced conflicts. Nonetheless, the existing international environmental agreements, in particular the conventions on climate, biodiversity and desertification adopted in Rio, do offer appropriate tools for preventing such conflicts. Wulffen thus argues in favor of a stronger integration of the issue of 'environment and security' in the Rio process, but at the same time underscores the importance of fora specific to the individual problematique in preventing crises that can be caused by environmental degradation.

Volker Quante addresses the issue from the aspect of the problem-solving capacities available to existing security organizations to minimize security risks presented by environmentally induced conflicts. Proceeding from the growing importance of non-military security threats (here, in particular, environmental change), the author discusses the possibilities and instruments of NATO in particular to respond to these new challenges.

The final part of the volume centers on the environmental and development policy challenges and approaches that result from the analysis of the conflictual effects of environmental stress (Part E). In his introductory overview, *Sebastian Oberthür* discusses the potential of environmental and development policy to prevent environmentally induced conflicts. He analyzes in particular the possibilities of preempting potentially violent conflict resolution by means of timely regulation of environmental problems in the context of international agreements. He further identifies approaches for reforms by which the organizations and conventions of international environmental policy could be placed in a position to discharge these tasks more successfully.

The following contribution by *Evita Schmieg* concentrates on the specific role of development cooperation in preventing crisis. With particular reference to German development cooperation, she presents the prevention of the causes of crisis and the strengthening of peaceful conflict resolution mechanisms as already existing instruments of crisis prevention. However, in light of the high frequency of crises in developing countries and the resultant rising expenditure for emergency relief and humanitarian aid, she posits the necessity of a targeted, crisis-preventing orientation of bilateral and multilateral development cooperation in the future.

Sascha Müller-Krenner discusses the role of non-governmental organizations (NGOs) in the context of this debate. He suggests that a main reason for NGOs to debate the security aspects of environmental degradation is to underscore the necessity of more far-reaching international cooperation in environmental affairs. He goes on to outline the specific possibilities for NGOs to contribute to preventing environmentally induced conflicts.

This book section examining the issue of environment and security from the environment and development policy perspective concludes with *Irene Freudenschuß-Reichl*, who discusses options for preventive action in the context of the debate on United Nations reform. Due to its status of an international (not supranational) organization, the capacity of the UN to exert a preventive influence upon conflicts is subject to the will of its member states. It thus depends essentially upon the willingness of its members to place the United Nations in a position, through financial commitments and through supporting UN institutional reform, to do justice to the role that devolves upon it in conflict and crisis prevention. Moreover, the author views a "sincere effort" on the part of governments and their citizens to implement the principle of sustainable development as a prime opportunity to prevent environmentally induced conflicts effectively and durably.

The editors wish to thank the German Environment Ministry for the financial support it provided to the organization of the conference and the preparation of the present volume. We owe a debt of gratitude to the authors, who wrote their papers specially for this volume and have integrated many of our suggestions. We are particularly grateful to Ecologic staff members Sebastian Oberthür, Ralph Piotrowski and Matthias Paustian, who commented on and revised, together with the editors, the greater part of the text. Kerstin Imbusch played a major part in the structuring and organization of the conference from which this volume stems. Eileen Petzold-Bradley undertook a substantial part of the editorial work for individual papers of this volume. Her linguistic and textual revision of the manuscript provided a valuable input. Thomas Leppert was responsible for the excellent technical support and layout, and coordinated the preparation of the book manuscript with the publishers Springer Verlag All of these contributions have been vital to bringing this project to fruition.

Berlin and Bonn, July 1999

Alexander Carius Kurt M. Lietzmann

Table of Contents

Part C: Modeling

Part D: Foreign and Security Policy Approaches

Part E: Environment and Development Policy Approaches

Annex

List of Figures and Tables

Figures

Tables

List of Authors

Dr. Günther Baechler
Schweizerische Friedensstiftung Bern
Gerechtigkeitsgasse 12
Postfach
CH-3000 Bern 8
e-mail: baechler@swisspeace.unibe.ch

Dr. Frank Biermann
Belfer Center for Science and International Affairs
John F. Kennedy School of Government
Harvard University
79 John F. Kennedy St
USA-02138 Cambridge, Mass.
e-mail: fbiermann@awi-bremerhaven.de

Prof. Dr. Lothar Brock
Center for European Studies
Harvard University
27 Kirkland St.
Cambridge, MA 02138
e-mail: lbrock@fas.harvard.edu

Alexander Carius
Ecologic
Gesellschaft für Internationale und
Europäische Umweltforschung
Friedrichstraße 165
D-10117 Berlin
e-mail: carius@ecologic.de

Priv. Doz. Dr. Wolf-Dieter Eberwein
Wissenschaftszentrum Berlin
für Sozialforschung (WZB)
Reichpietschufer 50
D-10787 Berlin
e-mail: eberwein@medea.wz-berlin.de

Dr. Irene Freudenschuß-Reichl
Ständige Vertretung der Republik Österreich
bei den Vereinten Nationen
Andromeda-Turm, 11. Stock
Donau-City-Straße 6
A-1220 Wien
e-mail: irene.freudenschuss-reichl@wien.bmaa.gv.at

Kerstin Imbusch
Osteuropa-Institut an der Freien Universität Berlin
Garystraße 55
D-14195 Berlin
e-mail: kimbusch@zedat.fu-berlin.de

Kurt M. Lietzmann
Bundesministerium für Umwelt, Naturschutz und Reaktorsicherheit
Postfach 12 06 29
D-53048 Bonn
e-mail: lietzmann.kurt@bmu.de

Sascha Müller-Kraenner
Heinrich Böll Foundation
Chelsea Gardens
1638 R Street, NW, Suite 120
USA-Washington, DC 20009
e-mail: sascha.washington@boell.de

Dr. Sebastian Oberthür
Ecologic
Gesellschaft für Internationale und
Europäische Umweltforschung
Friedrichstraße 165
D-10117 Berlin
e-mail: oberthuer@ecologic.de

Oberstleutnant i.G. Volker R. Quante
Ulica Krolowej Marysienki
PL-02-954 Warschau
e-mail: vquante@aol.com

Christoph Rohloff
Institut für Entwicklung und Frieden
Geibelstraße 41

D-47057 Duisburg
e-mail: christoph.rohloff@uni-duisburg.de

Dr. Jürgen Scheffran
Interdisziplinäre Arbeitsgruppe
Naturwissenschaft, Technik und Sicherheit
(IANUS)
Schloßgartenstr. 9
D-64289 Darmstadt
e-mail: scheffran@hrzpub.th-darmstadt.de

Evita Schmieg
Bundesministerium für wirtschaftliche Zusammenarbeit und Entwicklung
Planungsreferat (04)
D-53045 Bonn
e-mail: schmieg@bmz.bmz.bund400.de

Detlef F. Sprinz, Ph. D
PIK - Potsdam-Institut für Klimafolgenforschung e.V.
Postfach 60 12 03
D-14412 Potsdam
e-mail: dsprinz@pik-potsdam.de

Michael Windfuhr
Heidelberger Institut für internationale
Konfliktforschung e.V.
Universität Heidelberg
Marstallstraße 6
D-69117 Heidelberg
e-mail: jh1@ix.urz.uni-heidelberg.de

Dr. habil. Manfred Wöhlcke
Stiftung Wissenschaft und Politik (SWP)
Forschungsinstitut für Internationale Politik und Sicherheit
D-82067 Ebenhausen/Isar
e-mail: woe@swp.extern.lrz-muenchen.de

Dr. Bernd Wulffen
Auswärtiges Amt
Postfach 1148
D-53001 Bonn

Glossary of Abbreviations and Acronyms

AKUF	Arbeitsgemeinschaft Kriegsursachenforschung (A Peace and Conflict Research Group in Hamburg/Germany)
BMZ	Bundesministerium für wirtschaftliche Zusammenarbeit und Entwicklung (German Federal Ministry for Economic Cooperation and Development)
BUND	Bund für Umwelt und Naturschutz Deutschland (Friends of the Earth Germany)
CCMS	Committee on the Challenges of Modern Society (of NATO)
CFCs	Chlorofluorocarbons
CFSP	Common Foreign and Security Policy (of the EU)
CRED	Center for Research on the Epidemiology of Disasters
CRG	Collaborative Research Grants
CSCE	Conference on Security and Cooperation in Europe
CSD	UN Commission on Sustainable Development
DAC	OECD Development Assistance Committee
DC	Development Cooperation
DGFK	Deutsche Gesellschaft für Friedens- und Konfliktforschung (German Society for Peace and Conflict Research)
DNR	Deutscher Naturschutzring (German League for Nature and Environment)
ECE	UN Economic Commission for Europe
ECHS	Environmental Clearinghouse System (of NATO)
ECOSOC	UN Economic and Social Council
ECOWAS	Economic Community of West African States
ECRP	Environmental Change and Security Project (of the Woodrow Wilson Center, USA)
ENCOP	Environment and Conflicts Project
ENMOD	Convention on the Prohibition of Military or Any Other Hostile Use of Environmental Modification Techniques
ETH	Eidgenössische Technische Hochschule (Swiss Federal Institute of Technology, Zurich/Switzerland)
EU	European Union
FAO	UN Food and Agriculture Organization
FC	financial cooperation
GDP	Gross Domestic Product
GEF	Global Environment Facility
GEO	Global Environmental Organization

GNP	Gross National Product
HABITAT	UN Conference on Human Settlements
HDI	Human Development Index
HIIK	Heidelberger Institut für Internationale Konfliktforschung (Heidelberg Institute for International Conflict Research)
HSFK	Hessische Stiftung Friedens- und Konfliktforschung (Hessian foundation for peace and conflict research, Germany)
ICPD	International Conference on Population and Development
IGBP	International Geosphere-Biosphere Programme
IHDP	International Human Dimensions Programme on Global Environmental Change
JI	Joint Implementation
KOSIMO	Konflikt-Simulationsmodell (a conflict simulation model)
MCDA	Military Civilian Defense Assets
NATO	North Atlantic Treaty Organization
NGO	non-governmental organization
OAU	Organization of African Unity
ODA	Official Development Assistance
OECD	Organization for Economic Cooperation and Development
OSCE	Organization for Security and Cooperation in Europe
PIK	Potsdamer Institut für Klimafolgenforschung (Potsdam Institute for Climate Impact Research, Germany)
PPP	purchasing power parity
SEF	Stiftung Entwicklung und Frieden (Development and Peace Foundation, Germany)
SWP	Stiftung Wissenschaft und Politik, Ebenhausen (Foundation for Science and Politics, Ebenhausen/Germany)
TC	technical cooperation
UNCED	United Nations Conference on Environment and Development
UNCHE	United Nations Conference on the Human Environment
UNDP	United Nations Development Programme
UNDRO	United Nations Disaster Relief Organization
UNECE	United Nations Economic Commission for Europe
UNEP	United Nations Environment Programme
UNFPA	United Nations Population Fund
UNIDO	United Nations Industrial Development Organization
UNO	United Nations Organization
WBGU	Wissenschaftlicher Beirat der Bundesregierung Globale Umweltveränderungen (German Advisory Council on Global Change)
WEU	Western European Union
WTO	World Trade Organization
WZB	Wissenschaftszentrum Berlin für Sozialforschung (Science Centre Berlin)

Opening Address

Walter Hirche

Ladies and gentlemen,

It is a great honor for me to welcome you to this workshop.

The international environmental debate is being influenced anew by the special session of the UN General Assembly held in New York in recent weeks. I am afraid I have to admit that we fell short of the aim we set for ourselves, which was to re-ignite the momentum created by the Rio Summit of June 1992.

Environmental problems of local, national, regional and world-wide scope are posing as new challenges to the international community. This was the major realization in Rio. Only when the international community acknowledges its responsibility to future generations to commit itself to binding environmental objectives and sustainable economies, can Planet Earth remain viable.

This is quite obvious with the bitter wrangling over the tenths of percentage points of gross national product which the industrialized countries are supposed to be making available for the technology transfer required for the sustainable development of the developing countries. Also the bickering over reduction schedules for greenhouse gases have distracted the delegations and prevented them from recognizing and countering the continuing trends of global environmental problems, such as the loss of biological diversity, global warming, soil loss, water shortage and water pollution.

Any environmental policy which is expected to have a lasting effect must take account of the interests of society, whether of the state and its administration, industry, environmental groups or the people who are affected by environmental problems. We must also bear in mind, with regard to development policy, that the link between environment and development is particularly close in developing countries. People in the developing countries depend directly on the availability and sustainable use of natural resources such as water, soil and forests for their survival. There was much debate on this subject in New York but unfortunately, once again, it was impossible to reach a satisfactory conclusion.

It is up to politicians who appreciate the problem to identify the obstacles and set new goals. Because of the huge impact of the problem and above all because of growing awareness of it, the prospects are not as gloomy as the media in particular would like us to think. The project to set up a worldwide environmental partnership for the next century is a highly symbolic initiative, especially considering that

Brazil, Singapore, South Africa and Germany are among its main supporters. This means that four different states with entirely different histories and environmental traditions in four different parts of the world are taking a leading role. Please forgive me for referring constantly back to New York but I hope you will acknowledge that this provides an obvious introduction to a theme which should be raised during this workshop.

In particular the joint use of bodies of water, surface water or groundwater for drinking water or fisheries, the over-use of soils and the attendant widespread loss of valuable agricultural land compounded by extensive deforestation are causing increasing tension which, alas all too often, leads in turn to social and political crises and even to armed clashes.

Illustrations of this are provided by the most recent conflicts in Central Africa, in Rwanda and Burundi, which have spread like wildfire into eastern Zaire. I am sure there is no need here for me to cite more such examples from the recent past.

Those of you who are directly concerned with peace and conflict research know much better than I that the focus of military conflict has shifted to the South and remains for the time being at intrastate level. However, in future we will be faced increasingly with interstate conflicts centering, for example, on the joint use of water resources. Along the dividing lines between certain cultural differences or diverging economic interests, we are witnessing distributional conflicts influenced by extremely difficult political situations.

In view of the situation, it now seems appropriate to think about new approaches to the overall environmental debate. We should also be asking ourselves the question, or rather letting others ask us, whether we have always negotiated with effective partners in the past. It would also be an idea for us to look into the question of the effectiveness of institutions that have been set up for the practical implementation of internationally agreed-upon environmental objectives.

Environmental destruction and ecological mistakes are cited ever more frequently and unequivocally in lists of factors which lead to the outbreak of violent or military conflicts. I note with some satisfaction that, although only a short time ago the subject of the environment was not even regarded as a marginal factor in security policy and peace research, it has now come to be seen as a major influence whose importance is practically no longer disputed.

I hope you will make full use of this meeting to establish lasting cooperation between the academic and political communities in the field. Through your joint efforts I anticipate that you will produce answers to the following questions in particular:

- whether and to what extent conclusions can be drawn from evaluations of security risks about the need for action to prevent regional and global environmental risks;
- what other political and institutional questions or options for action are linked with the theme.

It is possible that the former proposals for a UN Environmental Security Council will be voiced again. At the very least, the recommendations for action that you produce should be applied to environment and security policy and submitted by the authorities to the relevant international bodies. In this connection I imagine that foreign and security policy in the field of conflict prevention in particular can provide the instruments with which these recommendations can be applied.

I have already mentioned that the question would have to be raised as to who the most effective partners might be to bring about a successful environmental policy. It is no secret that the environmental lobby alone is too weak to convey its concerns and put them into action. With conflict situations intensifying throughout the world it seems almost suicidal to continue running things as they have been up until now. If the 'security lobby' can be successfully persuaded to commit itself to the environmental cause, then we will have a partner with far more efficient means of implementing effective prevention strategies to appease conflicting parties. It is well worth getting these 'strategic partners' involved.

Whether this might lead to knock-on effects in the environmental efficiency debate or to binding agreements on the international exploitation of natural resources, including the introduction of effective conflict settlement mechanisms such as arbitration services, depends on broad public awareness of the consequences of human activities and management.

It is worth establishing this awareness on an academically sound basis, making it possible to foresee consequences in adequate time and take the appropriate preventive action. This is why you have come to Berlin for these two days.

I hope the workshop produces positive results and would be delighted if it could give rise to the setting up of a permanent network for the exchange of experience, future developments and recommendations to the authorities.

Dr. Walter Hirche

Former Parliamentary State Secretary (Parlamentarischer Staatssekretär) at the German Ministry for the Environment, Nature Conservation and Reactor Safety

Part A:

The Environment and Security Nexus

Environment and Security in International Politics – An Introduction

Alexander Carius, Kerstin Imbusch

1 Points of departure[1]

At the end of the millennium, the causal linkages among global, regional and local environmental change on the one hand and violent conflict on the other have emerged as an important topic of the social and political sciences. The debate is concerned with understanding the conditions under which environmental changes in different regions of the world lead to violent conflicts. What are the connections between environmental degradation, resource scarcity and violent conflict? Which regions of the world are particularly susceptible to conflict? Which demands do these new security policy challenges place upon polity and society? How can environmentally-induced conflicts be prevented?

These are the crucial questions that politicians and scientists are debating under the headings of "ecological security", "environmental security" or "environment and security". The present contribution aims to give an introduction to the debate on the environment-security nexus, to summarize the various lines of debate and to examine the underlying linkages among environmental changes and factors that condition, trigger or exacerbate conflicts. We will briefly outline the way in which this issue is treated politically on the agendas of national governments and international organizations, and will provide a synopsis of the present state of discourse in the social sciences. Finally, we will highlight approaches for political action to prevent environmentally-induced conflict in various policy arenas.

Within both the academic and policy communities, the debate on the links between environment and security has brought together two different families that had, in the past, only few points of contact and approach the issue with quite different objectives. Within the academic community, there is the security, peace and conflict research family on the one hand, and the environmental and development policy research family on the other. Peace and conflict research is concerned

1 We are grateful to Frank Biermann, R. Andreas Kraemer, Andreas March, Sebastian Oberthür, Ralph Piotrowski and Detlef Sprinz for valuable criticism and suggestions.

primarily with the causes of conflicts and the conditions under which these lead to violent conflict resolution. Here environmental degradation, heightening resource scarcities and their complex interactions with other socio-economic problems figure among many other factors studied. Environmental policy research, by contrast, proceeds from environmental changes and examines their socio-economic causes and consequences – recently extended to include their foreign and security policy dimensions – and derives recommendations for action from this perspective.

At the policy level, too, we find different interests and approaches. While security policy actors are mainly concerned in this context with new, non-military challenges to security policy, environmental policy actors employ foreign and security policy arguments to stress the urgency of solving environmental problems, particularly those at the global level. This is partly for reasons of political rhetoric, but also because of the real conflict potential presented by environmental changes, which joins other environmental policy exigencies as an independent motivation for action and thus should enhance the enforceability of efficient instruments of precautionary environmental action by grounding them in broader societal approval.

2 The political discourse

At a very early point in the scientific debate on the links between environment and security, a high-level political discourse was already under way.

The Club of Rome's 1972 "The Limits to Growth" report and the "Global 2000" report (Council on Environmental Quality/Department of State 1981) already called attention in urgent terms to the risks associated with natural resource scarcities and continuing deterioration of environmental quality and the connection with an array of socio-economic problems (population growth, urbanization, migration etc.), which, over the long-term, and particularly in Third World regions, could lead to security-relevant threats or even to the outbreak of violent conflicts.

At the international level, the Palme Commission stressed in 1982, in its first report, titled "Common Security", the connection between security and environment, and developed in its 1989 report a concept of "comprehensive (international) security", moving from the concept of war prevention to a comprehensive concept of global peace, social justice, economic development and responsibility for the environment. While the Palme Commission noted the environmental component, it did not further derive from this any political mandate for the United Nations.

In the 1987 Brundtland Report, the World Commission for Environment and Development was the first international institution to explicitly refer to the connection between environmental degradation and conflict. The Commission developed an expanded concept of security:

> The whole notion of security as traditionally understood – in terms of political and military threats to national sovereignty – must be expanded to include the growing impacts of environmental stress – locally, nationally, regionally, and globally. (WCED 1987: 19)
> Environmental stress is seldom the only cause of major conflicts within or among nations. [...] Environmental stress can thus be an important part of the web of causality associated with any conflict and can in some cases be catalytic. (WCED 1987: 291)

Since then, numerous international organizations and national governments have addressed this issue, with a diverse array of approaches. A study process commenced in 1995 with a Pilot Study commissioned by NATO's Committee on the Challenges of Modern Society (CCMS), which is examining the complex of "Environment and Security in an International Context"[2] (see also the contribution of Lietzmann in this volume). In the OECD context, two bodies, the Environmental Policy Committee and the Development Assistance Committee, discussed relevant studies in 1997, each body from different aspects.[3] As early as 1988, the United Nations Environment Programme (UNEP) initiated a study on environmental degradation and conflict together with the International Peace Research Institute, Oslo (PRIO), Norway.[4]

Particularly in the USA, the discourse on environment and security has evolved under the heading of "environmental security" to become almost a branch of research in its own right (ECSP Report 1995, 1996, 1997; for a number of the 'classics' see Brown 1977; Ullman 1983; Mathews 1989; Myers 1989, Kaplan 1994). This theme is met with considerable interest in US government circles. The debate centers here on threats to *national* security resulting from environmental

---

2 The Pilot Study is examining the linkages among environmental changes, resource scarcities and conflict and is developing appropriate approaches for preventive action for environmental and development policy and foreign and security policy. Almost 30 NATO states and European Partnership Council States are participating in the Pilot Study, which is being carried out under the joint leadership of the German Federal Ministry for the Environment, Nature Conservation and Nuclear Safety and the US Department of Defense. The authors of the present contribution are entrusted with conceiving and coordinating the work on this study in cooperation with the German Ministry, the US Department of Defense and Evidence Based Research (EBR), Inc. Its results shall be presented to the public in 1999 (for an interim report see Carius et al. 1997a).

3 At the end of 1997, the Environment Committee discussed a scoping paper on the economic dimensions of environment and security (this topic is not, however, being further pursued by this body at present). The Development Assistance Committee (DAC), being the body coordinating bilateral and multilateral donors in the OECD, is currently discussing the connection between the environment and conflict prevention from the development policy perspective. This is based on a discussion paper prepared by IUCN on behalf of DAC (see DAC 1997).

4 UNEP and PRIO agreed in 1988 to carry out "Studies in Environmental Security" at PRIO. A joint UNEP/PRIO program on "Military Activities and the Human Environment" comprised empirical research projects which were largely conceived and implemented by PRIO. From this initiative, a firm research focus on environment and security evolved at PRIO.

stress and resource problems, as expressed since the 1980s in numerous political statements and programs.

The strategic interest of the US administration is exemplified by the 1990 initiative of the then Senator Al Gore, in which he used the conceptual metaphor of "national security" to underscore the urgent need for action by the US government on global warming (Gore 1990, 1993).[5] At the beginning of the second term of President Clinton, the list of keynote addresses (including those by Baker, Christopher, Perry and Albright), statements and memorandums documenting the various nuances of the environment and security nexus has reached a length almost beyond enumeration.[6] In April 1997, the Department of State presented a first, now annual report on "Environmental Diplomacy" (Department of State 1997), which elevates environmental policy to one of the prime concerns of US foreign policy. This illustrated that in the USA the integration of environmental policy concerns into the foreign and security policy spheres had become a guiding vision of policy, even if policy remains at a very general level.

In 1989, the then Soviet Foreign Minister Eduard Shevardnadze called at the 46th General Assembly of the United Nations for the establishment of an Environmental Security Council entrusted with issues of ecological security (Shevardnadze 1990). In Western Europe, too, framing environmental issues in terms of security policy would appear to be gaining currency. Thus in 1989, the then Norwegian Defense Minister Johan Jørgen Holst (1989) already pointed out that environmental problems can become an important factor in the development of violent conflicts. British Foreign Minister Malcolm Rifkind (1997) has also stated that maintaining peace must proceed on the basis of equitable resource use and sustainable development. In June 1997, French President Chirac stressed at the Special Session of the UN General Assembly the threat of war can be sparked by water resource conflicts.

In 1998, the debate on cooperative approaches to resolving transboundary water conflicts formed the focal issue of the international dialogue forum "Global Water Politics – Cooperation for Transboundary Water Management" sponsored jointly by the German Federal Ministry for Economic Cooperation and Development (BMZ), the Federal Environment Ministry (BMU), the German Foreign Office, the World Bank and the German Foundation for International Development (DSE). The event, convened in March 1998 in Bonn, culminated in the adoption of the

---

5 On a different tack, it must be noted here that in the negotiations on a climate change protocol in Kyoto in December 1997 (the Kyoto Protocol), unacceptable constraints upon the national defense capacities of the USA resulting from climate policy commitments were put forward as an argument against far-reaching greenhouse gas reduction commitments, which ultimately led to the Kyoto Protocol containing extensive exemptions for the armed forces. This dimension of the environment-security nexus is not further examined here.

6 The various issues of the Environmental Change and Security Report published by the Woodrow Wilson Center (ECSP Report 1995, 1996, 1997, 1998) provide a good overview of the relevant government initiatives and agreements.

Petersberg Declaration, which sets out principles for cooperative international water politics.

However, this build-up of political pronouncements on the security policy relevance of environmental degradation has yet to be matched by a similar intensity of concrete political action. Despite regular exhortations of the dangers of environmentally-induced conflict, governmental expenditure for preventive environmental protection has risen just as little as has the commitment to international environmental agreements, such as on climate protection. The Rio pledge to commit at least 0.7% of GDP to development cooperation also has yet to be honored (on the role of development cooperation in crisis prevention, see Schmieg's contribution in this volume).

3 Links between environment and security

Having outlined the political debate on the linkage between environment and security and the way in which it has been addressed by various actors, the following two sections examine the systematic connections between environment and security.

3.1 Environment, security, conflict: Clarifying the concepts

For a long time, the scientific debate on environment and security concentrated mainly upon the concept of "security" and its political implications. In this debate, authors aimed at "redefining" the concept of security so as to include social, economic and ecological factors (see e.g. Mathews 1989, Myers 1989, Ullmann 1983). In contrast to the narrower concept of security (protection of national sovereignty and territorial integrity against external and in particular military threats), the extended concept proceeds from a differentiation of security objects (individual, national, regional and international security) and a differentiation of the factors impacting on security (poverty, environmental degradation, illegal arms trade, international drug trafficking etc.). Within this debate, environmental degradation and its security relevance is debated in a general fashion.

One segment of the debate addresses the framework conditions under which environmental changes and environmental degradation lead to violent conflict. This conceptualization of the relationship between environment and security forms the basis of two major empirical research projects being carried out at the University of Toronto, and at the Swiss Federal Institute of Technology (ETH Zürich) in cooperation with the Swiss Peace Foundation (see Baechler et al. 1997, Homer-Dixon 1994; for a discussion of the various approaches see Rohloff's contribution in the present volume). Section 4 of the present contribution takes up this conceptualization in the characterization and discussion of the conditions determining environmentally-induced conflict. "Conflict" can be understood as a continuum ranging from mere differences in the positions of actors, over sporadic use of violence, through to armed conflict. In this contribution we use the term "conflict" to designate a situation in which the use of violence appears at least probable.

Ecological factors connected to security or conflict include above all environmental degradation and scarcity of renewable natural resources.[7] Scarcity of resources refers both to the inadequate availability of natural resources, such as cropland, potable water and fisheries, and to the asymmetrical distribution of these resources. Environmental degradation refers to anthropogenic environmental problems such as climate change, ozone depletion, species loss and environmental pollution. As resource scarcity and environmental degradation are inextricably linked – scarcity can lead to further over-exploitation and thus degradation, degradation leads to scarcity – the two concepts are subsumed in the following under the term "environmental stress".

3.2 Dimensions of environment and security

The debate on the links between environmental changes and security can essentially be broken down into four dimensions (Figure 1). These are: (1) the impacts of military activities upon the natural environment in times of peace and of conflict; (2) the direct and indirect influence of a) environmental changes upon local, national, regional and international security but also b) their function of delivering causes for cooperation and thus building confidence; (3) the impacts of environmental changes upon social conflicts and their indirect consequences for security and, finally, (4) the instrumentalization of deliberate environmental changes as a means of warfare.

7 Non-renewable natural resources such as oil, coal, iron or gold have always been the subject of distributional and utilization conflicts between various groups. While non-renewable resources, as opposed to renewable resources, have an immanently strategic dimension due to their limited availability, they are not the subject of this discussion.

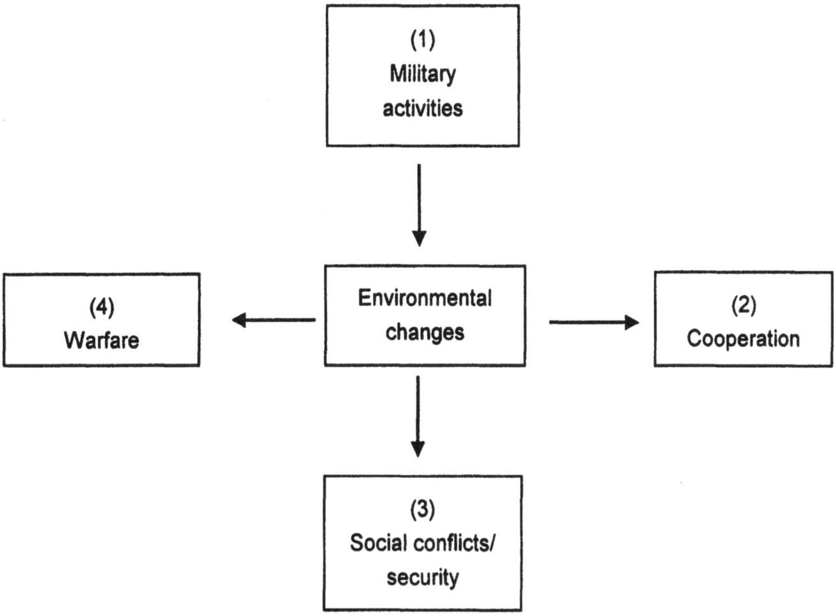

Figure 1: Four dimensions of the environment-security nexus

(1) *Environmental consequences of warfare and of military activities in times of peace and conflict.* Every kind of warfare entails an at least partial degradation of the natural environment. Furthermore, the consequences for the civilian population in war and conflict zones are immense, such as through the devastation of agriculturally utilizable areas or the contamination of freshwater sources. In times of peace, too, military activities have negative impacts upon the natural environment (Westing 1988). Despite a growing awareness within the armed forces of the environmental impacts of military activities, considerable areas continue to be used for maneuvers and military exercises, plant and animal communities are destroyed in the training areas, soil and drinking water are contaminated and the high levels of energy consumption contribute to air pollution.

(2) *Environmental cooperation as an instrument for confidence building and maintaining peace.* The access to and use of renewable natural resources and the pollution of the natural environment are the subject of several hundred international environmental agreements, international environmental regimes and countless regional and bilateral agreements on environmental protection and conservation. Thus, for instance, there are some 2000 agreements concerned with water-related issues alone, regulating the equitable distribution of water resources and their sustainable use at the regional, national or international levels (McCaffrey 1993). In the context of the two major regional "water conflicts" in the

Jordan River Basin and the Euphrates and Tigris river basins, water-related agreements play an important role in offering approaches towards or components for a peace process that is otherwise fraught with difficulty.

(3) *Environmental degradation and resource scarcities as a source of social and violent conflict* and thus an indirect threat to national and international security. Geographically contained disputes over the access to, equitable distribution and use of limited non-renewable resources (ores, oil) have always been a subject of violent conflict. The sustained degradation and depletion of the natural environment and renewable resources, in particular the social and economic impacts of climate change (IPCC 1997), indicate a new dimension of human intervention in the balance of nature. This new dimension has global impacts and can only be mitigated over very long periods. Here environmental changes become one element in a complex web of causality that can trigger a series of socio-economic problems such as overpopulation, poverty, mass migration and population displacement, hunger and starvation, political instability and ethno-political tensions. Under certain conditions, environmental degradation and social conflicts can lead to violent conflict.

(4) *Warfare through the deliberate modification or instrumentalization of the natural environment.* This includes the modification of micro-climate, the alteration of the electromagnetic and acoustic properties of water or of the atmosphere and the deliberate destruction of dams in areas that are below the sea level or are exposed to flood hazards (Westing 1984, 1997). The second Gulf War delivers a further example, in which oil wells were set on fire, thus intentionally instrumentalizing an environmental medium in the pursuit of violent conflict.

The environmental consequences of military activities in times of peace and times of conflict were already a focus of peace and conflict research in the 1970s (Albrecht 1986). In international law, these issues were addressed under the 1977 Protocol I addition to the 1949 Geneva Conventions, which contains provisions on environmentally destructive methods of warfare, and under the 1977 Convention on the Prohibition of Military or Any Other Hostile Use of Environmental Modification Techniques (ENMOD). The Protocol bans all forms of warfare that damage the natural environment in a sustained and comprehensive manner (see on this Biermann 1995: 77 ff.). ENMOD bans environmental warfare using environmental modification techniques, but does not prohibit the development of such techniques (Brock 1991).

The scientific and political debate on the environmental impacts of armed conflict and peacetime military activities and the discussion on the modification of the natural environment as an instrument of warfare is, as such, largely concluded. Nonetheless, despite relevant international agreements and their widespread acceptance, the need for political action remains. The environmental protection measures taken by military establishments also need to be further expanded, in particular the cleanup of contaminated sites.

The following discussion in this contribution focuses on environmental changes as a cause of violent conflict and as an occasion for cooperation and confidence-building, examining relations (2) and (3) in Figure 1. This is followed by an outline of available options for action to prevent environmentally-induced conflicts.

3.3 Challenges for political actors

Peace and conflict research has already been tackling the security impacts of global change and environmental degradation as a cause of violent conflict since the mid 1980s. Environmental policy research only took up this topic more recently. Against the backdrop of these debates, different challenges ensue for political actors and institutions.

(1) *The identification of "non-traditional security concerns".* After the end of the East-West stand-off, the geostrategic bipolar security architecture has undergone radical change. So have also the tasks of NATO, the OSCE, the WEU and the Council of Europe in their functions as pillars of the European security architecture. In addition to military threats, increasing attention is being devoted to a series of non-military risks such as transboundary waste trafficking, organized crime, international terrorism, and also renewable resource scarcities and environmental degradation.

(2) *The reorientation of security policy institutions and actors at the national and international levels.* After the political sea change of 1989, NATO, in particular, has underscored its self-perception as an alliance with not only military but also political dimensions. Thus its 1991 Strategic Concept states in paragraph 10:

> Risks to Allied security are less likely to result from calculated aggression against the territory of the Allies, but rather from the adverse consequences of instabilities that may arise from the serious economic, social and political difficulties including ethnic rivalries and territorial disputes [...].

The broadened security policy approach of NATO embraces political, economic, social and environmental elements. New challenges derived from this include, in addition to maintaining defense capabilities, above all the intensification of dialogue and cooperative mechanisms. The Conference on Security and Cooperation in Europe (CSCE) used from its establishment in 1973 onwards an extended security concept that embraces environmental problems as a security-relevant factor. However, no concrete measures have yet emerged from this inclusion of environmental factors. After the end of the East-West stand-off, the Organization for Security and Cooperation in Europe (OSCE) into which the CSCE has been converted is increasingly harking back to this extended security concept. It thus views intensified, but environmentally sound and non-inflationary economic growth, full employment and the removal of protectionist measures as being essential preconditions to promoting security in Europe. It is from this understanding that it derives its role as a promoter of rapprochement between Western and Eastern Europe through – albeit environmentally sound – market economies and pluralism (Broadhead 1997).

Propositions aimed at enshrining environmental policy aspects in the Common Foreign and Security Policy (CFSP) of the European Union in the context of the Amsterdam Treaty negotiations have not yet borne fruit (Robins 1996; Carius 1997b). This applies similarly to security policy institutions at the national level, in particular the national defense ministries and their subordinate bodies. The exception is the USA, whose government increasingly views environmental risks in states outside of the USA as having security policy relevance and is approaching environmental policy priorities at the national level against the background of their geostrategic security dimension. In other countries, however, the security policy aspect of environmental stress has rarely been taken up yet in national foreign and security policies. Therefore – particularly in the European context – the specter of a militarization of environmental policy would not seem to be looming as large as some critics might suggest (see Brock 1992, Deudney 1991, Daase 1991; see on this the discussion in Carius 1998, here section 3). Rather, we find a growing trend towards the 'civilianization' of the military, with commitments to humanitarian and disaster relief. Another aspect is that hardly any other policy sphere, apart from development policy, is as strongly influenced by non-governmental actors as environmental policy. Decisions on tackling and preventing environmentally-induced conflicts are thus not only taken by (inter)governmental actors, but also by non-governmental organizations (NGOs).

(3) *The foreign and security policy dimension of environmental degradation and environmental policy measures.* Although international environmental policy can look back on a successful development over the past 25 years (for a comparatively young policy area), implementation of the ambitious goals set out at the Rio Earth Summit in 1992 would seem to be losing steam. In the run-up to the Kyoto Climate Summit in 1997, the limits to the willingness to engage in joint environmental policy action became all too apparent. It is thus a concern of environmental policy to stress in future more clearly the negative impacts of non-sustainable economic practices upon other policy spheres, too, including foreign and security policy. Environmentally related causes of conflict hinge on gradual, potentially global environmental changes and their consequences which can only be reversed over a long period. Solving these global environmental problems and their local impacts all at the same time overtaxes the presently available environmental policy instruments. In the quest for preventive approaches to surmounting these problems, it is thus important to stress the foreign and security policy dimensions of environmental degradation and to involve specific actors at an early stage.

Some authors, including Deudney (1991) in particular, take a critical view of this conclusion. He and others label the recent concentration on the foreign and security policy dimension of environmental policy as "illegitimate" and merely a rhetorical construct. Voiced from the perspective of a natural scientist, this objection is understandable. However, from the environmental and development policy perspective it is not primarily a matter of assigning such conflicts their place within existing classifications and typologies of peace and conflict research (see Rohloff in this volume), which indeed does present methodological and theoretical problems (Daase 1992). Peace and conflict research itself is above all interested in developing an integrated perspective on the concerns of environmental degradation, development and armed conflict, wherever environmental degradation represents a significant underlying variable conditioning current and potential armed conflicts (Baechler et al. 1996: 9). At the political level, finally, the goal is to find solutions to the environmental and developmental problems that underlie violent conflicts, or to prevent these conflicts through an integrated strategy cutting across the various policy spheres. Discussing a problem in its full complexity by identifying its economic and social, but also foreign and security policy dimensions is neither an impermissible reduction nor an illegitimate line of argument. Environmental policy is an arena in which it is particularly necessary to refer to and make good use of structures and (cooperative) approaches for action in other policy areas.

(4) *Comprehensive crisis prevention as a task of development policy.* A considerable proportion of multilateral and bilateral spending on development assistance is expended for mitigating the consequences of wars and environmental disasters. Bilateral emergency aid alone now consumes 10% of development aid (see Schmieg in this volume). This trend is undermining development policy projects aimed at long-term, sustainable structures in developing countries. Although the currently available data-based early warning systems do not yet permit any statements as to the eruption of violent conflicts and thus can not provide concrete indications for political measures, empirical studies show the necessity of proactive development and environment policy moves to prevent conflict.

4 Environmentally-induced conflicts

Following on from the general introduction given above to the various facets of the debates on environment and security, and the presentation of the dimensions under which the issue area is currently being debated in the academic and policy communities, we discuss in the following the contextual variables and specific characteristics of environmentally-induced conflicts.

4.1 Characteristics of environmentally-induced conflicts

As environmental conflict research has shown, environmental changes lead by no means directly to violent conflict; they are rather one strand within a complex web of causality in which are intertwined a series of socio-economic problems such as overpopulation, poverty, forced mass migration, refugee movements, hunger and starvation, political instability and ethno-political tensions. Environmental degradation and natural resource scarcity are both causes and outcomes of these socio-economic problems or are intensified by these. The increasing scarcity of freshwater resources, the loss of ground cover vegetation, desertification, global climate change and rising sea levels are primarily the outcome of anthropogenic processes. These negative environmental changes are the result of resource-intensive, partially resource-wasting patterns of production and consumption and of inadequate agricultural practices which, in combination with the above-mentioned socio-economic problems, can expose national and international security to substantial risks (see for a general overview in particular Brock 1991; Homer-Dixon 1991, 1994; Baechler et al. 1993, 1996; Dabelko/Dabelko 1995; Käkönen 1994; Dokken/Graeger 1995, Gleditsch 1997a, 1997b).

The civil wars in Rwanda and Sudan, mining conflicts in the Southern Pacific, the water conflicts in the Jordan River Basin and the Euphrates and Tigris River Basins, or the intra- and interstate tensions on the Indian subcontinent bear testimony to the political volatility of environmentally-induced conflicts. These are, however, mostly limited to underdeveloped regions with a lack of development policy alternatives, to regions whose history makes them prone to conflict and where crises and conflicts are evidently an inherent part of their development (see Baechler in this volume). The conflicts are mainly intrastate in nature, with varying conflict intensities over long periods of time (ranging from latent crisis through to outright civil war – partly with transboundary dimensions).

Environmental degradation is by no means the sole and direct cause of violent conflicts. Rather the various forms of environmental degradation and resource depletion have a triggering or accelerating effect on conflicts, are at least interpreted as such or are classified as environmental conflicts. Thus it is commonplace to speak of "water conflicts" in reference to the conflicts over freshwater in the Jordan River Basin or in the catchments of the Euphrates and Tigris Rivers. In fact, however, the dispute over water resources is a manifestation of conflicts whose roots go deeper and are the expression of historical power policy, strategic, geographical and ethnic struggles. Crises of environmental degradation and resource scarcity thus exacerbate and accentuate already existing development crises.

Moreover, it would be too narrow to examine only violent conflicts. The issue of linkages between environmental degradation and security is also concerned with conflicts below the threshold of violent, armed struggles. This level is particularly important to the prevention of environmentally-induced conflicts, as it permits an exploration of forms of conflict prevention in a broad array of policy spheres. Furthermore, conflicts that at least remain below the violence threshold offer a forum for cooperation and confidence-building – an aspect discussed elsewhere in this contribution. Viewing all and every environmental change from a security policy perspective should, however, be avoided.

In the following, the term "conflict" is used to designate a situation in which the use of physical violence appears at least probable. The simplifying term "environmentally-induced conflict" is used, in awareness that this type of conflict neither develops linearly nor is based on a monocausal connection; it arises from complex causal connections. Environmentally-induced conflicts display the following characteristics:

First, environmental changes and the increasing scarcity of natural resources play a *decisive* role in the emergence of conflicts – namely by decisively accelerating or triggering social problems. As yet, the causative capacity of environmental stress as a conflict accelerating or triggering factor can only be ascertained in individual case studies.

Second, environmentally-induced conflicts generally operate below the violence threshold, their volatility is thus less visible at first.

Third, they are based on societal and economic problems or non-sustainable human-environment relations in general, this making them all the more complex.

Fourth, environmentally-induced conflicts arise mainly in regions with emerging economies and societies undergoing transformation, although in varying manifestations. They are not the result of singular influencing factors but the outcome of (complex) societal aberrations.

Fifth, despite the complexity of the sources of conflict, the environmental components provide a potential for preventive measures. Global environmental policy, in particular, relies on the principles of cooperation and reconciliation of interests. Using this experience, environmental policy instruments or negotiations on environmental problems can be used to peacefully resolve conflicts in which a peaceful solution to other, perhaps more decisive sources of conflict is not possible.

4.2 Causes of environmentally-induced conflicts

As opposed to resource scarcity, environmental degradation and depletion generally only lead indirectly to conflicts, by bringing about negative socio-economic consequences. Thus climatic changes can lead to regional reductions in water availability, causing parts of the population to migrate to other, already over-populated regions, generating further scarcities there, e.g. of agriculturally utilizable soil, food or firewood. They can lead to intensified agriculture and thus land use and erosion, in turn causing flooding and migration. Figure 2 illustrates schematically the interactions between environmental stress and resulting socio-economic effects, such as poverty, migration and refugee movements. The relationships shown in the figure are reciprocal, so that the socio-economic factors can in turn lead to environmental stress. The context variables are conflict accelerating or triggering factors.

4.3 Context variables of environmentally-induced conflicts

Whether environmental stress indeed harbors conflict or leads to violence depends upon a series of socio-economic context variables, primary conflict factors and cognitive processes. These include cultural circumstances and traditions, ethno-political factors, civil society mechanisms of peaceful conflict resolution, the stability of the interior policy system and, finally, societal, institutional, economic and technological capabilities. It follows that – in addition to regulating consumption and ensuring an equitable distribution of renewable resources – the identified context variables offer points of leverage to prevent such environmentally-induced conflicts.

Regarding propensity to generate conflict, disparate *(1) cognitive patterns* initially determine whether an environmental change, be it alone or in conjunction with socio-economic problems, represents a curtailment of sovereignty or of military, economic or public-welfare security and stability. The cognitive patterns of both social groups and state elites are based upon security policy assumptions, values and definitions and accordingly also determine the perception of being affected by non-military threats. Disparate cognitive patterns further influence the other seven factors listed in the following.

A second decisive criterion is the economic dependence of industry and agriculture upon renewable natural resources and thus the *(2) vulnerability* of a state or its elite. Marginalized agricultural societies, in particular, depend greatly upon the availability and equitable distribution of natural resources. In many cases, the absence of economic alternatives means that an increasing scarcity of resources leaves scarcely any other source of income and thus development options.

The vulnerability of an economy or a state as a whole depends strongly upon *(3) economic, technological and institutional capabilities*. These variables pertain to the treatment of both environmental problems and the consequential socio-economic problems. This is a matter of whether or not there are appropriate environmental policy institutions at the national, regional and local levels, such as a functioning environmental administration, legal and economic instruments by which to regulate resource consumption and a monitoring system. It is also a matter of the potential for creating economic alternatives for developing societies, the ability to engage in long-term planning processes, the ability to adopt strategic policies and the integration of state and non-state resources and capabilities.

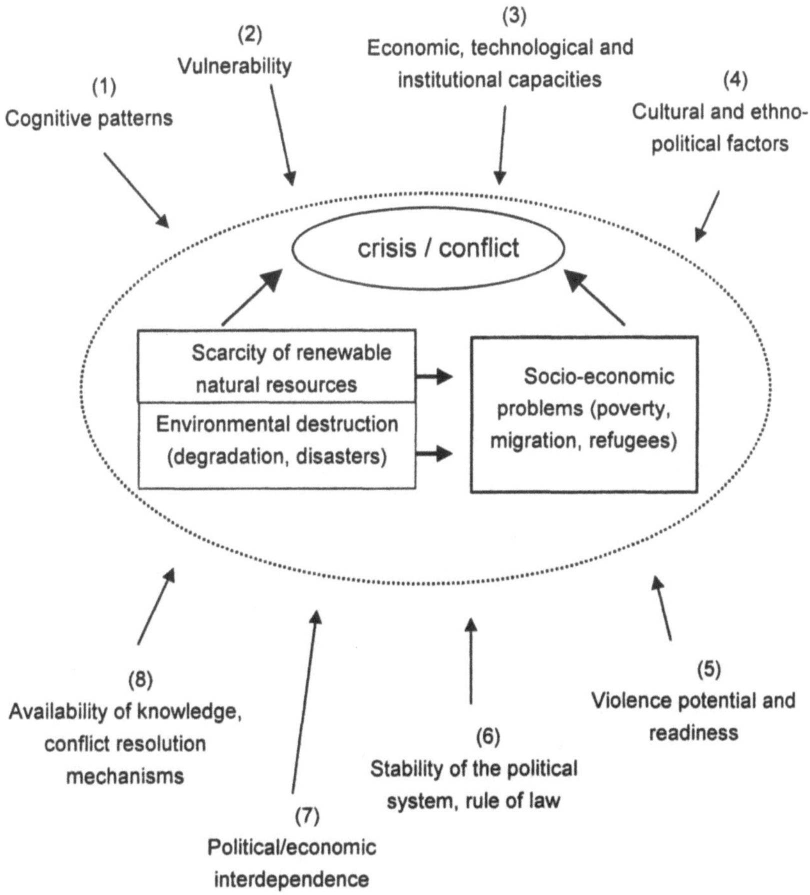

Figure 2: Context variables of environmentally-induced conflicts

(4) Cultural and ethno-political factors have a more direct link to conflict. Different ethnic groups are not per se conflict-oriented (meaning conflicts involving a substantial probability of violence), even under constrained geographic conditions, but do then represent a conflict potential if these ethnic differences are also perceived politically as a problem. The magnitude of conflict potentials of the ethno-political type results from the relative sizes of the various ethnic groups, from disparities in their positions in society, and from the historical process of their integration and representation in legitimated state institutions (Baechler 1998: 91). In addition, ethno-political tensions or crises can generally be traced to culturally and historically conditioned conflicts which are the outcome of power struggles or experiences of domination.

(5) The violence potential and *readiness to exercise violence* of the actors in conflict decisively determine whether the boundary lines between political and diplomatic tensions, political crises and violent conflicts are crossed. The capacity and readiness to exercise violence depend to a great degree upon whether actors are in a position to convince a critical number of individuals of the correctness of their position in the conflict. A precondition to violence is thus a sufficiently large, strategic group of individuals acting purposively. This is why migration need not necessarily lead to violent conflict. It is a characteristic of migratory peoples that they adapt their ways of life and husbandry to the momentary, changing location, and that, due to the absence of long-term links, they can not develop the potential for violent conflict resolution (with weapons). Readiness to exercise violence thus requires the presence of corresponding institutionalized structures for exercising violence, such as state military structures or the existence of armed units outside of the state military.

Ethno-political tensions and intrastate crises further depend upon *(6)* the *stability of a political system* and the *rule of law* as a whole. Governments rendered unstable, e.g. through frequent change of government or through a lack of democratic legitimization of decision-makers and elites, weaken the political stability of a state and increase the susceptibility to crisis, both internal and external. From the perspective of susceptibility to crisis, the notion of "rule of law" embraces all conceivable participatory mechanisms. For the criterion of participation, it is decisive that all groups in society have equal access to participating in social and political consensus-building and decision-making, and that these rules are recognized by the groups.

A variable of importance in cross-border conflicts, in particular, is the political and economic *(7) interdependence,* both regional and international. Integration in the narrower sense in regional security partnerships, but also integration in the broader sense in the world market, of developing countries in particular, decisively influences the susceptibility of states to crisis. The interdependence of markets can contribute both to stabilizing trade relations and to the direct transfer of technological and financial support and indirect transfer of capital and know-how.

A final component that merits attention is *(8)* the *availability of knowledge* and the availability of proven *conflict resolution mechanisms.* Contrary to the potential of and readiness to exercise violence, the decisive criterion here is the capacity to resolve conflicts, more precisely to resolve them by peaceful means. Such means can include institutionalized dispute settlement procedures, arbitrage tribunals and mediation procedures, but also forms of dispute settlement without the involvement of third parties (see Baechler 1998).

4.4 The explanatory value of this approach

The specific structure of the stated context variables contributes decisively to whether and when a conflict over environmental degradation and resource scarcity escalates to a violent conflict, or can instead be settled or solved by peaceful means. However, the explanatory power of this model is subject to a number of restrictions. Due to their complex causal structure, "environmental conflicts" are hard to identify as such. In their complexity, they are not fundamentally distinguished from other forms of conflict. Thus there are no practical examples of violent conflicts triggered exclusively by environmental changes.

Environmental changes only lead to conflict in their interactions with other social problems. Furthermore, these social problems, such as poverty, famine, refugees and mass migration already are potential causes of violent conflict when considered on their own. The decisive criterion for the characterization of environmentally-induced conflicts are the environmental stress factors underlying these conflicts, be they an essential determinant of the emergence of the conflict or be they merely an accelerating or triggering factor.

The discussion to date on environmentally-induced conflicts has been criticized for having produced nothing more concerning the environmental stress factor than that, in a general way, it plays a role (Levy 1995). This criticism is basically justified. Indeed, if no further explanations are provided, the finding would be banal. What has initially been demonstrated is the complexity of conflict causation. It has not, however, yet been possible to carry out a quantifiable weighting of the various components in the web of relationships (for some approaches towards this see Goodrich/Brecke 1997 and Sprinz in this volume). It is thus not yet possible to make reliable statements on the emergence of environmental conflicts after a certain, yet to be defined threshold has been crossed.

However, Levy's objection overestimates the scientific capacities and the availability of relevant data at the present point in time. To date, a comparatively small number of researchers and research groups are working worldwide on the connection between environment and security. Compared to the research input devoted to the social and economic consequences of global climate change which are tackled in various research networks, large-scale research facilities and the specially established worldwide expert panel, the Intergovernmental Panel on Climate Change (IPCC)[8], it is striking how little importance is currently accorded to research on environmentally-induced conflicts despite the enormous need for research and action that prevails in this area.

---

8 In addition to examining environmental problems as such, these research structures offer opportunities to address social and economic causes and consequences. The IPCC (1997) for instance has thus presented an assessment of the regional impacts of climate change upon ecosystems and social systems in particularly sensitive regions of the world.

A further objection concerning the explanatory power of the approach put forward here has been made variously and in particular by Gleditsch (1997b). These authors argue that previous studies – in particular empirical case studies – have predominantly concentrated on conflicts resolved violently, to the exclusion of cooperatively resolved conflicts. This has restricted the dependent variable "conflict" a priori to violent conflict resolution. This objection may apply (in retrospect) to the research project headed by Homer-Dixon, but only applies to a limited extent to the ENCOP study. Since the early 1990s, both projects have delivered to date only systematic empirical examinations of this issue, and indeed initially examined (violent) conflicts. In terms of a pragmatic approach to the research process, this is a purposeful way to proceed before identifying peace-promoting factors in a second step based on the conflict conditioning factors identified before (and on the basis of empirical studies of cases of environmental cooperation). The research processes based on the two quoted major empirical projects (Homer-Dixon and Baechler/Spillmann) have indeed already addressed the issue of to what extent conflicts induced by environmental change or environmental stress can lead to cooperation and can thus have a confidence-building effect. However, the issue of cooperation and confidence-building in the context of international environmental cooperation largely needs to be examined on an empirical basis and requires further research inputs.

There is currently a lack of adequate methods by which to quantitatively compare the conflictual effects of environmental changes. This presents methodological difficulties because changes in environmental media – water, air, soil – and fauna cannot be described as natural processes subject solely to human intervention. Even just as one category within the web of sources of conflict illustrated above, environmental changes are the outcome of complex interactions that require an integrated approach in analysis that gives equal attention to biological and physical processes and socio-economic factors. Numerous of such approaches have been proposed, but require further development in order to apply them to the description and modeling of complex human-environment relations and their conflictual relevance. The syndrome approach developed jointly by the German Advisory Council on Global Change (WBGU) and the Potsdam Institute for Climate Impact Research does justice to this complexity and has proven itself to be a useful modeling approach (see WBGU 1997 and the conceptual presentation of the syndrome approach by Biermann in this volume).

Nonetheless, there are still numerous points requiring further research on the path towards identifying and thus resolving or preventing environmentally-induced conflicts. For one thing, there are gaps in empirical environmental and conflict research, insofar as to date hardly any comparisons of case studies have been carried out that might give an understanding of the specific conditions under which environmental stress leads to violent conflict or to peaceful conflict resolution. In particular, there is as yet a lack of an inter-regional comparison of case studies of environmentally-induced conflicts that could deliver information on regional specifics. Data availability is a further complexity requiring improvement. There is a lack of disaggregated data on environmental stress factors and socio-economic influencing factors that are necessary to carry out a detailed analysis of environmentally-induced conflicts at the regional and local levels and to perform an estimation of the future risks associated with environmental changes. Most of the data used to estimate risk potentials (such as GDP, literacy rate, population growth) are currently exclusively available for the national level. Previous studies have shown, however, that environmentally-induced conflicts arise less between states than rather at the local or regional level (although they certainly have cross-border impacts). Similarly, there is a need for research on suitable indicators required to appraise the risk potential of environmental stress situations. The development of integrated socio-ecological indicators utilizable for the research issues outlined in this contribution is an important task for research efforts in the coming years.

5 Preventive policy approaches

We have pointed out the existence of research gaps in the identification of environmentally-induced conflicts. This should not, however, be taken to mean that the timely development and debate of political options for their peaceful solution or prevention can be dispensed with. These options can only be sketched in the present contribution. The available research results already give adequate indications of the conflict potential inherent in environmental stress under unfavorable socio-economic framework conditions and illustrate the necessity of political responses. The complexity of conflict factors, causes and respective framework conditions (environmental stress factors and socio-economic factors) suggests that patterns of reaction or preventive measures should be similarly diverse and complex. Efforts to prevent crises or safeguard peace thus need not primarily call for security policy approaches. The conflict-determining context variables illustrated in figure 2 make environmental, development, foreign and security policy approaches all similarly indispensable.

It is a particular concern of environmental policy to place the solution of environmental problems in the broader context of development, foreign and security policy, with the aim of highlighting the challenges of international and global environmental problems and vigorously demanding and developing solutions. Foreign and security policy actors command over more unutilized resources and options to take action to prevent environmentally-induced conflicts than the classic environmental policy actors alone. Conversely, the complexity of causal linkages calls for integrated approaches that go beyond the classic instruments of conflict prevention and containment. An integrated consideration of environmentally-induced conflicts and, above all, a practical integration of the capacities for action of the actors in the various policy arenas is therefore the most important guiding principle for action in comprehensive conflict and crisis prevention.

A further aspect is that approaches for action need to take into consideration the various levels at which environmentally-induced conflicts arise (local, regional, national, international). Regional conflicts, for instance over the use of increasingly scarce cropland, require other approaches (e.g. regionally and locally developed and applied dispute settlement mechanisms, see Baechler 1998) than the mitigation of the negative and conflict-laden consequences of global climate change (here strategies must address the level of global environmental policy).

Finally, appropriate approaches must make use of the specific potentials of the various actors. Local or regional conflicts call for the involvement of local and regional governmental organizations and NGOs which are in a position to develop specific problem resolution strategies adapted to the local or regional conditions, and command over more information on the causes of conflicts than central state and international actors.

From the *environmental policy* perspective, it is essential to intensify international cooperation on environmental concerns within the context of international environmental regimes, particularly in pursuit of solving global environmental problems with local and regional impacts. Intensified international cooperation should also be used as a long-term strategy to solve or prevent the environmental problems that underlie environmentally-induced conflicts. In existing environmentally-induced conflicts that have not yet escalated into violence, the relevant institutions of intergovernmental cooperation can serve as conflict resolution arenas. Institutions with well established routines facilitate communication among the conflicting parties and generally command over experts who can provide advisory support in the conflict resolution process. As already outlined above, it is the availability of a broad array of institutional structures geared to regulating and solving international environmental problems that makes the environmental component of environmentally-induced conflicts such a valuable point of departure for long-term solutions.

A further important approach towards preventing conflict is to more consistently integrate environmental concerns in other relevant policy sectors, in particular in development, foreign and security policy, but also in agricultural, energy and social policy. This further requires the continuous review of the realization of these concerns, e.g. in the context of comprehensive (strategic) environmental impact assessments for political programs and projects. These measures need to be implemented at the international and national levels equally. An internationally and nationally conceived policy oriented to sustainability aspects is thus the necessary strategy by which to prevent environmentally-induced conflicts over the long-term.

The conceptual discussion in this contribution has shown that environmental stress is not so much a direct cause of conflict but rather a catalyst that leads to conflicts or violent disputes through the agency of negative consequential socio-economic and political effects. The development of options for action thus also needs to include these socio-economic and political influencing factors. The outbreak of violent environmentally-induced conflicts is influenced by the specific structures of the above-mentioned context variables, which are the focus of a comprehensive preventive strategy. However, cushioning the negative socio-economic consequences of environmental stress or conversely, positively structuring the social, economic, political and cultural context factors falls within the remit of other policy spheres. Development and foreign policy have specific contributions to make to prevent violence arising from environmental stress, as does security policy in the event of concrete conflicts.

The prime *development policy* measures that can be applied to minimize or prevent the conflict-generating and potentially violent impacts of environmental degradation and resource scarcity include economic cooperation, the promotion of participation and democracy, cooperation in population policy matters, support for the establishment of the rule of law and the protection of human rights (DAC 1997). In view of the fact that environmentally-induced conflicts have occurred mainly in developing and transformation societies, development policy in the context of environment and security should place a focus on stabilizing the socio-economic and political framework conditions in these countries. This includes efforts to promote the involvement of segments of the population in decisions that concern them, and to promote a balanced representation of the various ethnic or cultural groups in decision-making bodies in general. A further point to be kept in mind is that many of the environmental problems confronting these societies can only be reversed over very long time scales. The necessary consequence of this is that development policy approaches need to focus on supporting the formation of problem-solving or -managing capacities. This includes creating a knowledge of the emergence and prevention of environmental degradation, knowledge of ways to cooperatively utilize increasingly scarce resources and knowledge for utilizing peaceful dispute settlement mechanisms (Baechler 1998).

Development policy measures can thus exert leverage in a quite concrete fashion at the level of direct conflict prevention or mitigation, often locally and involving local actors.

Foreign and security policy measures by which to address and prevent environmentally-induced conflicts can serve to stabilize the political environment in which local and regional conflicts are situated and which plays a decisive part in determining whether conflicts are resolved cooperatively or violently. Specific foreign policy instruments, such as dispute mediation and the initiation of political dialogs between states bear particular relevance to the channeling of conflicts. Foreign policy should be "deployed as a promoter of socio-structural strategies of change" and should deliver "substantial innovations" in terms of its peacemaking function, above and beyond its past role of representing national interests (Czempiel 1997: 47). Foreign and security policy actors should be involved well in advance in the prevention of such conflicts. Security policy organizations command over operative capacities can be utilized to monitor critical environmental changes and to appraise the risks entailed by environmental problems and their consequential effects. This last point once again underscores the supreme principle in preventing and peacefully resolving environmentally-induced conflicts: it is essential that environmental, development, foreign and security policy actors cooperate closely and that their specific instruments are integrated.

References

Albrecht, Ulrich 1986: "Weltweite Rüstung als Problem der Internationalen Umweltpolitik", in: Mayer-Tasch, Cornelius (Ed.): *Die Luft hat keine Grenzen. Internationale Umweltpolitik – Fakten und Trends.* Frankfurt am Main: Suhrkamp: 257-270.

Baechler, Günther et al. 1993: *Umweltzerstörung: Krieg oder Kooperation? Ökologische Konflikte im internationalen System und Möglichkeiten der friedlichen Bearbeitung.* Münster: agenda.

Baechler, Günther et al. 1996: *Kriegsursache Umweltzerstörung: Ökologische Konflikte in der Dritten Welt und Wege ihrer friedlichen Bearbeitung.* Band 1. Chur, Zürich: Rüegger.

Baechler, Günther 1998: *Zivile Konfliktbearbeitung in Afrika. Grundelemente für Friedensförderungspolitik der Schweiz.* Working paper No. 27. Schweizerische Friedensstiftung, Institut für Konfliktlösung.

Biermann, Frank 1995: *Saving the Atmosphere. International Law, Developing Countries and Air Pollution.* Frankfurt am Main.

Biermann, Frank et al. 1998: *Umweltzerstörung als Konfliktursache?* Beitrag zur NATO CCMS Pilot Study. Unpublished manuscript. Bremerhaven etc.

Broadhead, Lee-Anne 1997: "Security and the Environment: Taking the OSCE Approach Further?", in: Nyholm, Lars (Ed.): *OSCE – A Need for Co-operation. Towards the OSCE's Common and Comprehensive Security Model for Europe for the Twenty-first Century.*

Brock, Lothar 1991: "Peace through Parks: The Environment on the Peace Research Agenda", in: *Journal of Peace Research*, Vol. 28(4): 407-423.

Brock, Lothar 1992: "Security through defending the environment: An illusion?", in: Boulding, Edward (Ed.): *New Agendas for Peace Research: Conflict and Security Reexamined.* Boulder: Lynne Rienner.

Carius, Alexander et al. 1997a: "NATO CCMS Pilot Study: Environment and Security in an International Context. State of the Art and Perspectives". Interim Report, in: *Environmental Change and Security Project.* Report No. 3. Spring 1997. Woodrow Wilson Center: 55-65.

Carius, Alexander 1997b: *Environment and Security: A New Agenda for European Foreign Policies?* Paper presented at Europe's Footprints Workshop, London. October 1997 (Organized by the International Institute for Environment and Development and the Goethe Institute London).

Carius, Alexander 1998: "Umweltzerstörung: Sicherheitsproblem und Kriegsursache?", in: Bündnis90/Die Grünen im Deutschen Bundestag, Referat Öffentlichkeitsarbeit (Ed.): *Umweltzerstörung als Kriegsursache?*, Dokumentation, Reihe "lang & schlüssig" Nr. 1390, Juli 1998, Fachgespräch vom 8. Mai 1998 im Wasserwerk des Deutschen Bundestages.

Chalmers, Martin, David Mepham 1997: *British Security Policy: Broadening the Agenda.* London: Saferworld.

Council on Environmental Quality / Department of State 1981: *The Global 2000 Report to the President. Entering the Twenty-first Century.* Vol. 1. Charlottesville: Blue Angel.

Czempiel, Ernst-Otto 1997: "Alle Macht dem Frieden", in: Dieter Senghaas (Ed.): *Frieden machen*, Frankfurt/M.: Surhkamp: 31-47

Daase, Christopher 1992: "Ökologische Sicherheit: Konzept oder Leerformel?", in: Meyer, Berthold, Christian Wellmann (Eds.): *Umweltzerstörung: Kriegsfolge und Kriegsursache*. Friedensanalysen Nr. 27. Frankfurt am Main: Suhrkamp: 21-52.

Dabelko, Geoffrey, David D. Dabelko 1995: "Environmental Security: Issues of Conflict and Redefinition", in: *Environmental Change and Security Project*. Report No. 1. Spring 1995. Woodrow Wilson Center: 3-13.

DAC – Development Assistance Committee 1997: *Conflict, Peace and Development Co-operation on the Threshold of the 21st Century*, Policy Statement by the Development Assistance Committee of the Organisation for Economic Co-operation and Development (OECD), Paris.

Deudney, Daniel 1991: "Environment and Security: Muddled Thinking", in: *Bulletin of the Atomic Scientist*, Vol. 47(3): 16-22.

Department of State of the United States 1997: *Environmental Diplomacy. The Environment and U.S. Foreign Policy.* Washington D.C.: www.state.gov/www/global/oes/envir.html.

Dokken, Karin, Nina Græger 1995: "The Concept of Environmental Security. Political Slogan or Analytical Tool?", in: *PRIO Report*, No. 2. June 1995. Oslo.

ECSP – Environmental Change and Security Project, Woodrow Wilson Center: 1995, 1996, 1997, 1998 Reports. Washington D.C.: Woodrow Wilson Center.

Gleditsch, Nils Petter (Ed.) 1997a: *Conflict and the Environment.* Dordrecht, Boston, London: Kluwer.

Gleditsch, Nils Petter 1997b: *Armed Conflict and the Environment. A Critique of the Literature.* Paper presented to the 1997 Open Meeting of the Human Dimensions of Global Environmental Change Research Community. 12-14 June 1997. International Institute for Applied Systems Analysis (IIASA), Laxenburg, Austria. mimeo.

Goodrich, Jill Wieder, Peter Brecke 1997: *The Paths from Environmental Change to Violent Conflict.* Manuscript to be published.

Gore, Al 1990: "SEI: Strategic Environment Initiative", in: *SAIS Review*, Vol. 10(1).

Gore, Al 1993: Earth in the balance. Ecology and the human spirit. New York: Plume.

Holst, Johan Jørgen 1989: "Security and the Environment: A Preliminary Exploration", in: *Bulletin of Peace Proposals*, Vol. 20(2): 123-128.

Homer-Dixon, Thomas F. 1991: "On the Threshold. Environmental Changes as Causes of Acute Conflict", in: *International Security*, Vol. 16 (2): 76-116.

Homer-Dixon, Thomas F. 1994: "Environmental Scarcities and Violent Conflict. Evidence from Cases", in: *International Security*, Vol. 19(1): 5-40.

IPCC – Intergovernmental Panel On Climate Change 1997: *The Regional Impacts of Climate Change. An Assessment of Vulnerability.* Geneva.

Käkönen, Jyrki 1994: *Green Security or Militarized Environment.* Aldershot etc.: Dartmouth.

Kaplan, Robert 1994: "The Coming Anarchy", in: *The Atlantic Monthly,* Issue 2, 44-76.

Levy, Marc A. 1995: "Is the Environment a National Security Issue?", in: *International Security*, Vol. 20(2): 35-62.

Mathews, Jessica Tuchman 1989: "Redefining Security", in: *Foreign Affairs*, Vol. 68(2): 162-177.

McCaffrey, Stephen C. 1993: "Water, Politics and International Law", in: Peter H. Gleick (Ed.): *Water in Crisis: A Guide to the World's Fresh Water Resources.* New York: Oxford University Press.

Myers, Norman 1989: "Environment and Security", in: *Foreign Policy*, No. 74: 23-41.

Rifkind, Malcolm 1997: *Foreign Policy and the Environment: No Country is an Island* – Secretary of State's Speech on the Environment. Speech on January 14th 1997.

Robins, Nick 1996: *Greening European Foreign Policy*, in: European Environment, Vol. 6: 173-182.

Shevardnadze, Eduard 1990: "Ekologiya i diplomatiya", in: *Literarturnaya gazeta*, 22 November 1989. Published in an English version as: "Ecology and Diplomacy", in: *Environmental Policy and Law*, Vol. 20(1/2): 20-24.

Ullman, Richard H. 1983: "Redefining Security", in: *International Security*, Vol. 8(1): 129-153.

WBGU – Wissenschaftlicher Beirat der Bundesregierung Globale Umweltveränderungen (German Advisory Council on Global Change) 1997: *World in Transition. The Research Challenge*. Annual Report 1996. Berlin.

WCED – World Commission on Environment and Development 1987: *Our Common Future*. Oxford etc.: Oxford University Press.

Westing, Arthur H. 1984: *Environmental Warfare: A Technical, Legal and Policy Appraisal*. London, Philadelphia, PA: Taylor and Francis.

Westing, Arthur H. 1988: The Military Sector vis-à-vis the Environment, in: *Journal of Peace Research*, Vol. 25(3): 257-264.

Westing, Arthur H. 1997: "Environmental Warfare. Manipulating the Environment for Hostile Purposes", in: *Environmental Change and Security Project*. Report No. 3. Spring 1997. Woodrow Wilson Center: 145-149.

Environment and Security in the Context of the NATO/CCMS

Kurt M. Lietzmann

1 Introduction

The work currently in progress in the area of the NATO/CCMS on a study on the subject of "Environment and Security in an International Context"[1] both provided the occasion for the workshop, the papers presented at which are compiled in this volume, and is the reason for the increased interest on the part of the German Federal Ministry of the Environment in the subject of "Environment and Security". The aim of this article is to explain why and in which context this subject is being discussed in the North Atlantic Treaty Organization (NATO). In this connection, firstly the "*Committee on the Challenges of Modern Society*" (CCMS) will be presented and secondly the genesis of the topic of "Environment and Security" within the framework of the CCMS and ongoing projects on the study will be explained.

NATO is perceived by the public primarily, if not even exclusively, as a defense alliance. However, even in the past this view was not entirely accurate since in addition to having the obligation to provide mutual protection and support, the Member States of the Alliance have the obligation "...to contribute to the further development of peaceful and friendly international relations..." (Art. 2 of the North Atlantic Treaty). The perception of NATO as a purely defensive organization is in particular need of correction seen against the backdrop of its re-orientation after the end of the Cold War and of the intensive East/West partnership that has developed since then. With the "Strategic Concept" passed in November 1991, the Alliance reformed its approach to security policy by placing greater emphasis on *political* rather than military means in pursuing the aims of maintaining peace and

---

1 Kurt M. Lietzmann and Gary D. Vest (eds.) 1999: Environment and Security in an International Context. NATO/CCMS Pilot Study Report No. 232, Brussels, Bonn, Washington DC: North Atlantic Treaty Organization/Committee on the Challenges of Modern Society. The Pilot Study (including an English, French, and German Executive Summary) has been published and is available free of charge upon request from the CCMS Secretariat: NATO Scientific Affairs Division (Fax: *32 2 707 4232, email: ccms@hq.nato.int).

preserving the security of its members. The consolidation of political dialogue and the support of reform strategies places the relationship with the countries of Central and Eastern Europe on a new foundation. The most recent testimony to the restructuring of relations with Central and Eastern Europe is the decision taken in Summer 1997 to extend NATO eastwards in future. This Strategic Concept highlights the protection of the environment as an element of security and stability in addition to political, social and economic aspects (Art. 25 of the Strategic Concept of the Alliance).[2]

2 The "Committee on the Challenges of Modern Society"

The establishment of the "*Committee on the Challenges of Modern Society*" in 1969 must be seen in the context of the obligation to develop friendly international relations. Alongside cooperation in the military and security field, the CCMS embodies civilian cooperation between the NATO Member States. With this in mind, environmental topics were given priority on the CCMS agenda, in particular at the instigation of the German government at the time and prior to the world-wide environment conference in Stockholm in 1972, thereby creating the first international body for the discussion of environmental topics. Thereafter the CCMS developed into what was principally a NATO environment committee, which today deals with a large number of questions from the field of environmental protection, which are in some cases highly specialized and mostly of a technical nature.

Recently the CCMS has become increasingly important as a forum for cooperation with partners outside NATO, and in particular in relations with the countries of Central and Eastern Europe as well as with countries from the former Soviet Union. In the new definition of the future policy of NATO in its relations with the countries of Central and Eastern Europe the NATO Summit in Rome in November 1991 explicitly emphasized the "Third Dimension" of NATO for science and environmental protection in addition to cooperation in military affairs and matters of security policy.

In realigning the NATO/CCMS towards cooperation with the states of Central and Eastern Europe, the need to define the topics to be dealt with as against those handled by other cooperation bodies, especially within the framework of the UN Economic Commission for Europe (UNECE), the Organization for Security and Co-operation in Europe (OSCE) and the Organization for Economic Cooperation and Development (OECD) and its *Task Force for the Implementation of the Environmental Action Program (EAP) for Central and Eastern Europe* became

---

2 Future developments within NATO will be interpreted under the umbrella of the Strategic Concept for the 21st Century at the Washington DC NATO Summit in April 1999.

increasingly important. At the conference of environment ministers in Sofia in 1993 attention was drawn to the special role of the NATO/CCMS in the area of defense-related environmental protection. Since then several pilot studies have been initiated by the CCMS on this topic complex.[3]

However, the question also arises as to whether the upheavals in Europe after the end of the Cold War may not well involve further-reaching challenges ("*Challenges of Modern Society*"), which the CCMS could usefully take up.

At any rate, no matter which ideas and projects are adopted in the CCMS, they should contribute towards maintaining of peace in the widest sense since this is the core of NATO's civilian mandate. This gearing towards maintaining peace also applies to the study presented in the following, entitled "Environment and Security in an International Context", the main content of which takes up the NATO mandate for the sector of environment policy in its objectives.

3 The Pilot Study "Environment and Security in an International Context"

The Pilot Study "Environment and Security in an International Context" has its roots in the research reports from the Environmental Conflict Project (ENCOP) by the ETH Zurich and the Swiss Peace Foundation,[4] made available to the German NATO/CCMS coordinator by the Institute for Peace and Conflict Research of the University of Heidelberg, Germany. The German NATO/CCMS coordinator in turn took up the topic and introduced it to the NATO/CCMS where the subject was met with widespread interest, especially on the part of the US representatives. The USA made "*environmental security*" the object of a large-scale round-table discussion with the aim of drawing conclusions for the further work of the NATO/CCMS. The German proposal was a welcome addition to this, especially since the topic of "environmental degradation as a cause of war" was an important aspect being widely debated within the political and scientific communities in the USA and Canada.

---

3 Here the following pilot studies should be mentioned, inter alia, on the subjects of: "Cross-Border Environmental Problems Emanating from Defence-Related Installations and Activities" (1992-1995), "Environmental Aspects of Reusing Former Military Lands" (start 1994), "Protection of Civil Populations from Toxic Material Spills During Movements of Military Goods" (start 1994) and "Environmental Management Systems in the Military Sector" (start 1995).

4 The "Environment and Conflicts Project" (ENCOP) was worked on by the Swiss Peace Foundation in Bern jointly with the Center for Security Studies and Conflict Research at the ETH Zurich from 1992 to 1995. The aim of this study was to systematically investigate causal connections between environmental degradation and armed conflicts within and between countries, above all in the Third World, on the basis of empirical case studies. The results of this study have been published in a three-volume Final Report (Baechler et al. 1996)

Hitherto, the subject of "environment and security" has been discussed principally in peace and conflict research. Against the background of increasing tension and security risks as a potential danger to peace due to environmental problems and world-wide resource scarcity, however, this subject is receiving increasing attention in the global debate on the environment.

The different facets of this topic were discussed on the occasion of the round-table discussion in Washington D.C. on 14 November 1995. Here, the debate centered on both the use of resources from the defense sector for the requirements of environmental protection and the area of "environmental protection in the armed forces in peace and conflict" (e.g. *"environmental performance" and "environmental warfare"*). In addition to this, a further aspect of the overall topic under discussion was the possible threat to security resulting from environmental degradation or, in essence, the topic of "environmental policy as a peace maintaining instrument." As a result, the representatives of the NATO Member States and Cooperation Partners agreed to initiate a Pilot Study on this subject with the title "Environment and Security in an International Context" to be placed under the joint responsibility of Germany and the USA. This Pilot Study began in April 1996 and the Executive Summary and Full Technical Report were finalized in February 1999.[5]

In addition to Germany and the USA, this was actively supported in the Alliance by Canada, Denmark, France, Turkey and its new Member States-Poland, the Czech Republic and Hungary. Important contributions were also made by the western countries of the *"Partnership for Peace"*, in particular Austria, Finland, Sweden, and Switzerland. The study found widespread interest amongst the eastern European partners; participants include Belarus, Bulgaria, Estonia, Kyrgyzstan, Latvia, Lithuania, Macedonia, Moldova, Romania, the Russian Federation and Slovenia. Overall coordination of the Pilot Study is the responsibility of the German Federal Ministry of the Environment, which is supported in planning and organization by *Ecologic- Centre for International and European Environmental Research, Berlin.* The study was being structured in plenary meetings of participants in the Pilot Study, prepared in workshops on focal topics and then compiled initially in sub-reports and finally in an overall report. In its final section it contains recommendations for action for environmental, development, foreign and security policy.

On the basis of a CCMS resolution of March 1996 within the framework of the study, a scientific and political examination was made to determine whether and to what extent conclusions can be drawn from the assessment of the security risk for the need for action to prevent regional and global environmental risks, and which

---

5 It is important to draw attention to the following with regard to the method of working of the CCMS. The CCMS itself has no project funds to draw on. Pilot studies are commenced when at least three pilot countries have declared their willingness to participate and there are sufficient partners to carry out the study. They are elaborated and formulated in a consensus driven process. The results of the pilot studies are published and made available to other international organizations and non-participating states.

further political and institutional questions/options for action are connected with this topic. One particular challenge in dealing with these questions results from the need to integrate questions relating to environment policy and security policy. In looking for political solutions, regional and sectoral peculiarities also have to be considered.

Furthermore, the Pilot Study elaborates a typology of environmental problems to demonstrate when and under which circumstances environmental degradation constitutes a security risk and when the domestic conflicts of a country as a result of environmental degradation can have effects on the security of other countries. This already indicates the principal objectives of the Pilot Study: the development of suitable preventive strategies to avoid conflicts as well as considerations on the design or reform of suitable institutions, to be given the corresponding mandate or necessary authority to be able to cope with the complex tasks facing them. The Pilot Study is divided into the following chapters:

- Chapter 1: NATO Security Context (provides an overview of the North Atlantic Treaty, CCMS, and the Changing Security Context);
- Chapter 2: Assessing the Links between Environment and Security (clarifies the concepts between environmental stress and security; examines the consequences of environmental stress and their potential impact on the incidence or escalation of conflict; discusses contextual factors which may impact the consequences of environmental stress);
- Chapter 3: Typology of Environmental Conflict Cases (describes further the relevance of environmental stress, socio-economic conditions, contextual factors and conflict; exhibits the results of empirical research including historical cases and inductively derived case studies);
- Chapter 4: Integrated Risk Assessment (analyses and compares the conflict potential or security risk of specific unfavorable socio-ecological patterns; identifies regions that are affected by environmental stress factors or syndromes);
- Chapter 5: Indicators, Data and Decision Support Systems (presents a set of practical options to support policy makers for the development of early warning indicator systems, data bases, and decision support systems);
- Chapter 6: Policy Responses (presents an integrated approach of all policy areas and policy responses for environmental, development, foreign and security policy). [6]

---

6 The structure and progress are documented in the report on the Pilot Study published in the *Environmental Change and Security Project Report* (Carius et al. 1997). Further information can be obtained via the NATO *Environmental Clearinghouse System (ECHS)* under (http://www.vm.ee/nato/ccms/welcome.html).

The Pilot Study is of particular importance within the framework of the NATO/CCMS seen against the backdrop of its avowed intention to develop political solutions for environment and security policy. It is precisely in the search for new potential solutions in international environment policy that innovative and unconventional approaches are called for. At the same time, great expectations were placed in the German participation in the Pilot Study with regard to presenting proposals for solutions in the field of international environment policy.

In view of the fact that amongst a large number of dangers to the environment affecting security, the profoundest changes relevant to security are to be expected on a global level, in particular as a result of changes in the world's climate, the proposed solutions must be globally effective. Although there is no reason why NATO should not give some thought to the possibilities existing to protect NATO itself against the threat to its security presented by environmental degradation, there can nevertheless be no question that it is not only NATO, perhaps not even principally NATO, that is affected - assuming that NATO tends to embody the "North", whereas environmental degradation relevant to security is possibly to be expected mainly in the "South" (Baechler et al. 1996). This makes it seem apposite to formulate political recommendations within NATO with appropriate restraint and, moreover, to co-ordinate with partner countries outside NATO and other international organizations at an early stage.

References

Baechler, Günther et al. 1996: *Kriegsursache Umweltzerstörung: Ökologische Konflikte in der Dritten Welt und Wege ihrer friedlichen Bearbeitung.* Zurich: Rüegger.

Carius, Alexander et al. 1997: "NATO CCMS Pilot Study: Environment and Security in an International Context." *Environmental Change and Security Project Report*, Issue 3, 55-65. Washington D.C.: The Woodrow Wilson Center.

Environmental Conflict Research - Paradigms and Perspectives

Lothar Brock

1 Introduction

In the course of the 1990s, building upon isolated initial approaches leading back to the 1970s, a new field of research has emerged in the social sciences – *environmental conflict research.* The development of this field is the outcome of a confluence of quite disparate research interests: Traditional security studies have viewed the focus on environmental conflict as an opportunity to outline new dimensions of national security, the nuclear threat having subsided. For peace research it was a matter of overcoming the classic notion of security, with its military perspective and focus on the individual state. For developing country research the inclusion of environmental issues made sense because it promised a new approach towards analyzing violent conflicts in the Third World, and reference to environmental issues as factors of conflict provided a new line of argument for raising old demands for restructuring the global economy. Despite the early plea of Lester Brown for a redefinition of the concept of security from an ecological perspective (Brown 1977), environmental research in the narrower sense played a rather secondary role in the emergence of environmental conflict research in the early 1990s.

The impulse given by the end of the East-West standoff to the scientific debate on *environment and conflict* has weakened greatly in the meantime. The question thus arises today of to what extent environmental conflict research is based on a substantial nexus or rather a chance confluence of concerns. In the former case a further development of the field is called for, while in the latter case environmental conflict research would already have passed its zenith.

The present essay discusses the fruitfulness of and prospects for development in environmental conflict research. It first critically appraises the linking of environmental and security problems under the heading of *environmental security.* Subsequently, it briefly reviews the findings of research to date concerning the conflict potential inherent in environmental degradation. The third part sets forth an approach towards placing in historical context the problematique subsumed under the heading of *environmental conflict*, thus identifying a perspective for further progress in environmental conflict research.

2 On linking environment and security

It has remained unclear to this day which research interest is dominant in the field of environmental conflict research – the conservation of the environment and the transition to sustainable development on the one side, or the mitigation of violence on the other. It would seem probable that the field is concerned with both. However, this answer immediately presents new questions: Is it expedient to focus on the environmental problematique when examining the mitigation of violence in social conflict? Would not other focuses be more fruitful? Conversely, does it make sense to take violent conflict as a starting point when debating the greening of the economy (see Levy 1995a and b)? And further: Do environmental protection and violence mitigation go hand in hand at all? Can there not also be at least short-term contradictions between the two?

A large part of the relevant research links environment and conflict under the heading of 'environmental security'. This term is now in widespread use, which has not necessarily made its meaning clearer.[1] The situation is now comparable to that in Douglas Adams' *The Hitchhiker's Guide to the Galaxy*. In Adams' novel, a supercomputer, after a computation process of cosmic proportions, gives the now much-cited answer "forty-two". The answer remained incomprehensible because the question was partly forgotten, partly formulated too carelessly. 'Environmental security' is a similar case. It is an answer in the quest for its underlying question.

This quest can be concerned with national security at one end of the spectrum, just as with securing the basic needs of all people at the other end. It can be concerned with expanding notions of threat that call for new military precautions, just as it can be concerned with shifting notions of threat to the non-military domain. It can focus on politically upgrading the environmental problematique just as on narrowing it down to a security problem. In the research, all of these aspects are addressed in one or the other form.

The term 'environmental security' emerged in the context of the 'New Thinking' that accompanied the transformation and finally the dissolution of the East-West conflict in the second half of the 1980s, and that ultimately boiled down to giving greater political weight to the 'global problems' that had already been debated for some time in the Western literature. In the language of games theory: *Relative* gain was to be transposed into *absolute* gain. On the same understanding, the Palme Commission had proposed a transition from particularistic to "common security". In the same period, other sides argued for extending the concept of security itself (Ullman 1983, leading on from Brown 1977). The efforts following this plea took

---

1 Deudney 1990 and 1991; Brock 1992; Levy 1995a and 1995b; Dabelko and Dabelko 1995; Homer-Dixon et al. 1996; Lipschutz 1996; Graeger 1996; Carius et al. 1997; Brock 1997.

the form of a veritable movement in the Western social sciences in the second half of the 1980s and the first half of the 1990s.[2] The notion of "common security" was thus joined by "comprehensive security" (Westing 1989), then "global human security" (Kaul and Savio 1993) and finally Norman Myers' "ultimate security" (Myers 1993).

Environmental aspects played an important role in this evolution, and it was a concern of peace research to give environmental issues a higher rating on the scale of political priorities by means of connecting them to security issues. This was based on the assumption that by defining issues as security problems they would be elevated above the day-to-day wrangling of party politics.[3]

The objection was soon raised against this opportunistic linking of the two areas of concern that defining environmental issues as a security problem ignores the fundamental disparity between the two realms and must rather be expected to lead to a militarization of environmental policy, not to a demilitarization of security policy (Deudney 1990 and 1991; see Brock 1992 and 1994). The proponents of a 'securitization' of environmental policy respond that framing the environmental problematique as a security problem also transforms the common understanding of security itself, thus transcending its fixation upon military questions by stressing dangers to human life and welfare resulting from environmental degradation (and not from military threats) (Westing 1989; Mathews 1989; Meyers 1993). However, this response harbors a dilemma. Environmental concerns can only be upgraded by defining them in categories of security as long as the understanding of security retains some of the central characteristics that have made its pursuit such a high-ranking good. These characteristics of the 'high-ranking' concept of security include its focus on *national* concerns, its fixation upon maintaining the status quo and its orientation to *relative* as opposed to *absolute* gains. These characteristics are all diametrically opposed to thinking in categories of sustainable development. If we really succeed in making environmental concerns a subject of national security policy, then we risk military security considerations gaining considerable weight in the definition of environmental policy concerns, and a part of the resources required for environmental policy might remain under military control or even be increasingly channeled to the military sector. This would lead not to 'civilianizing' military resources to the benefit of environmental policy, but rather

---

2 Westing 1989; Mathews 1989; Mische 1989; Kaul and Savio 1993; Baechler et al. 1992; Myers 1993; Buzan et al. 1995.

3 In recent publications, the Copenhagen Center for Peace and Conflict Research has attempted to show that security issues follow a separate logic of their own. Ole Waever writes on this:

> Security is about survival, it is when an issue is phrased as existential, and it is on this basis argued that 'we' should use extraordinary measures to handle it. By saying 'security' a state-representative moves the particular case into a specific area, claiming a special right to use the means necessary to block this development. 'Security' is the move that takes politics beyond the established rules of the game, and frames the issue within a special kind of politics. (Waever 1994: 6)

to partially subsuming environmental policy under military security thinking. On the other hand, if we create a non-military concept of security and demilitarize and denationalize security policy, the intended political upgrading of environmental concerns is lost.

This dilemma can scarcely be circumvented by extending the content of the concept of *environmental security* so as to transcend the concept of security, such as e.g. Arthur Westing has proposed. According to Westing, environmental security would embrace, in addition to an intact environment, also political participation, good governance, the rule of law, the recognition of political, economic, social and cultural human rights and economic security (Westing 1989; Myers 1993). Such a broad understanding of *environmental security* would render the concept too unspecific. As an analytical instrument, it would become largely useless.

It would however be quite reasonable to speak of environmental security as the need of each and every individual to secure his or her natural bases of existence, in analogy to *social security* or to the concept of *food security* which is increasingly being used. Here, i.e. at the level of low politics, the above danger of an inappropriate 'militarization' of environmental issues would arise just as little as the danger of a 'militarization' of *social security*. However, *environmental security*, thus defined, would be much more difficult to operationalize than *social* or *food security*. Moreover, as we have seen in the human rights debate, the danger of an inflation of terminology would remain. This would be detrimental to the issue itself. Just as little as the actual recognition of human rights is helped if more and more aspects of social relations are reformulated as legal entitlements, as little does it help the enforcement of the legitimate welfare interests of the individual if all conceivable everyday questions are declared to be security concerns.

A further option would be to use the composite term *environmental security* as an indication of the necessity to observe certain ecological thresholds when pursuing economic activities. Thus Katrina Rogers (1997) has recently suggested understanding *ecological security* as a state in which the physical environment of a community satisfies its needs without depleting natural capital.[4]

Detlef Sprinz (1997) defines environmental security in a similar manner as the protection of the welfare gains that proceed from a certain quality of the environment. In this understanding, the environment would be secure as long as environmental pollution (at the national and international levels) remains within the "maximum assimilative capacity" of a country. International environmental security could be defined similarly in this perspective. However, it would be recommendable to speak of a *secure environment* instead of environmental

4 In this understanding, the term *environmental security* should be reserved for national security policy (Rogers 1997). See the analysis of Eberwein (1997) on this, who expressly affirms the proposition that environmental degradation can become a national security problem.

security, in order to underline that this concept is focused on a certain environmental quality and not on the management of environmentally induced conflicts.

Nonetheless, the question remains whether the concept of security, with its inherent status quo orientation, may not be fundamentally unsuited to expressing environmental concerns. The environment is in a process of constant change even without human intervention: "everything flows" (Heraclitus). Ultimately, the concern cannot be to preserve a certain environmental status, but rather to reflexively handle environmental change, meaning to adapt to this change such that alterations are not accelerated by human behavior to a pace that renders adaptation impossible. This means that even the concept of a 'secure environment' should not be used unreservedly, but more as a working title, as, as said above, environmental problems ultimately cannot be modeled as security problems.

3 Environmental conflict research

The critique of an opportunistic linking of environment and security under the heading of 'environmental security' is by no means a criticism of the assumption that environmental degradation could and must become a subject of conflict research. Quite on the contrary: the intention of this critique is to contribute to removing conceptual barriers on the path towards an appropriate appraisal of environmental conflict research.

The pivotal question of empirical environmental conflict research to date is that of which role an anthropogenically induced or accelerated degradation of the natural bases of human existence plays in acute societal conflicts that are accompanied by violence. The findings to date can be summarized as follows:[5]

Firstly: Environmental degradation and resultant ecological scarcity is very rarely, if ever, a sufficient condition and indirect source of collective violence. Ecological scarcity interacts with a broad array of other political, economic, social and cultural factors. To talk of "environmentally conditioned" and "environmentally caused" conflicts[6] thus highlights *one aspect* of a critical social constellation, but does not yet explain the outbreak of violence in this constellation. Strictly speaking, this means that it would even be wrong to speak at all of environmentally conditioned conflicts, not to mention environmentally caused conflicts. If we do so in the following, using the even more pointed term 'environmental conflicts', then this is with all due reservation and with reference to the multi-faceted nature of conflictual relations. We then use the term 'environmental

---

5 The most comprehensive empirical work has been presented by the Toronto group around Homer-Dixon (1994) and the Swiss ENCOP group (Baechler et al. 1996; Baechler and Spillmann 1996). An instructive collection of studies on individual conflicts is given in Gleditsch 1997. See also Loodgard and Hort af Ornäs 1992.

6 Both adjectives are used by Günther Baechler in this volume.

conflicts' as a cipher for a problematique that, while delineated as a field of research by the use of the term, has not been conclusively conceptualized by this usage.

Secondly: The probability of militant conflict grows in step with the degree in which general environmental scarcity is not only experienced by the persons affected as a problem in itself (environmental marginalization), but as a consequence of illegitimate appropriation (Homer-Dixon 1994) and discrimination (Baechler in this volume) by identifiable actors, culturally embedded mediation techniques (inter alia precisely because of the environmental scarcity and its consequential socio-economic and cultural impacts) no longer function and alternative institutions and techniques of conflict resolution through legal-administrative channels or by way of negotiation are not available or do not function. Insofar, the use of violence in connection with environmental scarcity is a phenomenon of developing and transition societies.

Thirdly: The use of violence in the context of environmental degradation and of the environmental scarcity caused by it is more probable at the intra-societal level than at the international level. There are numerous transboundary, indeed global environmental scarcities. However, the danger of inter-state wars, not to mention inter-regional conflicts between North and South, for reasons of environmental scarcity is slight.[7] On the contrary – the pressure to arrive at cooperative or negotiated solutions appears to be growing in step with the increasing internationalization of scarcity problems.

Marc Levy (1995a) has accused empirical environmental conflict research – and in particular the Homer-Dixon group in Toronto – despite the considerable effort applied and the far-reaching aspirations of this research, of having produced little more knowledge about 'environmental conflicts' than was already conventional wisdom before this research commenced. His criticism assumes that environmental conflict research has proceeded from a null hypothesis, then to deliver the proof that the environment does play a role after all in militant conflicts. As no one has ever assumed that anything else would be the case, the findings of environmental conflict research are, thus Levy, trivial. Here it must be held against Levy that the development was rather the opposite. The research departed from wide-ranging assumptions on future environmental wars, which have been repeatedly expressed since the oil crisis of 1973 and the military contingency planning of the USA drawn up at the time for the event of a longer-term blockade of oil supplies. The violent conflicts that have broken out after the demise of the East-West confrontation have further fuelled such speculation.[8] As noted at the beginning of this essay, such assumptions were taken up by environmental conflict

7 See already Lipschutz and Holdren 1990, who point out the ideological content of using resource issues to explain international conflicts.

8 See the assumption of Senghaas and Zürn, formulated as a question: "Are poverty and ecology wars looming?" (Senghaas, Dieter and Zürn, Michael 1992: 388).

research – not least in the expectation to be able to show them to be justified. It was precisely this expectation and its dramatization by Robert D. Kaplan which (temporarily) gripped the attention of the Clinton administration (Kaplan 1994). If now environmental conflict research arrives at the result that the links between environmental degradation and conflicts involving a substantial probability of violence are very much more complicated and are linked with each other via much longer causal chains than was presumed in the original, alarmist view of the problem, then this is not a trivial finding (see on this controversy Homer-Dixon et al. 1996).

We might however accuse environmental conflict research of itself continuing to convey or arouse unfounded presumptions of violence where it only addresses violent conflicts or stresses particularly the danger of war resulting from environmental degradation.[9] We must therefore agree with Levy when he calls for a comparative research aimed at identifying the circumstances under which environmental degradation and ecological scarcity lead to violence and the circumstances under which they lead to cooperation. It has already been set out at length elsewhere (Brock 1992) that reactions to environmental problems can encompass a broad range of behavior, and that peace research must concern itself with both confrontation and cooperation in the environmental sector. Gleditsch (1997) underscores this aspect, pointing out that environmental problems may perhaps even be more amenable to cooperative solutions than other scarcity problems (Gleditsch 1997). Here, however, environmental conflict research teaches us that the contexts of environmental conflicts must not be neglected. It is to be expected that the probability of a cooperative solution will differ depending upon the state of development and the political infrastructure and political culture of the countries affected, although without being strictly determined by these.

The syndrome analysis approach explained by Frank Biermann in this volume provides a tool for the comparative analysis demanded above. The tool has been developed by the German Advisory Council on Global Change (WBGU). The approach can be used to identify, world-wide, prevailing critical constellations of interests and to distinguish these according to certain characteristics (forms of combination of natural-geographical and socio-economic circumstances) (e.g. the "Sahel Syndrome", "Dust Bowl Syndrome" or "Aral Sea Syndrome"). A systematic analysis of the probability of cooperation or refusal to cooperate under the conditions of each specific environmental syndrome has yet to be carried out. It remains to be seen what this typology of environmental problems ("clinical profiles") may yield in this respect.

Christoph Rohloff also pleads for systematic, comparative research in his contribution to the present volume. He takes the view that it is only expedient to a

---

9 See the title of the ENCOP project report "Environmental Degradation as a Cause of War" (Baechler et al 1996, Baechler and Spillmann 1996). However, the ENCOP group is not only concerned with analyzing ecologically related conflicts, but also and above all with identifying ways for their peaceful resolution and with contributing through research (as action research) directly to such peaceful resolution.

limited extent to treat environmental conflict as a class of conflict in its own right. Rohloff's text therefore concentrates on comparing general conflict typologies, whereby the specific connection to environmental concerns becomes tenuous. The question thus once again arises of what is specific about environmental conflict research or about the conflicts that are termed here as "environmentally caused", "ecological" or "environmental".

Concerning the above, I offer the following hypotheses.

1. A clear-cut demarcation of the field of environmental conflict research against other areas of conflict research is not possible.

It follows from the findings of empirical research concerning the complexity of factors impacting upon specific conflict scenarios reported above that environmental conflict research refers to a certain focus of conflict analysis, but not a separate field of conflict research. Whether a certain conflict is viewed as migration-related or as environmentally induced does not depend so much upon the matter itself as rather upon the research interest guiding the analysis. The typologies of environmental conflict offered in the literature could accordingly also be presented as typologies of development conflict or of modernization conflict.

2. The difficulty experienced in demarcating the field of environmental conflict research is an expression of the circumstance that ecological scarcity is societally mediated.

Environmental conflict research is concerned with those cases of conflict and of readiness to exercise violence that result from a deterioration of the natural basis of existence of human societies (i.e. the degradation of ecosystems). But 'the natural bases of human existence' does not imply that nature is given to human society as an unalterable datum. It is rather the product of a more or less lengthy process of political, socio-economic and cultural development, in the course of which the original natural physical-geographical conditions and their perception have undergone thousand-fold change. This means that environmental scarcity is not an 'objective' determinant of societal development, but is a part of this development. It does albeit make a difference whether one lives in the Sahel zone or in the US-American cornbelt. However, this difference lies not only in the natural physical-geographical conditions, but also in the societal forms of existence. In other words, we do not live in nature, but in specific societal relationships to nature (Jahn 1990).

Seen thus, environmental conflicts are the outcome of a differentiation of the leeway available for adaptation to environmental change. This differentiation is societally mediated (and also by international economic interpenetration). Such conflicts usually take the shape of a struggle over the distribution of specific resources ('resource conflict'). Resource conflicts can in turn lead to environmental conflicts, if the weaker parties see themselves forced to compensate for their lack of individual resources (water, soil) through intensifying environmentally damaging management practices (e.g. cultivating slopes prone to erosion).

3. 'Internally' caused environmental degradation leads to diffuse, 'externally' caused environmental degradation leads to relatively clear-cut conflictual constellations, whereby a distinction needs to be made between poverty-related and commercially generated exogenous environmental scarcity.

The problems at the heart of environmental conflict research could thus be viewed as the degeneration of ecosystems caused by the inhabitants of an ecoregion, i.e. *internally,* and the resultant *endogenous* ecological scarcity. The internal degeneration of ecosystems is a complex process. It runs its course as a lengthy process of interaction between a great number of determinants of societal development. While not all inhabitants of an ecoregion contribute equally to its degeneration, it can be presumed that, through the predominant ways of life and styles of conducting economic affairs, all inhabitants do ultimately participate in the transformation of the environment. Actors and responsible parties can thus not be clearly localized. As every individual tends to be a 'perpetrator' and 'victim' at the same time, diffuse conflictual positions arise in which the probability of a *sudden* escalation of conflict through to the use of violence is small. The situation is different where interventions in local resource use are external and clearly identifiable, such as through dam building, mining, commercial logging of forests or the displacement of small farmers by large landowners. As here the responsible parties are clear (or are perceived as being so), there is a greater danger of violence in such conflictual constellations than there is in use conflicts arising from endogenous scarcity. The violent resource use conflicts in the Brazilian Amazon are an example of this. Here the subsistence possibilities of small farmers are attenuated by external users (expanding large landowners, mining). However, the small farmers, for their part, are in conflict as external users with the indigenous peoples of the forests (and the rubber tappers). It is thus expedient to distinguish between poverty-related and commercially generated exogenous environmental scarcity.

With reference to the discrimination hypothesis set out by Baechler (see above), *proposition 3* can be summarized by stating that exogenous environmental scarcity is more likely to be experienced as discrimination than endogenous environmental scarcity, and that it thus poses the greater risk of escalation.

4. The 'higher' the level of analysis – from the local to the global – the more blurred the boundary becomes between internally and externally caused environmental degradation. This is a further reason for the low probability of 'direct' eco-wars (i.e. resulting directly from environmental problems) between states or even between global regions (North and South).

It is well known that the industrialized countries, with twenty percent of the world's population, consume eighty percent of the world's natural resources or contribute on this scale to the burdens placed upon renewable resources. Accordingly, the industrialized countries accepted in Rio in 1992 that they bear a particular responsibility for the global environmental problematique. So to speak in return for this, the developing countries have declared their willingness to view global environmental degradation not only as an external problem for which the industrialized countries are responsible, but also as an internal problem that concerns their own behavior. This is of course also an expression of the circumstance that the developing countries, being those most affected, must do something to contain global environmental degradation no matter who is responsible for it. But a further aspect is that the developing countries are themselves – in an increasingly differentiated way – involved in global degradation, and that their share in this is rising, driven by their own efforts. The further this process evolves, the more differentiated the specific interests of the various countries become, so that confusing lines of confrontation and cooperation emerge. This is one reason why the USA are not simply accused by the rest of the international community of ecological aggression, although they account for by far the greatest share of the emissions of climate-relevant pollutants.

4 From environmental conflicts to conflicts over the environment as such

In a historical perspective, the reference to the 'rising' level of analysis – from the local to the global – can also be understood as a gradual internalization of the environmental problematique. This would further mean that environmental conflicts are increasingly becoming conflicts over the environment as such – that is, conflicts over the question of how to handle the environmental changes that are under way. May we phrase the issue in these terms, and would this mean that, with the advance of global modernization, environmental conflict research of the classic mold will lose its object?

Before plunging into this question, we must first step back and reflect on what is specific about environmental conflicts if we proceed from the environmental issue rather than from the issue of violence. Antonio Hill (1997) recently made an interesting proposal on this, recurring to hierarchy theory (Giampietro 1994) and human needs theory (Burton 1990). Hill underscores the distinction between environmental conflicts and (traditional) conflicts over natural resources (soil, water) as economic resources. In his understanding, a conflict is then an "environmental conflict" if at least one of the parties to conflict elevates environmental aspects above particularistic economic interests. The environmental aspects function here as a mechanism of cultural control through which economic interests are filtered in favor of hierarchically higher needs and the commitments resulting from these needs.

In the terminology suggested here, this type of conflict would be a "conflict over the environment as such" (German: Konflikt über die Umwelt). We can assume that environmental conflicts in the sense discussed above, namely as differences in position that are brought about by environmental degradation and harbor a probability of violence, have predominated historically as compared to conflicts over the protection of the environment. My hypothesis is now that, in the course of modernization, conflicts over the environment as such are becoming the predominant form. However, this is not a process rendered irreversible by the inherent dynamism of modernization. Indeed, it is threatened by the possibility of failure to the same extent that modernity itself is.

The transformation of 'environmental conflicts' into 'conflicts over the environment as such' is a part of the increasing reflexivity of human enterprise. This development can be outlined as follows (see Giampietro 1994; Hill 1997): In pre-industrial history, unintended ecological changes have filtered away those collective attitudes and value-orientations from the totality of possible attitudes and value-orientations of human societies that led to an unconsidered exhaustion of nature while failing to give due regard to ecological connections (learning from the plague and cholera). In the course of the industrial revolution, human societies have found themselves more and more in a position to disregard ecological connections or to manipulate these without punishment. The vehicle for this was technological progress, which has been misconstrued as an increasing domination of nature; for the outcome of technological development through to its present climax in genetic manipulation is not that the 'natural' limits of human satisfaction of needs have been overcome, but rather that the limits have been transformed while creating ever new needs. The result has been a destabilization of the global ecosystem at an increasingly higher level. Thus human society is now threatened by new unintended environmental changes. This will either lead to behavior being adapted (out of necessity) according to the historical pattern again, namely through famine, epidemics and war, or it will provide the impetus for adapting behavior directly to the functional requirements of the ecosystem, through a transition to globally viable development.

Expressed in an even more reduced shorthand: In the beginning, life was *in* nature with inherent reflexivity. Remnants of this life are still to be found to this day in some tropical rainforests, where the people continue to have the leeway to pursue their environmentally adapted ways of life and husbandry, and where the cultural filters that attenuate detrimental attitudes and value-orientations are still intact. In pre-modern times, life *in* nature was transformed into a life *with* nature, in which on the one hand the forces of nature were increasingly harnessed, but on the other hand corrections to human behavior were forced by nature 'brutally' and recurrently. Many regions in the poorer developing countries are in this 'epoch' today. Under the aegis of the industrial revolution emerged a life *against* nature based on the above-mentioned transformation of the natural limits to socio-economic development. In contrast to this, first elements of a conscious reflexivity of societal development have recently emerged in the industrialized countries in the form of the environmental movement and environmental policy. These elements are by now also making themselves felt in Third World countries. However, it is all too apparent that these are extremely precarious. This applies not only to the developing countries, but also to the industrialized countries. Under the influence of world market oriented deregulation, the priority given to the provisions of environmental (and social) policy is dwindling. A return to adaptation forced by (human-induced) natural disasters is thus by no means out of the question.

The following table structures the environmental problematique in a manner that attempts to illustrate, in a historical perspective, the logic of the transformation of 'environmental conflicts' to 'conflicts over the environment as such' (or over the protection of the environment) (Table 1).

The table distinguishes between four different types of economic activity, thus addressing the historic shift in main activities (in the most advanced states) from the primary to the secondary and tertiary sectors. This shift of course does not mean a process of historical succession. The extractive activities (mining) remain; in absolute terms, they even grow. Paradoxically, the absolute physical throughput of materials has (until now) risen in step with the relative predominance of the non-material activities in total value added. The shift of economic activities towards the tertiary sector entails *dematerialization* of economic activity, but only in *relative*, not in absolute terms.

Table 1: Environment and conflict in historical perspective

Forms of adjustment / *Economic activities*	**Relations with nature**	**Relations within society**	**Conflict resolution**	**Strategies for reducing the consumption of 'environmental space'**
Extractive	Exploitation of non-renewable resources; local degradation	Resource conflicts and consequential environmental conflicts	Regulation of access and use	Miniaturization; habitat restoration
Agrarian	Cultivation of natural resources; geographically idespread degradation	Environmental and consequential resource conflicts	Regulation of access and use; redistribution of land	Productivity improvement; organic farming
Industrial	Transformation of natural resources; degradation across all dimensions	Environmental conflicts and emergent conflicts over the protection of the environment	Internalization and redistribution of environmental costs	Increasing eco-efficiency
Post-industrial	Relative dematerialization; artificial life worlds	Conflicts over the protection of the environment and a drop in environmental conflicts	Redistribution of environmental space	Complex decentralization; subsidiarity

Local environmental degradation is the typical ecological effect of the exploitation of non-renewable resources – and not only in the shape of deep holes, but also in the form of the contamination of the soil, air and water (for example: mining on Bougainville). The continuous increase in technical inputs to agriculture (machines, fertilizers, pesticides) means that the cultivation of natural resources goes hand in hand with their widespread ecological impoverishment (above all in the extreme form of large-scale monocultures). The industrial transformation of natural resources is the complement to this development, introducing an extensive and fundamentally global degradation of environmental space. The latter is increasingly substituted by artificial life worlds, which no longer order and structure nature (as gardens and parks previously did), but substitute nature, serving as fun landscapes that deliver higher economic yields than 'natural nature'.

Resource conflicts (as conflicts over discrete partitionable goods) are the predominant form of conflict in the realm of extractive economic activities. As explained above, environmental conflicts (as conflicts resulting from the degradation of ecosystems) emerge as a consequence of the scarcity of individual resources (environmentally damaging compensation). In the agrarian realm, environmental conflicts (as a consequence of widespread degradation) are the predominant type – however, most of these manifest themselves as resource conflicts (over land and water distribution). With advancing industrialization, resource conflicts diminish to the extent that resource availability is expanded through substitution and through exploiting marginal deposits. At the same time, *environmental conflicts* are increasingly joined by *conflicts over the environment as such* as first elements of a reflexive modernization. In a process connected to the relative dematerialization of economic activities (and the commensurate rise in importance of information and communication) *conflicts over the environment as such* become the predominant type of conflict.

At first, approaches towards resolving these conflicts revolve around the agreement of access and use rules. In the course of modernization, these are joined by redistributional activities: the redistribution of soil (agrarian reform), the redistribution of the consequential costs of environmental degradation (concretized under the international ozone regime and in the differentiation of the requirements placed upon the actions of industrialized and developing countries) and finally the redistribution of environmental space as a step towards globally viable development (concretized conceptually until now as joint implementation or as tradable emission entitlements).[10] Redistribution of the consequential costs of environmental degradation and of environmental space implies an at least partial internalization of environmental costs.

Regulatory and redistributional mechanisms do not, however, solve the underlying problem of growing environmental scarcity. Accordingly, approaches towards resolving conflicts are linked to activities aimed at reducing the consumption of environmental space. Principal strategies include in the extractive

---

10 For a critique see Altvater 1996, referring to BUND 1996.

realm the reduction of resource inputs and the reclamation or restoration of habitat, in the agrarian realm the increased use of organic farming methods, in the industrial realm an efficiency revolution and finally, in the process towards a virtual economy, the decentralization of economic activities in conjunction with the implementation of the subsidiarity principle in the structuring of these activities. A complex, networked decentralization of economic activities could deliver enormous savings in the environmental impacts caused by transportation. First moves in this direction are becoming apparent in the decentralization of jobs in the 'virtual factory'.

However, as yet, all of these activities have at best succeeded in yielding a relative reduction in environmental scarcity (e.g. through decoupling energy consumption from overall economic growth). This is not only due to the timeless discrepancy between aspirations and action. The wrangling that surrounded the adoption of emissions reduction commitments at the follow-up conferences to the 1992 Climate Convention demonstrates that status quo thinking based on prevailing power disparities sets narrow limits to a reflexive approach to natural resources.

5 Outlook

The hope placed in learning from history is underpinned by the circumstance that, today, improved efficiency can yield more resources than conquest can, and that every relative gain extracted at the cost of the environment is associated with an absolute loss in development options in the global economy which at some point will fall back on the winner. This leads to a logic in which cooperation ultimately yields more than power-based interest politics does.

Nonetheless, we know that the general gain offered by cooperation does not nullify the striving for particularistic advantage as long as the individual actors can hope (due to the prevailing power relations and development disparities) that they will gain more in a concrete case through refusing cooperation than through cooperating, and as long as the internalization of the environmental problematique only gives rise to diffuse conflict constellations (with numerous, overlapping lines of conflict). It follows that the interest-based approach (Sprinz and Vaahtoranta 1994) will not suffice to preserve the natural basis of existence for all. This is where the question of value hierarchies comes into play. It is a matter of establishing a scale of values on which the pursuit of economic interests is linked to the condition that the adaptability of the whole of humanity to environmental change is guaranteed or restored.

This is also because in addition to the two scenarios set out above (learning through disasters according to the historical pattern, learning *from* history while avoiding disasters) a third scenario is also conceivable – the increasing polarization of humanity between a part that shields itself against disastrous developments in artificial life worlds and indeed profits from bringing about such developments, and another part that is cast out into the wilderness. It is the danger of such a polarization of general living conditions that makes it necessary to think about establishing a hierarchy of the values represented in our societies – a hierarchy that upgrades the concerns of environmental and social policy vis-à-vis economic interests. However, such an endeavor harbors a further danger, namely that referring to the necessity of a hierarchy of values may tend to make societies forget that this is a pragmatic necessity (to secure the survival of humanity). This could lead to a new, environment-based dogmatism that no longer permits the continuous adaptation of the corrective mechanism itself. Accordingly, thinking about value hierarchies can only be fruitful in an institutionalized process of dialogue at both the intra- and inter-society levels (Payne 1996). First steps towards such a dialogue are to be found in those organizations, initiatives and movements that have been interpreted in recent years as an indication of the emergence of a transnational civil society. The euphoria of this interpretation has in the meantime been dampened, which, as such, is to be welcomed. However, there is no cause to let disillusionment swing into resignation. The historical perspective on the environmental problematique proposed here, oriented to changing economic activities, removes the ancillary character of the debate on sustainability with NGO participation (BUND/Misereor 1996) that has emerged in recent years. This perspective underscores the crucial importance of this debate for constructively resolving environmental conflicts. It furthermore highlights the necessity of moving environmental conflict research onward from its restriction to violent conflicts in the Third World and giving it a stronger focus on global conflicts over environmental scarcity and ways to address these.

References

Altvater, Elmar 1996: "Der Traum vom Umweltraum", in: *Blätter für deutsche und internationale Politik*, Issue 1, 82–91.

Baechler, Günther (Ed.) 1992: *Perspektiven der Friedens- und Konfliktforschung in Zeiten des Umbruchs*. Zürich: Rüegger.

Baechler, Günther et al. 1992: *Gewaltkonflikte, Sicherheitspolitik und Kooperation vor dem Hintergrund der weltweiten Umweltzerstörung*. Bonn: Projektstelle UNCED des Deutschen Naturschutzringes, Bund für Umwelt und Naturschutz Deutschland.

Baechler, Günther et al. 1996: *Kriegsursache Umweltzerstörung. Ökologische Konflikte in der Dritten Welt und Wege ihrer friedlichen Bearbeitung* [Ecological Conflicts in the Third World and Ways for their Resolution]. Vol. I. Zürich: Rüegger.

Baechler, Günther and Kurt R. Spillmann (Eds.) 1996: *Environmental Degradation as a Cause of War*, Vols. II and III. Zürich: Rüegger.

Brock, Lothar 1992: "Peace Through Parks? The Environment on the Peace Research Agenda", in: *Journal of Peace Research*, 28/40, 407–423.

Brock, Lothar 1994: "Ökologische Sicherheit. Zur Problematik einer naheliegenden Verknüpfung", in: Wolfgang Hein (Ed.): *Umbrüche in der Weltpolitik*, 443–458. Hamburg: Deutsches Übersee-institut.

Brock, Lothar 1997: "The Environment and Security: Conceptual and Theoretical Issues", in: Nils Petter Gleditsch (Ed.): *Conflict and the Environment*, 17–34. Dordrecht: Kluwer.

Brown, Lester 1977: *Redefining Security*. Worldwatch Paper 14. Washington D.C.: Worldwatch Institute.

BUND, Misereor (Eds.) 1996: *Zukunftsfähiges Deutschland – Ein Beitrag zu einer global nachhaltigen Entwicklung*. Studie des Wuppertal-Instituts für Klima, Umwelt, Energie. Basel: Birkhäuser.

Burton, John W. (Ed.) 1990: *Conflict: Human Needs Theory*. London: MacMillan.

Buzan, Barry, Ole Waever and Jaap de Wilde 1995: *Environmental, Economic and Societal Security*. Working Papers 10. Center for Peace and Conflict Research.

Carius, Alexander et al. 1997: "NATO Pilot Study: Environment and Security in an International Context." *Environmental Change and Security Project Report*, Issue 3, 55–65. Washington, DC: The Woodrow Wilson Center.

Dabelko, Geoffrey and David D. Dabelko 1995: "Environmental Security: Issues of Conflict and Redefinition", in: *Environmental Change and Security Project Report*, Issue 1, 3–13. Washington, DC: The Woodrow Wilson Center.

Deudney, Daniel 1990: "The Case Against Linking Environmental Degradation and National Security", in: *Millennium*, 19/3, 461–476.

Deudney, Daniel 1991: "Environment and Security: Muddled Thinking", in: *Bulletin of the Atomic Scientist*, 47/3.

Eberwein, Wolf-Dieter 1997: *Umwelt – Sicherheit – Konflikt. Eine theoretische Analyse*. Papers P 97 - 303. Berlin: Wissenschaftszentrum Berlin für Sozialforschung (WZB).

Giampietro, Mario 1994: "Using Hierarchy Theory to Explore the Concept of Sustainable Development", in: *Futures*, 26/6, 616–625.

Gleditsch, Nils Petter 1997: "Environmental Conflict and the Democratic Peace", in: Nils Petter Gleditsch (Ed.): *Conflict and the Environment*, 91–106. Dordrecht: Kluwer.

Gleditsch, Nils Petter (Ed.) 1997: *Conflict and the Environment*. Dordrecht: Kluwer.

Gleick, Nils P.1993: "Water and Conflict. Fresh Water Resources and International Security", in: *International Security*, 18/1, 79–112.

Graeger, Nina 1996: "Environmental Security?", in: *Journal of Peace Research*, 33/1, 109–116.

Hill, Antonio 1997: "Environmental Conflict: A Values-oriented Approach", in: Nils Petter Gleditsch (Ed.): *Conflict and the Environment*, 51–70. Dordrecht: Kluwer.

Homer-Dixon, Thomas, Marc Levy, Gareth Porter, and Jack Goldstone 1996: "Environmental Scarcity and Violent Conflict: A Debate", in: *Environmental Change and Security Project Report*. Issue 2, 49–71. Washington, DC: The Woodrow Wilson Center.

Jahn, Thomas 1990: "Das Problemverständnis sozialökologischer Forschung. Umrisse einer kritischen Theorie gesellschaftlicher Natur-Verhältnisse", in: Egon Becker (Ed.): *Jahrbuch für sozialökologische Forschung*. Frankfurt.

Kaplan, Robert 1994: "The Coming Anarchy", in: *The Atlantic Monthly*, Issue 2, 44–76.

Kaul, Inge and Roberto Savio 1993: *Global Human Security. A New Political Framework for North-South Relations*. Rome: UNDP

Levy, Marc A. 1995a: "Time for a Third Wave of Environment and Security Scholarship?", in: *Environmental Change and Security Project Report*, Issue 1, 44–46. Washington, DC: The Woodrow Wilson Center.

Levy, Marc A. 1995b: "Is the Environment a National Security Issue?", in: *International Security*, 20/2, 35–62.

Lipschutz, Ronnie and John P. Holdren 1990: "Crossing Borders: Resource Flows, the Global Environment, and International Security", in: *Bulletin of Peace Proposals*, 21/2, 119–22.

Lipschutz, Ronnie D. 1996: *Environmental Security and Environmental Determinism: the Relative Importance of Social and Natural Factors*. Draft Paper prepared for the "NATO Advanced Research Workshop on Conflict and the Environment", Bolkesjo, 12–16 June 1996.

Loodgard, Sverre 1990: *Environmental Conflict Resolution*. Paper presented to the UNEP meeting on 'Environmental Conflict Resolution'. Nairobi, 30 March 1990.

Loodgard, Sverre and Anders Hjort af Ornäs 1992 (Eds.): *The Environment and International Security*. PRIO-Report Issue 3. Oslo: Peace Research Institute Oslo.

Matthew, Richard A. 1995: "Environmental Security: Demystifying the Concept, Clarifying the Stakes", in: *Environmental Change and Security Project Report*. Issue 1, 14–23. Washington, DC: The Woodrow Wilson Center.

Mathews, Jessica Tuchman 1989: "Redefining Security", in: *Foreign Affairs 68*, 162–177.

Moran, Alan, Andrew Chrisholm and Michael Porter (Eds.) 1991: *Markets, Resources and the Environment*. Sydney: Alan & Unwin.

Myers, Norman 1993: *Ultimate Security: The Environmental Basis of Political Stability*. New York: W.W. Norton

Payne, Rodger A. 1996: "Deliberating Global Environmental Politics", in: *Journal of Peace Research*, 33/2, 129–136.

Rogers, Katrina S. 1997: "Pre-empting Violent Conflict: Learning from Environmental Cooperation", in: Nils Petter Gleditsch (Ed.): *Conflict and the Environment*, 503–518. Dordrecht: Kluwer.

Senghaas, Dieter and Michael Zürn 1992: "Kernfragen der Friedensforschung der Neunziger Jahre", in: Baechler, Günther (Ed.) 1992: *Perspektiven der Friedens- und Konfliktforschung in Zeiten des Umbruchs*, 379–389. Zürich: Rüegger.

Sprinz, Detlef F. 1997: "Environmental Security and Instrument Choice", in: Nils Petter Gleditsch (Ed.): *Conflict and the Environment*, 483–502. Dordrecht: Kluwer.

Sprinz, Detlef and Tapani Vaahtoranta 1994: "The Interest-Based Explanation of International Environmental Policy", in: *International Organization*, 48/1, 77–106.

Tennberg, Monica 1995: "Risky Business: Defining the Concept of Environmental Security", in: *Cooperation and Conflict*, 30/3, 239–258.

Ullman, Richard 1983: "Redefining Security", in: *International Security*, 8/1, 129–153.

Westing, Arthur 1989: "The Environmental Component of Comprehensive Security", in: *Bulletin of Peace Proposals*, 20/2, 129–134.

Waever, Ole 1994: *Insecurity and Identity Unlimited*. Working paper 14. Copenhagen: Center for Peace and Conflict Research.

The Role of Environmental Policy in Peace and Conflict Research

Michael Windfuhr

1 Introduction

In recent years, the issue of 'environment and security' has gained remarkable prominence. Since 1989, a great amount of essays and studies have been published, illuminating the topic from the most varied aspects. These have ranged from more conceptual discussions of the link between environmental degradation and security (see e.g. Brock 1991 and 1994; Eberwein 1997; Wöhlke 1997) to the findings of more comprehensive research projects (see e.g. Homer-Dixon 1994, Baechler et al. 1996) involving extensive case studies. Rarely do any of these authors doubt that there is a link between environmental degradation and security. However, how this link is to be characterized and where a substantive connection is seen depends upon many different prior assumptions, e.g. upon the questions of what is understood by the term 'security', which types of environmental degradation are taken into consideration and so forth. Due to the great range of meanings attached to it, the term 'environmental security' used in the debate is hard to operationalize. Some authors define this very broadly as the absence of serious ecological problems, i.e. as the realization of 'sustainable development'[1], while, conversely, other authors define it very narrowly as the protection of the military against environmental threats.

Some have explained the remarkable popularity enjoyed by the topic – particularly in recent years – by pointing to the motivation behind this research, claiming, for instance, that the military is looking for new tasks after the end of the East-West conflict. Much the same could be said of peace and conflict researchers. Similarly, the environmentalists who have tackled this topic have been accused of using it to upgrade their own 'low politics' issue by construing a security nexus. This dispute over motivation, as interesting as it is for explaining the genesis of the

---

1 Understood here in the sense of the conceptual tradition of "sustainable development" that has emerged in the follow-up process of the 1992 UNCED conference in Rio de Janeiro (see Harborth 1991).

field of study, shall not be a further topic of this essay, as it contributes little to clarifying the actual matter at hand.

At the substantive level, we shall therefore clarify which types of link between environmental degradation and security are to be expected at which points; for the existence of transboundary or global environmental problems is indubitable. At the same time, we know, and not only from environmental policy research, that the connections between the ecological and the social system are highly complex. These must nonetheless be identified precisely enough to distinguish between those possible environmental conflicts that can exert security relevance and those that cannot. The notion of conflict used here embraces all conflicts, i.e. also those that can be resolved by peaceful avenues. By contrast, the notion of security used here is limited to situations in which there is a possibility of groups or states using organized violence, at an intrastate or international level.

The neutral term 'link' is used intentionally, as a link can be both positive and negative. Without further concretization, it at first merely expresses interdependence between the two domains.[2] The literature formulates highly disparate types of links between environmental problems and security, with differing relevance to security studies or action scenarios. Depending upon the type of link to which an author refers, very different assessments of the conflictual nature of environmental problems emerge. For the analytical purposes of this essay, these types of link are first organized heuristically according to three classes. The aim of this approach is to explore whether it is possible to distinguish between acute, medium-term and long-term security threats emanating from ecological problematiques, and thus to offer a tool for assessing the potential of individual phenomena of environmental degradation to give rise to conflict. The proposed typology is examined in the course of the essay and modified or supplemented where appropriate.

A: A very direct link can be made between environment and security if conflicts over discrete environmental resources are assessed as directly promoting conflict or causing war, as some authors do for the case of water resources.[3] This perspective on the link shall be termed 'environmentally-induced conflict' (or environmentally-induced war) in the following. The condition for classifying a conflict in this category is the direct causal effect of an environmental problem upon the security situation within a state or in a cross-border context.

B: Most empirical studies have shown that environmental degradation is not generally a sole cause of war or conflict, but only causes or exacerbates conflicts in the interplay with other existing social and economic factors. This type of link is classified in the following under the heading of 'extended environmental security'.

---

2 It was Lothar Brock who first pointed out that international cooperation on environmental problems may also help to build confidence and trust in international relations (see Brock 1991: 408f.).

3 For one example among many, see Barandat 1997.

The analyses belonging to this type focus on the potential of environmental problems to promote or exacerbate conflicts.

C: The literature sees a third link between the general impacts of global environmental problems and possible future conflict scenarios (scarcities of raw materials, food, utilizable soil and water due to climatic changes etc.). In this perspective, every form of non-sustainable production and consumption essentially becomes a future security problem. This type of link shall be categorized under the heading of 'ecological interdependence'. This can be used to identify future problem horizons which, while presently generating no or only limited security concerns, are expediently examined from security aspects for early warning reasons.

The broader the link is formulated (C), the more speculative its security relevance becomes. The more strictly the link is defined, e.g. between specific forms of environmental degradation and the outbreak of armed conflict or war (A), the better the cause-effect relationships need to be defined in the argumentation, as an unequivocal causal relationship is then being sought. Research on the causes of war has shown that the identification of such cause-effect relationships is a difficult and complex task, as a large array of variables is involved in the process leading from a conflict to an armed struggle. No simple line can be drawn from the environmental problem to the armed conflict. This is why it is important for a comprehensive perspective on the environment and security nexus to keep all three types of causal explanation in view – from a very broad awareness of basic connections to a very tight causal explanation of violent conflicts.

In the following, I shall briefly summarize the findings to date of the research carried out in the various disciplines on the link between environmental degradation and security. First an overview is given of the literature that has tackled the issue of environment and security at the interface between peace and conflict research on the one hand and environmental policy research on the other. The purpose of this overview is to gain methodological indications for the identification of the link. Although various studies have delivered exceedingly valuable preliminary work, no convincing, hard and fast concept has yet been provided, both as regards the typology of environmental conflicts and as regards the appraisal of the propensity of conflicts to escalate. Nor can such a hard and fast concept be elaborated within the scope of the present essay. The aim here is merely to develop important building blocks for such a concept. After the review of the findings of peace and conflict research and of environmental policy research, the fourth section of this essay gives a résumé, proposing, on the basis of the literature presented, a typology of environmental problems that can lead to or can exacerbate conflicts. The final section gives methodological recommendations for the identification of links between environmental degradation and security, whereby an effort is made to relate these proposals to the research design that has been pursued until now by the NATO CCMS Pilot Study "Environment and

Security in an International Context" (Carius et al. 1997). At the end of each section, interim résumés are provided.

2 Environmental problems and conflicts as a theme of peace and conflict research

2.1 The expanding concept of security

Peace and conflict research has only recently turned to environmental problems as a cause of conflict. The discipline has traditionally concentrated on military clashes between states over issues such as political power and ideological value differences. Conventionally, international relations have been examined from the perspective of political or military power. The realization of the growing importance of economic interdependence in the global system brought about first changes in research focuses. North-South conflict, access to scarce resources etc. are catchwords in this debate. In the meantime, the concept of interdependence has itself been extended. At the latest since ecological issues are also debated in terms of global interdependence – in a quite classic mode as the rising sensitivity and vulnerability of states and individuals to the activities of other states or individuals[4] – it was a logical step to take up this theme in peace and conflict research, too. From the early and mid 1960s onwards, environmental problems were initially understood as local phenomena (in Germany an important catchphrase at the time was "Blue sky over the Ruhr river").[5] In many cases, tackling such local phenomena led directly to the need to tackle transboundary environmental threats, which were addressed more systematically for the first time at the 1972 UN Conference on the Human Environment in Stockholm.[6] Since the 1980s, the understanding of bi- or trilaterally attributable transboundary environmental problems has been further extended to embrace the dimension of global environmental threats, i.e. problems in whose causation all states play a part, irrespective of the question of who has to suffer most under the consequences. The realization of the dangers posed to the 'global commons' (e.g. climate or biological diversity) has given rise to terms such as the 'global community of fate'. In this understanding, every economic activity (of every individual) potentially has long-term environmental relevance. Peace and conflict research has accompanied this

---

4 Keohane and Nye 1977.

5 Von Weizsäcker 1989 gives an overview of the conceptualization of environmental problems.

6 It was no coincidence that this conference took place in Sweden. Sweden had made intensive efforts to have the conference convened, in view of numerous water-related problems in the country caused by the long-range transport of air-borne pollutants from Germany, Poland and elsewhere.

expansion of the concept of environmental threat by a parallel extension of the concept of security. The extended concept of security[7] often used today embraces economic and social elements such as poverty and injustice, in both the national and international context, and recently also aspects of environmental security. The United Nations have played an active role in this process of extending the concept of security. Both the Brandt Commission and the Palme Commission had already stressed the necessity to extend the concept of security to embrace further elements concerned with creating the fundamental conditions for peaceful relations between nations and the solution of non-military problems.[8] The Brundtland Commission then introduced the extension to the concept of environmental security in 1987[9] arguing in particular for a concept of comprehensive security. The United Nations Development Programme now even uses since 1990 the term "human security", which can ultimately be viewed as the realization of sustainable development for the individual.[10] Arthur H. Westing has been arguing for years for a very extensive concept of security – in his view, comprehensive security has environmental and political security strands, political security being a combination of military, economic and social security (Westing 1986: 183–200).

A concept of security that is extended in this fashion harbors the danger of muddled categories. If every state and non-state action, whether intentional or not, has the potential to be relevant to security interests, it becomes almost impossible to precisely identify when behavior crosses the threshold to organized violent action by collective actors (states and intrastate groups), or at which point this is to be expected. Daase therefore urgently warns against a conceptual understanding of ecological security in which "nature itself is threatened by humankind and its destruction in turn represents a threat to humankind". If the concept is used in such an extended manner that environmental degradation is understood in its totality as a security threat, then, thus Daase, responsibility for environmental degradation can no longer be assigned to anyone, as there are no longer any identifiable actors

7 Christopher Daase describes the development of the concept of security in three phases, (1) from the negative (security against certain threats) to the positive security of e.g. justice and peace), (2) from hazard containment to precautionary action, i.e. the extension of the concept of security by the precautionary aspect, with the consequence of decoupling threats and actors (not only intentional actions of an actor are relevant, but also every kind of action by other actors that restrict the quality of life in a state or the leeway for action of a state), and (3) from national security to global security, calling for a holistic security policy as a response to the global, transboundary threat, particularly in the environmental sector. Daase notes the risks attached to a concept of security extended in this manner: "It is precisely the holistic definition of the concept of security, which, while essentially correct, disguises the connection between personal responsibility and collective fate which is so central to environmental degradation; ..." (Daase 1992: 32f.).

8 Brandt Commission (Independent Commission on International Development Issues) 1980; Palme Commission (Independent Commission on Disarmament and Security Issues) 1982.

9 Brundtland Commission (World Commission on Environment and Development) 1987.

10 The concept has been expanded upon since 1990 in various issues of the *Human Development Report* (UNDP 1990).

(Daase 1992:37). Eberwein notes in a similar vein that the concept of security becomes analytically meaningless if all conceivable levels of security and affectedness (individuals, groups etc.) are included, as then "the complexity of the phenomenon would approach infinity" (Eberwein 1997b:4).

Interim résumé: If environmental threats are to be meaningfully operationalized for the identification of security problems, then a precise terminology is essential. A very broad concept of security that abandons a link to violent conflict resolution does not serve this purpose, as then all environmental changes would effectively have to be considered as relevant topics of research, even if they have scarcely any potential to develop security relevance.

2.2 Phases of 'environmental conflict' scholarship

The linkages among environmental degradation, security and war have already been a subject of debate for some time. The ENCOP research group dates the emergence of interest in this issue to 1972, when environmental problems generally became a subject of public and political debate in the wake of the Stockholm Conference on the Human Environment.[11] However, the initial interest in these linkages concentrated on environmental damage caused by war or by forms of warfare involving deliberate environmental damage; this interest was fueled in the USA above all by the environmental warfare in the Vietnam War. Debate on environmental issues within peace and conflict research has gone through three phases.

After the first phase, which focused on environmental warfare and the ecological consequences of wars, in the 1980s the environmental and peace movements spotlighted the environmental consequences of security policy in peacetime, too. This debate has been concerned with noise pollution by low-flying aircraft, environmental contamination at military sites and environmental pollution caused by the production of armaments (Krusewitz 1985). Examples of these two thematic linkages (phases I and II) have been empirically identified and described from the outset. Both forms of link between environmental degradation and security remain relevant topics of analysis to this day.

In the context of the present discussion, the third phase is of particular relevance – the perception in peace and conflict studies of environmental problems as a potential cause of conflict. The analysis of the issue of environmentally-induced violent conflict proved to be considerably more difficult than the other two phases, as there have been very few conflicts for which it can be proven without doubt that environmental degradation was the most important or exclusive cause of conflict. This has created considerable difficulties in theory formation, all the more so in a field of research – the study of the causes of war – of which most

[11] ENCOP (Environment and Conflicts Project) in: Baechler 1996: 10.

of its own representatives say that in view of the fact that wars are highly complex subjects which can only be explained in a multi-factor approach, "the generalizable findings on the causes of war that might form the basis for a body of theory are extremely meager to this day".[12]

Interim résumé: The most varied factors need to be examined in order to explain the outbreak of armed conflicts and wars. Environmental degradation can never be identified as a sole cause of war. Environmental degradation as a cause of conflict does not lead monocausally and directly to armed conflicts, but only generates together with other, mainly social, factors a web of causality that can lead under certain conditions to armed conflict. Two different forms of social process are fundamental to the understanding of the causal structure of conflicts and their escalation. Gantzel points to societal constellations and their importance to the causation of conflict. These are key categories in the development of modern conflicts. Baechler et al. stress the social factors that are of importance to the process in which a conflict crosses the threshold to armed struggle.

2.3 The propensity of environmental conflicts to escalate and their amenability to resolution

There is far less literature on 'environmentally-induced conflicts' that take a violent course. We shall therefore first examine here whether and how *research on the causes of war* and *empirical studies of conflict* have addressed environmentally-induced violent conflicts until now.

Various empirical studies have made important contributions to the topic of environmentally-induced conflicts. These have used case studies of environmentally related violently developing conflicts to determine the courses that these conflicts took and which factors played a role in their inception and escalation. Various groups of researchers have in the meantime presented the findings of such case studies.[13] The research group around Homer-Dixon has identified four principal social effects of ecological degradation that could lead to armed conflict, above all in developing countries: (1) falling agricultural production, (2) economic decline, (3) migration and (4) collapse of legitimate political institutions and movements. On this basis, resource scarcity and environmental degradation can produce three types of conflict: Simple scarcity conflicts, ethnic and national

---

12 Quoted from and representative: Krell and Müller 1994, 133–156 (quoted from p. 133).

13 Space does not permit a detailed presentation of the findings of these research projects. They have included specifically: Findings of the International Peace Research Institute, Oslo (PRIO), (see Dokken and Graeger 1995), a research group at the University of Uppsala in Sweden (see Hjort af Ornäs 1992), the group around Homer-Dixon at Toronto University (Homer-Dixon 1991 and 1994) and the ENCOP research group (see Baechler et al. 1996). See the good overview of the findings of these groups in Baechler et al. 1996, 10–17.

conflicts and relative deprivation conflicts. The studies show clearly that the socio-structural framework conditions are decisive to the explanation of the course taken by a conflict. Homer-Dixon refers above all to the link between environmentally-induced mass migration and ethnic clashes.

The ENCOP project has used case studies to analyze violent conflicts with environmental links in a manner comparable to the approach taken by the group around Homer-Dixon. The ENCOP group arrives at the result that environmentally caused conflicts develop due to the degradation of renewable resources (water, land, forests, vegetation) above all in regions of developing and transition societies susceptible to socio-ecological crisis "if, due to the prevailing stratification, social cleavages exist which can be instrumentalized in such a fashion that through these – partially violent – social, ethno-political, power-political and international conflicts arise or are heightened" (Baechler et al. 1996:292). However, their analysis suggests that armed conflict only arises where there is a confluence of several of the following five factors: If (a) the situation is unavoidable for the actors, (b) there is a lack of mechanisms by which to resolve armed conflict, (c) environmental degradation is instrumentalized by state or social groups, (d) conflicting parties have opportunities to organize, form alliances and arm themselves or (e) there is already an ongoing armed conflict. Baechler et al. illustrate with their analysis that the phenomenon of environmental conflict cannot be explained in a monocausal fashion. Instead, it only becomes virulent through the agency of catalysts in complex cause-effect relationships, and that it particularly tends to do so in developing regions.

All research groups largely agree in their findings inasmuch as that scarcity problems caused by environmental degradation contribute mainly to intrastate conflicts, while interstate conflicts are highly improbable. They further underscore that the development and dynamism of conflicts can only be explained through identifying the linkages among environmental, social and economic determinants. A problematic aspect of these case studies is that they focus solely on 'realized cases of conflict'. The environmentally-induced conflicts are thus not compared with – possibly similar – conflict constellations in which it has proven possible to resolve the conflict. They offer an important heuristic basis for statements on 'eligibility for conflict', i.e. the propensity of problem constellations to escalate into violence, although they do not permit any prediction of the probability of a specific dispute escalating into armed, violent conflict.

It would therefore also be important to carry out case studies on positive conflict developments, in order to be able to assess the propensity of environmentally-induced conflicts to escalate or, conversely, their amenability to resolution. No major research projects have yet been carried out on this aspect, although the findings of environmental policy research may be used for first answers. Before further pursuing this in the following section, we shall first review the general findings of empirical conflict research and peace studies as to whether they can provide answers concerning the propensity of conflicts to escalate.

Empirical conflict research, a branch of study that emerged in German-speaking countries above all after 1970, attempts, on the basis of as comprehensive databases as possible, to formulate typical, characteristic variables with which to classify and, in the best case, to interpret violent conflicts and their development.[14] These studies thus aim at drawing up typologies of conflict – an approach that can serve as a first step in theory formation but does not have the ambition to describe general causes of war. A problematic aspect of this approach is that central concepts, such as war, armed conflict, latent crisis etc., are not defined uniformly and are used differently by the various researcher groups.[15] Nonetheless, registering these variables makes it possible to make statements on important actors, structures and main issues of violent conflicts, and thus to formulate or check new hypotheses. Environmentally-induced conflicts can be integrated in this analytical structure by perceiving and classifying resource scarcity as one of the countless conceivable issues of dispute.

The classification of types of war drawn up by the AKUF (Arbeitsgemeinschaft Kriegsursachenforschung) research group around Gantzel in Hamburg is organized around the relationships of actors (parties to conflict) among each other, distinguishing four types: anti-regime wars, other internal wars, interstate wars and decolonization wars. These types can further be linked with a classification of conflicts according to main causes. Gantzel distinguishes several main classes here: (a) system conflicts, (b) internal power conflicts, (c) participation conflicts, (d) self-determination/awareness/identity conflicts, (e) territorial conflicts in the proper sense and (f) regional supremacy struggles.[16] In this taxonomy, conflicts induced by environmental degradation would mainly belong to the class of 'territorial conflicts', as Gantzel groups under this heading both territorial boundary conflicts and conflicts over resource access, i.e. in particular international conflicts. Gantzel's classification does not yet cover intrastate conflicts in which a deterioration of the environmental situation (soil degradation, water scarcity) can exacerbate existing or generate new potentials for social conflict (migration).[17]

---

14 A good overview of empirical conflict research is offered by the two seminal works: Gantzel and Meyer-Stamer 1986; Pfetsch and Billing 1994. The Heidelberg Institute for International Conflict Research (Heidelberger Institut für Internationale Konfliktforschung, HIIK) draws up an annual conflict barometer registering the ongoing and new conflicts of each year, and showing successful peace or mediation efforts (see the essay by Rohloff in this volume).

15 Pfetsch and Billing refer to the most varied war lists that have been drawn up since the middle of the 1960s. Important databases include: Small and Singer 1982, Gantzel and Meyer-Stamer 1986, Gantzel and Schwinghammer 1994 and the KOSIMO (Konflikt-Simulations-Modell) database created at the HIIK institute in Heidelberg (see the essay by Rohloff in this volume).

16 Each of these classes of conflict is broken down into several further subclasses, see on this Gantzel's overview of wars waged from 1985 to 1990 (Gantzel 1994: 148ff).

17 These might possibly be integrated in the AKUF classification through inclusion in the class of 'participation conflicts', if this were extended to cover not only socio-political forms of discrimination as is presently the case, but also socio-economic forms.

The Heidelberg Institute for International Conflict Research has created a "KOSIMO" database distinguishing between seven different conflict goods or values in a manner similar to the classification of types of conflict carried out by the AKUF group in Hamburg. Some of these categories are identical to the Hamburg classification. They include: (a) territory, (b) decolonization and national independence, (c) ethnic, religious or regional autonomy, (d) ideological or system conflict, (e) national power struggle, (f) international, geostrategic power struggle and (g) conflict over resources. This taxonomy could similarly integrate environmental conflicts as conflicts over access to resources. The KOSIMO database contains a further differentiation according to type of conflict resolution and level of intensity of conflicts.[18] These two additional levels of classification make it possible to distinguish between phases of intensity of a conflict in its historic development, and to cover conflicts resolved without recourse to violence (here labeled as latent conflict and as crisis). However, none of the present databases covers conflicts peacefully resolved on the basis of constitutional rights or international agreements. In the World Trade Organization (WTO) regimes alone, numerous disputes between states are resolved every year within the WTO dispute settlement procedure without escalation occurring. Since some years now, this mechanism has also tackled disputes involving environmental goods.[19]

Interim résumé: Approaches that fail to cover conflicts resolved without recourse to violence will scarcely be able to provide an understanding of the probability of conflicts escalating into violent resolution. Environmental degradation is a sphere in which it would be particularly important to include non-escalating international conflicts, as international conflicts over environmental goods frequently take place in arenas in which the parties to the conflict are members in international organizations or are participants in international regimes, i.e. where there are legitimized methods and codified procedures of conflict resolution. Similarly, it is essential to study the factors that have favored resolution in non-escalating intrastate conflicts.

Peace research is a field of study that has given even less systematic attention to environmentally-induced conflicts than empirical conflict research. Moves in this direction have only just begun and shall be briefly outlined here. Conflicts over environmental degradation tend rather to be conflicts over goods than over values. Peace research therefore tells us that they are relatively more amenable to

---

18 Types of conflict resolution are distinguished as: (a) negotiations, (b) authoritarian or judicial resolution, (c) pressure or threat and (d) use of force (sporadic or systematic). This differentiation permits an assessment of conflict escalation on the basis of the instruments of conflict resolution used. Conflict intensity is differentiated according to four levels (a) latent conflict, (b) crisis (without the use of violent instruments), (c) violent crisis and (d) war (Pfetsch and Billing 1994).

19 Among others, Kulessa 1995 and Helm 1995 give an overview of environmentally relevant GATT/WTO disputes.

peaceful resolution.[20] Registering resource conflicts whose prospects for resolution are particularly poor if at least one or several parties suffer resource scarcity would be a first step, but this does not suffice on its own. Baechler points out that

> ...modern environmental conflicts are increasingly assuming a character that is different from that of historical resource conflicts over land, water and mineral resources (....) While resource conflicts have been and continue to be concerned with a struggle over the material wealth of peoples and states, environmental conflicts of the more recent stamp unfold against the backdrop of the elementary threat to biological wealth caused by precisely these peoples and states. We might say that the issue is no longer one of conquering the oasis, but of preventing its devastation (Baechler 1997:377).

Research on factors conducive to peace – itself a recent branch of study – can not yet tell us much about the amenability of environmental conflicts to peaceful resolution. This branch has mainly studied international conflict resolution mechanisms and crisis management. There has been a growing interest in recent years in negotiation and mediation strategies. According to Rohloff, conflicts with a high potential for escalation include above all those where struggles over state power arise. Disputes over improved integration, autonomy or secession of minorities and international disputes over territories or resources also have a high propensity to escalate into violent conflict.[21] Volker Matthies notes that in view of the new qualities of intrastate conflicts in particular, new mechanisms of conflict resolution need to be found, as the deficits of state governance are becoming all too apparent here.

> The predominant forms of irregular civil wars and ethnically characterized conflicts, frequently associated with processes of state disintegration, the fragmentation of entire societies and widespread humanitarian disasters, would seem to escape the tools of peace policy and indeed be beyond the control of the actors and tools of classic state governance (Matthies 1997:529).

This is why the literature has now placed a strong focus on the potential of non-governmental actors in particular to resolve disputes (see Ropers 1995 and, typically: Calließ 1993), but with little specific attention to environmentally-induced conflicts. An Environmental Conflicts Management Project has emerged from the ENCOP project that is concerned with traditional methods of conflict resolution in the Horn of Africa. In a recent essay on the peaceful resolution of environmental conflicts, Baechler reports two examples of positive solutions that were based on the exchange of equivalents (e.g. land for peace). He draws from his findings to date the conclusion that constructive conflict resolution

> is inseparable from environmental management, and this in turn from the interests of the actor groups involved and their socio-economic, institutional and technical capacities to remodel societal relationships to nature (Baechler 1997:386).

---

[20] Recent publications giving an overview of the work of peace research in this field include: Klotz 1996; Matthies 1997; Rohloff 1996; Senghaas 1997; Wall and Lynn 1993; Zürn 1997.

[21] See Rohloff 1996. A considerable body of literature has been produced in the meantime on peaceful conflict resolution. Ropers and Debiel 1995 gives a good overview.

Baechler points out that solutions tackling the underlying dynamics of conflict – "namely the anthropogenic transformation of the environment" – need to "understand, stop and re-channel" this dynamism, and that in so doing the instruments chosen need to act in a "pincer movement" at both the local level (of internal causes) and international level (of international webs of causation). However, the question of how such local and international approaches should be designed in order to have positive impact has as yet been studied more by environmental policy and regime research than by peace and conflict research – the uptake by the latter of the findings of other fields of study is only just beginning. The findings of research to date on the conditions of success of international regimes in international environmental law are summarized in the third section of this essay.

Interim résumé: Environmentally-induced conflicts are not monocausal. Research carried out to date on environmentally-induced conflicts has shown that the precise identification of possible security impacts of environmental problems requires a consideration of national and international causal factors, and careful study of the processes by which social, economic and political factors mutually influence the course and dynamism of conflict. A characterization of environmentally-induced conflicts should yield an understanding of whether conflictual situations have the potential to escalate (have 'eligibility for conflict'). Empirical conflict research has created systematic databases recording conflicts, in which environmental conflicts have as yet – if at all – been registered as conflicts over access to scarce resources. It would be of great value to extend their coverage to include both environmental issues and peacefully resolved conflicts.

3 Security as a theme of environmental policy

Building on the overview of peace and conflict research given above, I shall summarize in the following the findings of environmental policy research pertinent to the nexus between environment and security. The goal here is to identify the most important research findings for the four principal levels of analysis of security relevance of environmental problems. These four levels comprise:

(1) Taxonomies of prime environmental problems that have the potential to cause new or exacerbate existing conflicts;
(2) Assessment of the eligibility for conflict of these various environmental problems;
(3) Taxonomies of the factors influencing the development of conflicts, i.e. promoting their escalation or attenuation;
(4) Assessment of the efficiency of envisaged policy responses, i.e. the degree in which constellations can be tackled so as to reduce or regulate conflict.

For the typological classification of environmental problems, we must first clarify which types of environmental problems are classified in the literature as having security relevance. In a second step, we shall examine how webs of causality are identified for these environmental problems – for the more precisely these chains of causation can be identified, the more specifically strategies for their resolution can be framed. With regard to the second and third levels of analysis, this section shall review the approaches taken in past research to evaluate the structurally inherent hazard potential and propensity to escalate of environmental conflicts, i.e. to determine when severe conflicts can emerge from environmental problems. Can criticality measurements be made in order to identify, using indicators, the points at which the threshold to security-relevant conflict is crossed?

3.1 Typological classifications of ecological problematiques

We may generally say that environmental problems can be organized according to quite different aspects. Depending upon the interests pursued and the classificatory scheme of an author, problems can be organized according to environmental media (soil, water, air), according to regions or geographic distribution, according to potential severity of damage (emissions, wastes, radiation etc.), according to causal relationships, environmental policy tools etc. In the following I shall briefly present some of the organizing models that are to be found in the literature on environment and security.

Manfred Wöhlcke classifies ecological damage and conflict potentials according to seven issue areas. He follows a breakdown according to environmental media (plus resources), complementing this with issues relating to the urban environment and natural disasters: (1) damage to the atmosphere, (2) damage to the biosphere, (3) scarcity of mineral and fossil resources, (4) scarcity of freshwater, (5) scarcity of soil, (6) environmental pollution and contamination and degradation of a urban environment and, finally, (7) 'classic' natural disasters and their increasing anthropogenic causation (Wöhlcke 1997). He then goes on to organize these issue areas in two groups: divisible collective goods such as mineral resources, freshwater and land on the one hand, and non-divisible collective goods such as the atmosphere, 'CO_2 sinks' and ocean fish stocks on the other. The security relevance in the group of divisible goods arises from the circumstance that the combination of economic and demographic growth is leading to demand for resources overstepping their availability or regenerative capacity. While historic conflicts over resources have been mainly concerned with tapping additional resources, their focus in the future will rather be concerned with securing basic existential needs, i.e. avoiding ills. According to Wöhlcke, these conflicts will initially escalate internally, but can then, if internal solutions are inadequate or unenforceable, lead also to external conflicts. In contrast, the security relevance of non-divisible collective goods is, according to Wöhlcke, that they belong to all and their overexploitation by individuals can damage all others. This entails the freerider problem; for even if individuals try their best, a destructive ecological dynamism can emerge if not all involved take ecological restrictions into consideration.

> In such constellations, the temptation to pursue a strategy of maximizing own benefit is particularly great, which accelerates the overexploitation of non-divisible collective goods (Wöhlcke, 1997:88).

Wöhlcke views these environmental problems as not having all that high propensity for conflict if not compounded by other conflict-exacerbating variables. Wöhlcke further identifies two further problematiques, firstly, indirect conflict exacerbation through environmental refugees, and, secondly, environmental warfare.

This first typological classification can be complemented by a further classification of environmental problems according to their scope of causation, i.e. according to whether they are local, transboundary bilateral, trilateral (i.e. regional) or global.[22] Traditional concepts of security policy (Haftendorn 1992) contain similar distinctions according to national, international (i.e. also regional) and global security. These classifications use the same regional breakdown, but are not identical. While global security problems can be structured by, for instance, clear foe images and a limited number of actors, global environmental problems are characterized by precisely the opposite: many, often not individually identifiable generators. Such a classification remains necessary, on the one hand in order to be able to identify actors involved and on the other hand with respect to the choice of tools by which to address problems (see for example: Görrissen 1993).

Harald Müller usefully combines both elements in his typological classification of environmental conflicts (Müller 1992: 82ff), although he excludes nationally limited conflicts or environmental problems from his scheme. He similarly distinguishes cross-border environmental problems according to conflicts over 'divisible' and over 'quasi-public' (generalized damage) goods. In this scheme, conflicts over divisible goods include utilization conflicts over environmental resources that represent an economic good (e.g. fish), conflicts triggered by transboundary environmental pollution (e.g. air pollution export) and contamination export (e.g. toxic waste export). In a manner similar to Wöhlcke, Müller distinguishes between generators and victims. In Müller's scheme, generators and

---

22 Some environmental researchers tend in the meantime to call such a geographical classification into question, pointing out that it is precisely the sum of many local environmental problems that leads to global complexes of causation (see Brown 1997). As justified as this view is over a long-term perspective, it curtails opportunities for differentiation and leads to such a broad concept of environmental problems that an appropriate assessment of e.g. security hazards becomes impossible.

victims are identifiable and different in the category of conflicts over divisible goods.

The differentiation according to benefit-damage distributions is made similarly for conflicts over quasi-public goods. Here Müller distinguishes between conflicts with asymmetrical benefit distribution (some few) and symmetrical damage distribution (all, e.g. species extinction or tropical forest degradation), and, secondly, conflicts where both damage and benefit are distributed asymmetrically (as is e.g. the case for ozone depletion or global warming).[23] With this characterization according to benefit and damage distribution, Müller has created a structural classification of problems that helps to identify the 'eligibility of conflict' of constellations, and would, moreover, permit a partial assessment of the amenability of conflicts to resolution if sufficient empirical case studies were available.[24]

Two further approaches that I will present in the following take a different route to classifying environmental problems. These not only consider general aspects of environmental problems, but attempt to integrate from the very outset causal relations and the potential for escalation of conflicts.

The ENCOP study "elaborates three dimensions of the quantitative and qualitative transformation of the environment that decisively characterize conflictual behavior in the international system" (Baechler et al. 1996:58). Baechler et al. therefore distinguish between: (1) the transformation of the environmental goods in their physical form and material substance, (2) ecoregional changes, i.e. the spatially effective degradation of water resources, soil and vegetation, and (3) the global effects of the human exploitation of nature, among which the authors count both the overexploitation of resources in developing countries caused by industrialized countries, and the long-range impacts caused by changes in the composition of gases in the atmosphere. With this scheme, the ENCOP authors aim to arrive at a more precise identification of those problems that are based more on 'ecoregional' causal relations or, on the other hand, more on global 'inter-ecoregional' relations – whereby they are aware that there are many and diverse interconnections between these spheres (impacts of global warming upon the water resources of a region etc.). They thus initially perform a geographical classification of environmental problems, then combining this with a social attribution of causation. The authors aim to thus identify the socio-ecological heterogeneity of environmental conflicts, i.e. asymmetries in responsibility for problem generation:

---

23 Müller further distinguishes between an asymmetrical damage-benefit distribution (e.g. ozone) and an asymmetrical benefit-damage distribution (e.g. global warming), whereby both categories signify the same, a small number of states reap benefits while many states reap damage, whereby benefits and damages do no accrue symmetrically to the same states.

24 This scheme is structured in a manner parallel to that of the findings of regime research, which analyzes structural aspects of conflicts in order to differentiate the opportunities for addressing these (see Rittberger 1993). Regime research has by now delivered a considerable number of case studies that could be used for such an assessment.

> It is particularly in such regions where, due to energy and material dependence upon renewable resources, societal relationships to nature have developed such that they tend towards crisis or are becoming non-sustainable that environmental transformation has assumed the shape of qualitative and quantitative limits to socio-economic development which let us expect a considerable potential for conflict.

The German Advisory Council on Global Change (WBGU) has succeeded in creating an even more detailed identification of conflict potentials in the interplay between environmental degradation and anthropogenic causation. In collaboration with the Potsdam Institute for Climate Impact Research, the WBGU has developed over the past years a concept under the heading of "Syndromes of Global Change" with the aim of structuring the convoluted dynamism of people-environment relations across the local, regional and global dimensions. The concept was not specifically developed with the aim of analyzing environmentally-induced conflicts, but can be extremely useful for the further analysis of this type of conflict.[25] The syndrome approach distinguishes 16 causal complexes organized in three groups: (1) utilization syndromes, relating above all to the overexploitation of fragile natural physical-geographical areas, (2) development syndromes, relating mainly to the overexploitation of natural factors caused by rapid economic development and (3) 'sink' syndromes, meaning mainly manifestations of degradation linked to excessive stress upon the buffering capacity of ecosystems (waste, emissions, contaminated land). The classification abandons a systematic distinction according to local, transboundary or global environmental problems, aiming rather to focus on highly complex dynamics of change with many mutually intervening variables. The assessment of the damage potential of individual syndromes is not always simple or clear, as, for instance, while economic development can lead to certain syndromes it can also provide the means by which to address them. However, the WBGU has made comprehensive attempts to identify the damage potential in more detail for the example of soil and of water resources (WBGU 1995 and 1998). As opposed to Baechler et al., the syndrome approach refers not only to the socio-ecological heterogeneity of conflict scenarios, but further aims to identify the prime dynamics of people-environment relations, i.e. also locally limited ecological consequences within the industrialized countries.

The various typological classifications of environmental problems all have good reasons for their various schemes. Some concentrate on classifying environmental problems according to environmental media and their development over time, others upon the geographic scope of environmental problems, a third group upon the structural characteristics of problems that can emerge from case to case, a fourth upon cause-effect complexes in the emergence of environmental problems and a fifth upon the prime dynamics of people-environment relations at a global scale. Various elements of the different classifications can be utilized to analyze

---

25 The concept has been set out in detail in the various annual reports of the WBGU since 1993.

environmentally-induced conflict in order to move beyond identifying potential problems to gain an improved understanding of the expected development, the propensity to escalate and the amenability to resolution of environmental problems.

A general heuristics of environmental problems can help to concentrate analysis upon pivotal problems. This heuristics must then be geographically assigned in the double sense of ecoregional localization (i.e. according to critical regions) and of the scope of causation (local, transboundary, global). The scope of causation can be integrated in the (eco)regional distribution, if causation is differentiated according to local (internal) and external causal relationships. This appears expedient in order to appropriately reflect the second main variable, problem structure. A local utilization conflict (land access/tenure) can be changed by external intervening factors (climatic change etc.) so that structures of conflict are mingled. These elements must nonetheless be identified separately, as their resolution requires both national tools (tackling internal causation) and international tools (tackling external causation). This identification will not be easy in each situation, but empirical studies that consider a combination of quantitative and qualitative indicators will indeed by able to yield such differentiation. This is impressively shown by the findings of the ENCOP studies.[26] This typological classification can further be combined with information on benefit-damage distribution that forms a part of identifying the situational structure. This would make it possible to make trend statements on the amenability of conflicts to resolution.

Interim résumé: In sum, six levels can be distinguished that are essential to identifying environmental problems in order to be able to assess these with respect to their security impacts. The following levels of analysis are proposed for a typology:

(1) Heuristics of environmental problems;
(2) Regional distribution coupled with scope of causation (local, transboundary, global);
(3) Structural elements of constellations (benefit-damage distribution);
(4) Amenability of conflicts to resolution;
(5) Structures of causation;
(6) Types of linkage.

In the following Section (3.2) we shall examine in more detail how the causal structure of environmental conflicts can be identified appropriately, before considering in the subsequent sections how the criticality (Section 3.3) of environmental problems can be surveyed and which prospects (Section 3.4) the use of environmental policy tools can have to steer environmental conflicts.

---

[26] In addition to the outcomes of the ENCOP project already cited above, the three final report volumes contain highly instructive detailed studies (Baechler et al. 1996 Vols. 2+3).

3.2 Identifying webs of causality

Some of the typological classifications of environmental conflict presented above take possible causal linkages into consideration from the outset. The syndrome approach of the WBGU was explicitly developed in order to identify environmental problems within their web of causality. Typical cause-effect chains and connections lead to the interaction of different elements of global change. The approach lists 16 syndromes defined by characteristic interactions of different complexes of problems. The names given to the syndromes already illustrate that the approach always identifies environmental hazards along the lines of specific human patterns of activity – the economic activities of human societies are thus the basic reference of the classification. The syndrome scheme thus delivers a classification of environmental problems embedded in a qualitative environmental systems analysis whose causal relationships are a constituent part of the analysis (WBGU 1998). The approach can thus be expected to be superior to most other relevant classifications. In its assessment of environmental hazards, the WBGU proceeds from an assessment philosophy oriented to the notion of 'crash barriers'. Every conflict calls for specific crash barriers for tolerable environmental stresses, which need to define a corridor that is all the narrower the more irreversible the expected environmental damage may be. This concentration upon crash barriers as a tool of environmental policy underscores that while the complex web of causality underlying syndromes and thus also social factors are concentrated upon, therapy focuses nonetheless upon environmental policy measures and not upon social factors. If other factors (e.g. the structure of land distribution) are more important than environmental causes to the dynamism of conflicts and above all operate as catalysts, then it is essential to define these very precisely and to tailor therapies mainly to these factors (e.g. redistribution of land).

The authors of the ENCOP study go the furthest in this sense when they formulate that environmental degradation cannot lead directly as a cause of conflict (i.e. monocausally) to armed conflict. It only operates indirectly, mediated through social effects in a complex web of causality. Environmental degradation thus assumes the function of a catalyst, such as if, for instance, distributional conflicts over scarce land or water resources lead to the emergence of a conflictual constellation based on ethnopolitical exacerbation (deprivation). The NATO/CCMS Pilot Study "Environment and Security in an International Context"[27] partially takes up this idea. The authors group the main 'environment and security' problems that have a potential to trigger conflict in two classes, firstly, those

[27] CCMS is the NATO Committee on the Challenges of Modern Society, which has commissioned a study under the stated title. The authors, Alexander Carius, Sebastian Oberthür, Melanie Kemper (of Ecologic – Centre for International and European Environmental Research, Berlin) and Detlef Sprinz (of Postdam Institute for Climate Impact Research), have presented an interim report with the same title (Carius et al. 1996), which formed a basis of debate at a conference on this issue in July 1997.

triggered by 'resource scarcities' and, secondly, those caused by 'environmental degradation'. They assume that conflicts triggered by environmental degradation can only become virulent through the mediation of 'secondary social problems', while those caused by natural resource scarcities can also contribute directly to or can trigger violent conflict.[28] Among the secondary social problems, the authors list migration, poverty, food scarcity and health problems. In this classification – as in that of Baechler et al. – the secondary social problems are viewed as variables that determine the triggering of violent conflict and its dynamics. Thus in the case of environmental degradation a causal chain of conflict emergence and dynamics becomes the triggering variable for 'social effects', which in turn are decisive for the explanation of conflict.

I would like to note here that probably a large proportion of scarcity and of degradation phenomena are precisely triggered by social problems, i.e. environmental degradation is an outcome of preceding social processes. It would therefore be expedient to not only survey 'secondary social problems' as factors conditioning the development and dynamics of conflict, but also 'primary social and primary economic problems' underlying the processes of ecological change.[29] The ENCOP authors also point to this in the context of the poverty problem, by showing that environmental and economic aspects are closely interwoven. The "State of World Rural Poverty" study published in 1993 by IFAD is an excellent example of the analysis of poverty problems that can also lead to environmental degradation. This study describes the possible causes of poverty through 11 central processes. The process descriptions provide a good basis for understanding the interplay among causative social conditions, environmental degradation and secondary consequential social problems (IFAD 1993).

---

28 The authors list natural disasters under the environmental degradation class (Carius et al. 1997:6, Note 6), under which they also group contamination and local and global processes of environmental intervention. The resource scarcities class includes renewable (fish, water, forest) and non-renewable resources.

29 Two examples may serve to illustrate these primary social problems:
(1) The dynamics of the destruction of tropical coastal mangrove forests: In many countries, these are currently being cleared to make place for the production of shrimp cultures (which require brackish water). Considerations of economic exploitation are thus the prime reason for degradation in these countries (considerations that are due in the rarest of cases to population growth, as the shrimps are produced for export – not for local food supply). A consequence of the degradation of the mangrove forests is that the countries concerned become more vulnerable to the impacts of tropical cyclones and future sea-level rise.
(2) In many countries, soil degradation is primarily a consequence of inequitable land distribution. In El Salvador, land distribution has forced many leaseholders and small farmers to use high sensitive slope plots. The resultant soil degradation intensifies the original problem together with population growth. The original factor responsible for the scale of degradation now experienced was the historically uneven distribution which led to the overexploitation of sensitive ecoregions.

Interim résumé: Almost all authors take the view that environmental conflicts rarely lead directly to organized collective violence, but rather contribute to the genesis of conflict in a manner mediated through consequential social problems. The appropriate identification of causal linkages requires the identification of primary social and economic processes that can underlie environmental degradation and resource scarcities. Secondary social problems, on the other hand, are important to explain the escalation potential and the dynamics of conflicts. The causal linkage can therefore be analytically subdivided into two phases in which social problems assume a catalytic function. This differentiation is important because different solution strategies need to be considered for these – possibly quite different – social problems. Inequitable land distribution can lead to soil degradation, this in turn leading to mass migration. Addressing the issue of land distribution requires other tools than handling migration. Secondary social problems will be a decisive variable in the identification of the actual propensity of conflicts to escalate.

3.3 Measuring conflict criticality and assessing risk

The above review has mainly focused on methods for typologically identifying environmentally-induced conflicts, and has discussed which circumstances and processes may play a role in the emergence of such conflicts. The determination of the dangerousness of environmentally-induced conflicts and of their propensity to escalate is a further important step essential to making security-relevant statements concerning this type of conflict. The great diversity of factors that can play a role in the emergence and development of environmentally-induced conflicts does not make risk assessment any simpler. As in the previous sections, I shall give a brief review of proposals made elsewhere for measuring the hazard potential of environmental conflicts, and shall then present some ideas on the way to move forward.

Proceeding from their analysis of historical or ongoing ecological conflicts, Baechler et al. have distinguished five factors (see above) that are considered to trigger conflict if several of them come together: (1) the perception of actors that a situation is unavoidable, (2) the lack of societal conflict resolution mechanisms, (3) instrumentalization of environmental problems by state or social actors with the aim of enforcing their own interests, (4) the opportunities of parties to conflict to organize or to forge alliances and (5) the existence of an ongoing conflict. This system shows that according to Baechler et al. the decisive conflict-triggering parameters are to be sought in causal linkages at the national level, particularly as ecological conflicts rarely take on a cross-border dimension (Baechler et. al. 1996:308-317). The ENCOP authors thus offer a qualitative heuristics for risk analysis that takes into consideration both ecological criticalities and social criticalities. Harald Müller (Müller 1992) offers a comparable heuristics composed of ecological and social criteria. Under the heading of (1) "degree of threat"

Müller posits three categories of dangerous situations: (a) A threat through environmental hazards that severely endanger the economic system and trigger wide-ranging social and political instabilities (soil degradation, water), (b) mass migration and (c) large-scale loss of territory (climate change, Chernobyl) would be situations representing a sufficient threat potential. In Müller's system, the potential alone does not suffice to explain escalation. Actors must further (2) be convinced that they can successfully conclude the conflict[30] and they must (3) command over sufficient action competence. For Müller this action competence is a key category, and embraces institutional competence, economic and demographic flexibility, social control of the affected population and technological competence. Müller further includes in his system (4) the general conflictual constellation (relationships of the actors among each other) and (5) the institutional integration of parties to conflict in alliances. Both systems presented here rely on a bundle of qualitative indicators intrinsic to crises or latent conflicts to assess security risks (eligibility for conflict). A drawback of qualitative threshold indicators is that they always depend upon the interpretation of the observer.

Consequently, Detlef Sprinz (Sprinz 1997) has suggested determining danger potentials by means of a quantitative threshold value that should be calculated such that when a number of environmental problems exceed a certain threshold value of problem intensity, a reliable prediction of escalation is permitted.[31] Such a threshold value proposal is well suited to identifying critical levels of environmental stress. However, this can only capture to a limited extent the above-mentioned intervening social variables which according to Baechler et al. primarily control the propensity of conflicts to escalate (secondary social problems). It may be hypothetically assumed that for some conflicts there can also be an objective threshold value of environmental degradation that renders the outbreak of conflict probable, but this will largely only apply to scarcity problems. Such a threshold value will not cover adaptation strategies or capabilities of the political and social systems of actors involved. Particularly in disastrous

---

30 Müller points out that such weighing requires 'rational actors'. In regime analysis the attempt is made under the heading of situation structure analysis to assess such weighing processes using the analytical apparatus of game theory. Here generally four types of problematic situations are distinguished: Coordination games with distributional effects and coordination games without distributional effects, dilemma situations and Rambo situations. While in the first two situations regime formation is possible (and promotable), dilemma situations require long confidence-building processes. Rambo games are assessed as being scarcely capable of regime formation, i.e. of cooperation (see Rittberger 1993b).

31 In Sprinz's empirical-quantitative research approach, anthropogenic forces, particularly production and consumption, are understood as central drivers of environmental problems. The processes of 'primary social and economic' causation discussed above are thereby partially covered. However, the complexity of social processes of causation (e.g. of the emergence of poverty) will presumably only be measurable by means of weighted qualitative indicators.

constellations (problem overload) these can often assume unprecedented levels.[32] Nor does it cover irrational action preferences of actors.

To address global water resource problems, the German Advisory Council on Global Change (WBGU) proposes in its 1997 Annual Report a combination of quantitative state indicators and qualitative process indicators. The report first develops a "criticality index as a measure of the regional importance of the water crisis", which integrates water extraction and water availability as state indicators and takes into consideration the available problem-solving capacity. Problem-solving capacity is not determined in such a complex manner as proposed by, for instance, Müller with his five indicators. Nonetheless, the Council includes the following sub-indicators in its criticality index: an indicator for water-related know-how, an indicator for the existing infrastructure for supply, distribution and treatment of (waste) water, one for the economic strength of a location and one for the efficiency and stability of the relevant political institutions (WBGU 1998:121ff). The criticality index offers an indicator that considers all at once hydrological, climatological, demographic and economic factors and is surveyed in a spatially differentiated (cartographic) manner. This indicator can further be combined with the regional networks of interrelations identified in the form of the above-mentioned syndromes. This is the system that currently offers the most precise statements in risk assessment of critical ecological constellations.

Interim resume: Only a combination of quantitative and qualitative state and process indicators appears suitable to perform more reliable risk assessments of conflict escalation potentials. It is essential that an assessment takes into consideration the action capacity of actors under both normal conditions and conditions of crisis.[33] The assessment can further be complemented by statements on actor rationality in the context of specific problem and conflict structures. It follows that the most important influencing variables are to be sought at the national level. Here it is important to also appraise problem resolution capacities. It nonetheless remains necessary to also consider international influencing factors, both with regard to their contribution to the emergence of problems and with regard to their possible function in helping to solve these problems.

---

32 The recent literature on political control generally assumes that the action competence of regulative policy can as a rule be described assuming a reduced political control potential in view of recognizable limits to political control, but that when problem overload occurs it can considerably rise in the specific situation. Insofar, critical problems may release considerable conflict resolution potentials (see Mayntz and Scharpf 1995; Ulrich 1994).

33 Conditions of crisis can (not must) enhance the reaction and action capacities of actors involved.

3.4 Environmental policy tools and their application to environmental conflicts

A great number of different tools are under debate and can be applied at all levels of conflict (local, national, international) to resolve conflict dynamics. These can be political, economic or legal tools, whose appropriateness and efficiency depends above all upon the structure of a conflict and the action capacity of actors involved. The possible tools shall not be listed in detail here. The German Advisory Council on Global Change (WBGU) has discussed the scope of possible instruments for the example of water resource problems (WBGU 1998:262–332). The WBGU report considers in detail the influence of socio-cultural and individual framework conditions, of environmental education and of public discourse, and economic and legal tools at both the national and international levels. These are complemented by disaster prevention and control tools, and by dispute settlement mechanisms at the national and international levels.[34] It is of great importance to ensure that selected instruments have coherent effects and that these effects are not mutually defeating. The issue of coherence goes beyond the question of compiling the proper package of tools. While one policy arena concentrates on solving specific causal relationships (e.g. inappropriate forms of land management), it may well be that tools implemented in other policy arenas significantly exacerbate the emergence or dynamics of conflict. Points of incoherence can, in the best case, concern ineffectively coordinated political strategies, or can produce countermanding effects. In the worst case, they are the outcome of fundamental goal conflicts in the enforcement of certain political projects. Such goal conflicts have been a particular cause of concern in the field of species conservation. Some non-governmental organizations (NGOs) have pointed out that for particularly endangered animal species even the permission to engage in controlled trade compliant to the criteria of the international CITES conservation agreement effectively leads to a parallel expansion of the black market and thus nullifies the protective effect. NGOs therefore recommend a total trade ban for such species. For economic reasons, however, those countries that have, for instance in the case of ivory trade, been able to stabilize elephant populations by means of special control measures demand that limited trade be permitted.[35] In the sphere of development policy in particular, NGOs have succeeded in highlighting such points of incoherence among different policy arenas.[36] Policy incoherence

---

[34] Dispute settlements and mediation mechanisms were already touched on in Section 2 above. Bercovitch 1995 and 1996 gives a good overview of the state of research.

[35] See the detailed and very precise presentation of this issue given by Princen 1994.

[36] This has concentrated particularly on European agricultural policy. While the European Union or its Member States has attempted since years to strengthen cattle farming as a sustainable source of income in West African Sahel countries, the markets for the meat produced in Africa – generally located in West African coastal cities – have been severely

thus concerns not only the resolution phase of ecological conflicts, but can in fact also be a part of problem causation.

Interim résumé: The choice of tools depends upon the precise identification of the main chains of causation. A great diversity of tools is available:

a) In various policy arenas (environmental policy, foreign policy, development policy, financial policy, international economic policy), each of which offer specific inputs to resolving local, national, regional and international causal factors. Their capacity can be determined by examining the inputs already provided in similar constellations.

b) Systematic experience (i.e. an appraisal of tools e.g. according to command-and-control versus economic instruments etc.) in environmental policy, environmental economics and regime analysis can be used to assess the scope of the various tools (see for instance Jänicke and Weidner 1995; Rittberger 1993).

c) Possible points of incoherence between different strategies or tools need to be identified.

4 Overall résumé: Methodological recommendations for studying the nexus between environment and security

The above review of different approaches and research projects has shown that the studies approaching the nexus between environment and security from the peace and conflict research side or from the environmental policy research side have very different perspectives on the issue. They lead to a number of different conclusions and additional recommendations:

1. Proceeding from different concepts of security, the discussions of the nexus between environment and security arrive at very different thematically crystallized conclusions. A purposeful and operationalizable use of the term 'environmental security' can only be gained if the concept of 'security' is restricted so as to only include such conflicts that involve the organized use of violence by collective actors.

2. On the other hand, I propose here a broad concept of 'conflict', also including such conflicts that have the potential to develop security relevance within the meaning of the definition proposed above, but which are resolved peacefully. This is because firm statements on the propensity of conflicts to escalate can only be

disrupted by the highly subsidized sales of EU beef surpluses. In addition to the example of beef, NGOs have in the meantime documented further examples of such policy incoherence.

made if both conflict-heightening and conflict-resolving developments are analyzed. Empirical conflict research has developed systematic databases registering conflicts. However, if at all, these only register environmental conflicts as access conflicts over scarce resources. It would be desirable to extend these so as to embrace both environmental issues and peaceful conflict developments. The greatest problem will be to purposefully demarcate the definition of conflict 'downwards', i.e. to operationalize it such that only those conflicts are registered that had the potential to develop security relevance.

3. Proceeding from these preliminary assumptions, the objective is formulated to develop elements for a methodology that would permit distinguishing violent environmentally conditioned conflicts according to whether they tend to emerge acutely, over the medium term or over the long term. Such a classification would be useful for timely prevention of escalation, i.e. in order to engage in 'preventive action (diplomacy)'.

4. In order to elaborate such a classification, it is essential to develop an appropriate reference framework. Four main levels of analysis appear pivotal:

(1) Taxonomies of prime environmental problems that have the potential to cause new or exacerbate existing conflicts;
(2) Assessment of the eligibility for conflict of these various environmental problems;
(3) Taxonomies of the factors influencing the development of conflicts, i.e. promoting their escalation or attenuation;
(4) Assessment of the efficiency of envisaged policy responses, i.e. the degree in which constellations can be tackled so as to reduce or regulate conflicts.

For the typological classification of environmental problems, it needs to be clarified which types of environmental problems are classified in the literature as having security relevance. In a second step, it needs to be ascertained which causal relationships have been identified for these environmental problems; the more precisely these chains of causation can be stated, the more concretely prospective solution strategies can be identified. At the second and third levels of analysis, I have presented above the two main groups of approaches: firstly those aiming at assessing the danger potential and propensity to escalate inherent in the problem structure of environmental conflicts and, secondly, those that have attempted to perform criticality measurements in order to be able to state, using indicators, the points at which the threshold to security-relevant conflict is crossed.

5. The typological classification of critical ecological constellations shows that the reference framework of most authors differs considerably, each concentrating on different aspects in the identification of these constellations. In summary, six levels can be distinguished that are used to identify environmental problems and whose elements should be taken into consideration when determining security-relevant effects.

(1) Heuristics of environmental problems;
(2) Regional distribution coupled with scope of causation (local, transboundary, global);
(3) Structural elements of constellations (benefit-damage distribution);
(4) Amenability of conflicts to resolution;
(5) Structures of causation;
(6) Types of linkage.

Important analysis matrixes used in the various studies are synthesized in Table 1 together with a number of additional suggestions.

Table 1: Elements of a typological classification of environmental problems that can lead to or can exacerbate conflicts

(1) Heuristics of environmental problems

(1) Damage to the atmosphere
(2) Damage to the biosphere
(3) Depletion of mineral and fossil resources
(4) Depletion of freshwater resources
(5) Depletion of soil resources
(6) Environmental pollution and contamination, and degradation of the urban environment
(7) 'Classic' natural disasters

(2) Regional distribution / critical regions

A: National (local) problems
B: Cross-border problems
C: Global problems

(3) Situation structure – conflict structure

A: Conflicts over divisible goods:
- Utilization conflicts over environmental resources that represent an economic good
- Transboundary environmental pollution
- Exportation of environmental damage

B: Conflicts over quasi-public goods
- Asymmetrical benefit – symmetrical damage distribution
- Asymmetrical benefit – asymmetrical damage distribution

(4) Eligibility for conflict / amenability to resolution

Differentiated according to:
- Degree of threat
- Action competence
- Conflictual constellation (relationship of actors to each other)
- Institutional integration of actors

or according to:
- Criticality indexes
- Syndrome approach

(5) Process (of causation)

<table>
<tr><td colspan="2">Phase I: Causation</td><td colspan="4">Phase II: Development / dynamics of conflict</td></tr>
<tr><td colspan="2">Causation of environmental problems (primary social and economic problems)</td><td colspan="4">Causation of conflict development
Environment as influencing variable</td></tr>
<tr><td colspan="2"></td><td colspan="2">direct</td><td colspan="2">indirect (man-made)</td></tr>
<tr><td>Impact upon scarcity</td><td>Impact upon degradation</td><td>Scarcity</td><td>Degradation</td><td>Scarcity</td><td>Degradation</td></tr>
</table>

(6) Type of linkage

A: Environmentally-induced conflicts (direct link)
B: Extended concept of environmental security (medium-term)
C: Ecological interdependence (long-term)

6. The most important methodological indications concern the following elements:

– A great variety of factors need to be considered to explain the occurrence of armed conflict and war. Environmental degradation as such – even if very severe – will therefore presumably never be identifiable as the sole cause of war. Two different forms of social process are fundamental to determining the causal relationships of conflicts and of their escalation: firstly the social processes that play a role in the causation of conflict-promoting constellations, and secondly those that contribute to a dispute crossing the threshold to armed conflict.

– Environmentally-induced conflicts are not monocausal. Research carried out to date on environmentally-induced conflicts has shown that the precise identification of possible security impacts of environmental problems requires a consideration of national and international causal factors, and careful study of the processes by which social, economic and political factors mutually influence the course and dynamics of conflict.

– A general heuristics should situate environmental problems geographically, in the double sense of critical regions and of the scope of causation (local, transboundary, global). The scope of causation can be integrated in the (eco) regional distribution, if causation is differentiated according to local (internal) and external causal relationships. This typological classification can further be combined with information on benefit-damage distribution that forms a part of identifying the problem structure. This would make it possible to make trend statements on the amenability of conflicts to resolution. The entire causal dynamics of environmental problems needs to be considered in order to be able to make meaningful statements about the associated social, economic, political and cultural framework conditions relevant to identifying conflict dynamics and the propensity of conflicts to escalate.

– Environmental conflicts rarely lead directly to organized collective violence, but rather contribute to the genesis of conflict in a manner mediated through consequential social problems. The appropriate identification of causal linkages requires the identification of primary social and economic processes that can underlie environmental degradation and resource scarcities. Secondary social problems, on the other hand, are important to explain the escalation potential and the dynamics of conflicts.

– Only a combination of quantitative and qualitative state and process indicators appears suitable to perform more reliable risk assessments of the conflict escalation potentials of environmental conflicts. It is essential that an assessment takes into consideration the action capacity of actors under both normal conditions and conditions of crisis. The most important influencing variables are to be sought at the national level. Here it is important to also appraise problem resolution capacities. In addition, it is necessary to also consider international influencing factors, both with regard to their contribution to the emergence of problems and with regard to their possible function in helping to solve these problems.

7. The attempt to systematically structure the linkages between environment and security represented by the four main levels of analysis is similar to the study scenario proposed in the interim report of the NATO CCMS Pilot Study, according to which the variables to be studied are identified along the lines of "human driving forces (pressures on the environment), state of the environment (environmental performance), policy responses (e.g. instruments), and indicators of violence" (Carius et. al. 1997:10). The four levels of analysis proposed here aim to differentiate more strongly among various causal relationships, and to identify in more detail the problem structure and the eligibility for conflict of situations.

8. The determination of the factors proposed here presupposes the consideration of a whole set of indicators whose basic characteristics have been set out in the course of the discussion above. The indicators required range from qualitative process indicators through to (ecological) state indicators. These can then be integrated in a twofold respect in a model. The criticality assessment carried out by the German Advisory Council on Global Change (WBGU) for the example of water resources is a successful attempt at combining such different indicators. The "Integrated threat assessment for the NATO region as well as for other regions" proposed in the NATO/CCMS Pilot Study could readily be performed by means of integrating the findings in a geographic information system (GIS). Conflict developments (escalating and peaceful) could thereby be considered, with a number of extensions, within conventional conflict identification schemes in a conflict database such as the Heidelberg KOSIMO database. An extended database embracing both escalating and peaceful conflicts could then also be helpful in developing a decision support system for the use of tools if the historical prospects of success of mediation tools were also surveyed.

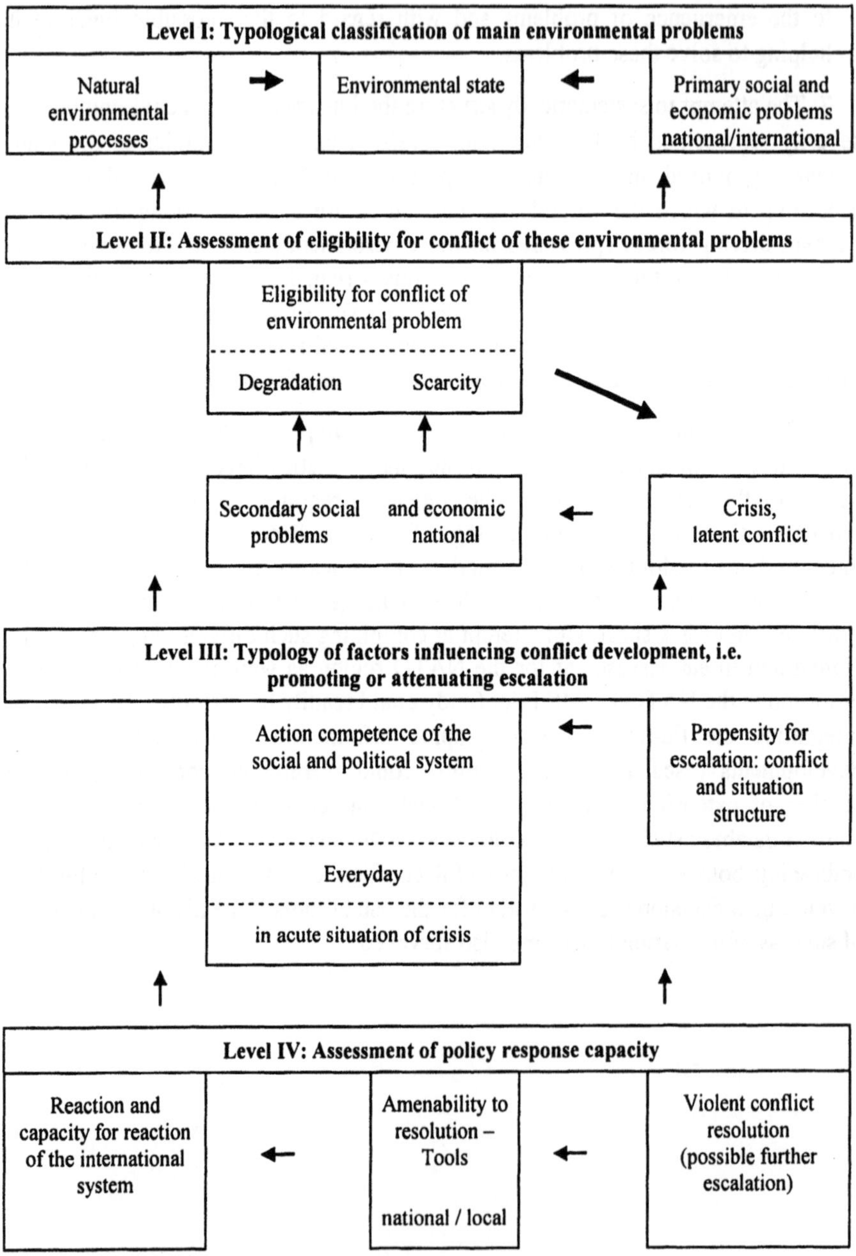

Figure 1: Main levels of analysis of environmental conflicts

References

Baechler, Günter et al 1996: *Kriegsursache Umweltzerstörung. Ökologische Konflikte in der Dritten Welt und Wege ihrer friedlichen Bearbeitung* [Ecological Conflicts in the Third World and Ways for their Resolution]. ENCOP Study Vol. 1. Chur / Zürich.

Baechler, Günter, 1997: "Wie sich Umweltkonflikte friedliche regeln lassen", in: Senghaas 1997b, 376–396.

Barandat, Jörg (Ed.) 1997 : *Wasser. Konfrontation oder Kooperation*, Baden Baden.

Bercovitch, Jacob (Ed.) 1996: *Resolving International Conflicts. The Theory and Practice of Mediation.* Colorado.

Bonacker, Thorsten and Imbusch, Peter (Eds.) 1996: *Friedens- und Konfliktforschung. Eine Einführung mit Quellen.* Opladen.

Brandt Commission (Independent Commission on International Development Issues) 1980: *North-South: A Programme for Survival.* London.

Brock, Lothar 1991: "Peace through Parks: The Environment on the Peace Research Agenda". *Journal of Peace Research,* Vol. 28, No. 4, 407–423.

Brock, Lothar 1994: "Ökologische Sicherheit. Zur Problematik einer naheliegenden Verknüpfung" in: Hein 1994, 443–458.

Brock, Lothar and Hessler, Stephan 1992: "Globaler Umweltschutz oder Öko-Imperialismus? Die ökologische Notwendigkeit neuer Nord-Süd-Beziehungen. Das Beispiel Amazonia" in: Meyer and Wellmann 1992, 194–228.

Brundtland Commission (World Commission on Environment and Development) (Ed.) 1987: *Our Common Future.* Oxford.

Calließ, Jörg and Merkel, Christine M. (Eds.) 1993: *Peaceful Settlement of Conflict – A Task for Civil Society,* Dokumentation einer Tagung der Ev. Akademie Loccum, Loccumer Protokolle 7/93.

Calließ, Jörg and Merkel, Christine M. (Eds.) 1994: *Peaceful Settlement of Conflict II – Third Party Intervention.* Dokumentation einer Tagung der Ev. Akademie Loccum, Loccumer Protokolle 9/94.

Carius, Alexander et al. 1996: *NATO CCMS Pilot Study. Environment and Security in an International Context. State of the Art and Perspectives. Interim Report (October 1996).*

Görrissen, Thorsten 1993: *Grenzüberschreitende Umweltprobleme in der internationalen Politik.* Baden-Baden.

Czempiel, Ernst-Otto 1981: *Internationale Politik.* Paderborn.

Czempiel, Ernst-Otto. 1997: *"Alle Macht dem Frieden"* in: Senghaas 1997b, 31–47.

Daase, Christopher 1991: "Der erweiterte Sicherheitsbegriff und die Diversifizierung amerikanischer Sicherheitsinteressen. Anmerkungen zu aktuellen Tendenzen in der sicherheitspolitischen Forschung", in: *Politische Vierteljahresschrift,* 32. Jg., H.3, 425–451.

Daase, Christopher 1992: "Ökologische Sicherheit: Konzept oder Leerformel?", in: Meyer and Wellmann 1992, 21–52.

Dicke, Klaus 1995: "Konfliktprävention durch Minderheitenschutz? Möglichkeiten und Grenzen multilateraler und innerstaatlicher Rechtsgarantien" in: Ropers and Debiel 1995, 233–256.

Dokken, Karin and Graeger, Nina 1995: *The Concept of Environmental Security – Political Slogan or Analytical Tool.* Report prepared for the Royal Ministry of Foreign Affairs. Oslo.

Eberwein, Wolf-Dieter, 1997a: *Umwelt-Sicherheit-Konflikt. Eine theoretische Analyse.* Wissenschaftszentrum Berlin für Sozialforschung, Paper P97-203, Berlin.

Eberwein, Wolf-Dieter, 1997b: *Umweltbedingte Konflikte. Methodologische Notizen, Tagungsbeitrag für den Workshop "Umwelt und Sicherheit"*, 3–4. Juli, Berlin.

Gantzel, Klaus-Jürgen 1994: "Kriegsursachen: Theoretische Konzeption und Forschungsfragen", in: Krell and Müller, 133–156.

Gantzel, Klaus-Jürgen and Meyer-Stamer, Jörg 1986: *Die Kriege nach dem zweiten Weltkrieg bis 1984*. München.

Gantzel, Klaus-Jürgen and Schwinhammer, Torsten 1995: *Die Kriege nach dem Zweiten Weltkrieg 1945–1992. Daten und Tendenzen.* Münster.

Gurr, Ted Robert 1994: "Peoples against States: Ethno-political Conflict and the Changing World System", *International Studies Quarterly*, 94/38, 347–377.

Haftendorn, Helga 1992: "The Security-Puzzle: Theory-Building and Discipline-Building in International Security", *International Studies Quarterly,* 35/1, 3–17.

Harborth, Hans-Jürgen 1991: Dauerhafte Entwicklung statt globaler Selbstzerstörung. Berlin.

Helm, Carsten, 1995: *Sind Freihandel und Umweltschutz vereinbar? Ökologischer Reformbedarf des GATT/WTO-Regimes.* Berlin.

Hjort af Ornäs, Anders and Krokfors, Christer 1992: "Environment and International Security", in: Lodgaared and Hjort af Ornäs 1992, 9–18.

Lodgaared, Sverre and Hjort af Ornäs, Andres (Eds.), 1992: *The Environment and International Security*. PRIO-Report No. 3, Oslo.

Holst, Johan Jorge, 1989: "Security and the Environment: A Preliminary Exploration", in: *Bulletin of Peace Proposals.* 20/2, 123–128.

Holsti, Kalevi, J. 1989: *Peace and War. Armed Conflicts and the International Order 1648–1989.* Cambridge.

Homer-Dixon, Thomas F. 1991: "On the Threshold: Environmental Change as Causes of Acute Conflict", in: *International Security*. 16/1, 76–116.

Homer-Dixon, Thomas F. 1994: "Across the Threshold: Empirical Evidence on Environmental Scarcities as Causes of Violent Conflict", in: *International Security*. 19/1, 5–40.

IFAD – International Fund for Agricultural Development (Ed.) 1993: *State of World Rural Poverty*. Rome.

Jänicke, Martin and Weidner, Helmut (Eds.) 1995: *Successful Environmental Policy*. Berlin.

Kende, István 1982: "Über die Kriege seit 1945" *DGVN- Issue* Nr. 16. Bonn.

Keohane, R.O. and Nye, J.S. 1977: *Power and Interdependence. World Politics in Transition*. Boston.

Klotz, Sabine 1996: "Gelungene Friedensschlüsse und Konfliktregulierungen in: Matthies et al. 1996, 33–48.

Krell, Gert, 1994: "Die Friedensforschung vor neuen Herausforderungen", in: Krell and Müller 1994, 61–95.

Krell, Gert and Müller, Harald (Eds.) 1994: *Frieden und Konflikt in den internationalen Beziehungen*. Frankfurt a.M.

Krusewitz, Knut 1985: *Umweltkrieg, Militär, Ökologie und Gesellschaft.* Königstein/Ts.

Kulessa, Margareta 1995: *Umweltschutz in einer offenen Volkswirtschaft. Zum Spannungsverhältnis von Freihandel und Umweltschutz*. Baden-Baden.

Matthies, Volker 1996: "Friedensforschung und Friedensursachenforschung", in: Matthies et al. 1996, 7–17.

Matthies, Volker 1997: "Zwischen Kriegsbeendigung und Friedenskonsolidierung", in: Senghaas 1997b, 527–559.

Matthies, Volker u.a.: "Frieden statt Krieg. Gelungene Aktionen der Friedenserhaltung und der Friedenssicherung 1945–1995", in: *Reihe Interdependenz 21*, Stiftung Entwicklung und Frieden. Bonn.

Mendler, M. and Schwegler-Rohmeis, Wolfgang 1989: *Weder Drachentöter noch Sicherheitsingeniur: Bilanz und kritische Analyse der sozialwissenschaftlichen Kriegsursachenforschung.* HFSK-Forschungsbericht 3/1989. Frankfurt a. M.

Meyer, Berthold and Wellmann, Christian (Eds.) 1992: *Umweltzerstörung: Kriegsfolge und Kriegsursache.* Friedensanalysen 27. Frankfurt a.M.

Meyer, Berthold (Ed.) 1996: *Eine Welt oder Chaos?* Friedensanalysen 25. Frankfurt a.M.

Meyer-Abich, Klaus Michael 1996: "Klimakrieg und Klimafrieden – Öko-Kolonialismus oder globale Umweltkooperation?", in: Meyer 1996.

Müller, Harald 1992: *Umwelt und Konflikt.* in: Meyer and Wellmann 1992, 72–99.

Müller, Harald 1993: "The Internationalization of Principles, Norms, and Rules by Governments. The Case of Security Regimes", in: Rittberger 1993, 361–388.

Myers, N. 1989: "Environment and Security", in: *Foreign Policy,* (74), 23–41.

Palme Commission (Independent Commission on Disarmament and Security Issues) 1982: *Common Security: A Blueprint for Survival.* With a Prologue by Cyrus Vance. New York.

Pfetsch, Frank R. 1994: *Internationale Politik.* Stuttgart.

Pfetsch, Frank R. and Billing, Peter 1994: *Datenhandbuch nationaler und internationaler Konflikte.* Baden-Baden.

Princen, Thomas 1994: "The ivory trade ban: NGOs and international conservation", in: Princen, Thomas and Finger. Matthies (Eds): Environmental NGOs in World Politics. London/New York.

Risse-Kappen, Thomas 1992: "Demokratischer Frieden? Unfriedliche Demokratien? Überlegungen zu einem theoretischen Puzzle", in: Krell and Müller 1992, 159–189.

Rittberger, Volker (Ed.) 1993a: *Regime Theory and International Relations.* Oxford.

Rittberger, Volker 1993b: "Research on International Regimes in Germany: The Adaptive Internationalization of an American Social Science Concept", in: Rittberger 1993, 3–22.

Rohloff, Christoph 1996: "Frieden sichtbar machen! Konzepte und empirische Befunde zu Friedenserhaltung und Friedenssicherung", in: Matthies et al. 1996, 18–33.

Ropers, Norbert 1995: "Die friedliche Bearbeitung ethno-politischer Konflikte: Eine Herausforderung für die Staaten- und die Gesellschaftswelt", in: Ropers and Debiel 1995, 197–232.

Ropers, Norbert 1997: "Prävention und Friedenskonsolidierung als Aufgabe für gesellschaftliche Akteure", in: Senghaas 1997b, 219–242.

Ropers, Norbert and Debiel, Tobias 1995: "Einleitung: Friedliche Konfliktbearbeitung in der Staaten- und Gesellschaftswelt", in: Ropers and Debiel 1995, 11–34.

Ropers, Norbert and Debiel, Tobias (Eds.) 1995: *Friedliche Konfliktbearbeitung in der Staaten und Gesellschaftwelt.* Stiftung Entwicklung und Frieden. Bonn.

Senghaas, Dieter 1997a: "Frieden – Ein mehrfaches Komplexprogramm", in: Senghaas 1997b, 560–575.

Senghaas, Dieter (Ed.) 1997b: *Frieden machen.* Frankfurt a.M.

Siegelberg, Jens 1994: *Kapitalismus und Krieg: eine Theorie des Kriegs in der Weltgesellschaft.* Münster.

Small, M. and Singer, J. D. 1982: *Resort to Arms.* Beverly Hill.

Sprinz, Detlef F. 1997: *Die Modellierung umweltbedingter Konflikte.* Vortragsskript präsentiert auf dem Internationalen Workshop "Umwelt und Sicherheit" am Wissenschaftszentrum Berlin für Sozialforschung, 3–4. Juli, Berlin.

UNDP – United Nations Development Program (Ed.) 1990: *Human Development Report 1990.* Oxford.

Vogt, Wolfgang R. (Ed.) 1995: *Frieden als Zivilisierungsprojekt – Neue Herausforderungen an die Friedens- und Konfliktforschung.* Baden-Baden.

Wall, James A./Lynn, Ann 1993: Mediation: "A Current Review" *Journal of Conflict Resolution*, Vol. XXXVII, No. 1, 160–194.

Weiss, Thomas G. (Ed.) 1995: *The United Nations and Civil Wars.* Boulder/London.

Weizsäcker, Ernst-Ulrich von 1994: *Earth Politics.* London / New Jersey.

WBGU (Wissenschaftlicher Beirat der Bundesregierung Globale Umweltveränderungen – German Advisory Council on Global Change) 1994: *World in Transition: Basic Structure of Global People-Environment Interactions.* Annual Report 1993. Bonn.

WBGU 1995: *World in Transition: The Threat to Soils.* Annual Report 1994. Bonn.

WBGU 1998: *World in Transition: Ways Towards Sustainable Management of Freshwater Resources.* Annual Report 1997. Berlin.

Environment and Security: The Demographic Dimension

Manfred Wöhlcke

1 Introduction

In the discussion of environment and security to date, the demographic dimension has been somewhat neglected, although it illuminates important aspects of the subject's framework conditions. As shown by the result of empirically comparative studies of "Environment and Security", the occurrence of (in part) environmentally induced conflicts has so far been restricted to the developing countries alone. This leads to the supposition that the economic, social and political conditions in the respective countries could have an influence on the occurrence and development of such conflicts. The following paper aims to illuminate the special significance of demographic developments for the possibility of (in part) environmentally induced conflicts.

Since the beginning of human history, population development over the millennia has been extremely slow and nearly linear. Not until the previous century did population start to grow exponentially. In 1900, the global population was 1.6 billion, in 1950 it was 2.5 billion and by 1987 it had already doubled, to 5 billion. Presently it totals nearly 6 billion and by 2010 it will pass the 7 billion mark. However, the *growth rates are not stable*, but have a *declining tendency*. Despite falling growth rates, there is a growing absolute base of population. Presently we are in the middle of the steepest part of the growth curve that began about 50 years ago and that will not flatten out for about 50 to 100 years to come.[1]

Against this background, it is relevant to ask how many humans can in fact inhabit the earth, under realistic economic and ecological assumptions and in conditions consistent with human dignity. Obviously, the key question here is not the population density in a given area, but rather the ability of the societies concerned to provide an appropriate quality of life for the population as a whole. From this perspective, the earth has long since been overpopulated, and it is doubt-

---

1 The present chapter is based on two working papers by the author: Wöhlcke 1997a and Wöhlcke 1997b. They contain detailed commentary and bibliographies.

ful whether scientific and technical innovations, such as genetically engineered food production, can change this in any decisive way.

At the same time, it should be noted that the global population is not evenly distributed, but concentrated in certain regions: these are primarily in the developing countries. Around three quarters of the world population live in them, and they contribute 90% of the global population growth. Only a few countries have succeeded in implementing a rigid birth control policy despite the obvious humane or religious considerations. In this connection, China in particular should be mentioned: it is estimated that, because of this policy, there are 200 million fewer people alive today than would otherwise be the case. Most countries deal with population problems half-heartedly, a tendency also found in the world population conferences held to date. Although this avoids the immediate problems arising from a rigid birth control policy - the need to justify this policy to religious authorities, the intervention it requires in the personal lives of citizens - it nevertheless creates grave problems in the long-term. These will affect not only each of the respective countries, but will also have global impacts. It will substantially affect the interests of, and scope for action by, industrial countries as well as supranational institutions.

This paper will examine the demographic dimension of the subject of environment and security. Starting from an analysis of current demographic tendencies (2), it will trace the special significance that population problems have both for international security and for the origin and handling of environmental problems. The central question here will be, to what extent demographic factors influence the inception or the course of (in part) environmentally induced conflicts (3). The paper concludes with a discussion of the prospects that result, based on the dynamics of population development, for the problem such conflicts represent (4).

2 Recent developments in population problems

2.1 Strategies for coping with population problems

In an appropriate strategy for coping with population problems, four key points can be discerned, which, however, have until now been implemented as practical measures only in exceptional cases, or without sweeping success:

- Direct population policy measures that curb population growth without taking the long "detour" through the development of the societies concerned.
- Increased efforts for the development of the "Third World", in order, on the one hand, to retard population growth *indirectly*, and, on the other, to achieve satisfactory socio-economic integration of the already existing, continually growing population.

- Substantial efforts in the direction of sustainable development, in the national as well as the international sphere. The present pattern of consumption and production in the industrial countries cannot, for environmental reasons, be adopted globally. In the long-term, it is out of the question for one quarter of the earth's population to consume three quarters of its resources while also causing three quarters of its environmental problems.
- Greater involvement of the elites in developing countries in a political process grounded in joint responsibility for dealing with global dangers. For all the justified self-criticism of the industrial countries, in the effort to achieve constructive results it must be stressed that many elites in developing countries do not assume enough of the national and international responsibilities devolving on them on the basis of their sovereignty, resources and scope for action.

However, as long as this multiple strategy does not lead to real solutions or fails altogether, it is necessary to prepare for a (probable) situation in which the continuing population growth *cannot* be satisfactorily integrated and sustainable development – both in the industrial and developing countries – *cannot* be realized, at least not in the next fifty to one hundred years. Regrettably, many problems can only be partially solved or not at all, and the ideally constituted global society still lies in the distant future. In the real world, population growth is a global problem that must be taken seriously. Instead of *dealing directly* with it– i.e., with a deliberate birth control policy – with few exceptions, countries prefer *indirect strategies*: i.e., the solution of the demographic problem is made dependent on solving a number of other problems. Under ideal conditions, this would indeed be the optimal course of action, but in reality it will not be possible to slow population growth significantly in this way.

2.2 Current developments

Current developments can be simply formulated as follows: the death rate is falling faster than the birth rate; more women are reaching maturity; although on the average they give birth to fewer and fewer children, the total number of their children in absolute terms continues to grow. To understand this, think of the mounting total returns from an invested capital fund and the interest and compound interest[2] on it *despite falling* interest rates. If the interest rate is stable, an increasingly steep growth curve results (what is known as the e-function). If the interest rate falls, an S-shaped growth curve results, which does not stop rising until the interest rate drops to 0%. With respect to global population growth, it is estimated that this point will be reached in about 100 years, when the total

2 "Kind und Kindeskind vermehren sich wie Zins und Zinseszins" (Jacobi, quoted in: Klüver 1993: 7).

population is between 10 and 14 billion, depending on changes in the birth rate. Figure 1 shows eight variations.

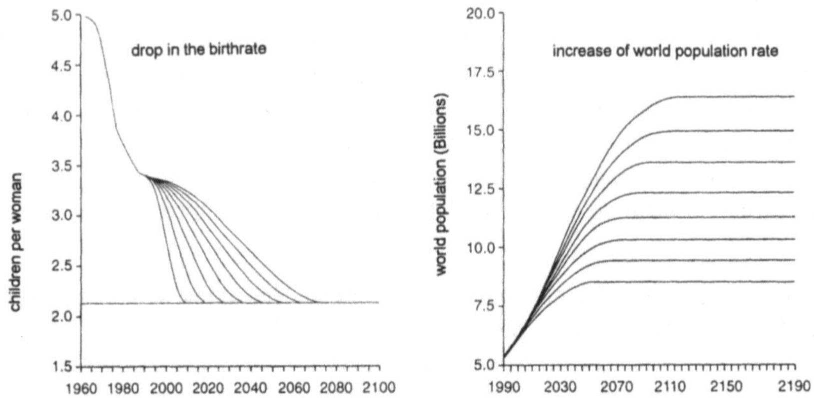

Figure 1: Increase in world population at varying rates of reduction in the birth rate (Source: Birg 1995: 110)

Prognoses are not only uncertain in regard to the assumptions underlying the model,[3] but also because many states do not have reliable population statistics at their disposal. Their demographic data are derived in part from estimates which are constantly brought up to date and revised. For this reason, United Nations population figures always appear under the heading "Devision". Differences from one revision to the next are due to methodological problems in data collecting and are often interpreted tendentiously. If the 1996 revision, for example, gives lower figures than that of 1994, this does not necessarily show that population growth has slowed correspondingly, but can mean only that more reliable data are now available. The 1996 revision was rashly interpreted in this regard as "a successful reduction of world population growth" and as "a success story of development cooperation" (Deutsche Stiftung Weltbevölkerung 1996) and the supposed slowing of population growth was assessed as a success for the United Nation Population Fund (UNFPA) (see Birg 1996b).

3 Nevertheless, prognoses to date have been astonishingly accurate. Thus, in the 1980 "Global 2000 Report" a population of 6.3 billion was predicted for the year 2000, which will come very close to the actual figure (Cf. Birg 1995: 31f.).

2.3 Population growth in developing countries

Population growth takes place in today's developing countries much faster than was the case during the "demographic transition" of today's industrial countries, for the following reasons:

- The birth rate in today's developing countries is considerably higher than in the industrial countries at the beginning of their "take-off".
- The death rate in today's developing countries is likewise falling much faster, because it is not geared to the respective social conditions, but is rather determined "from outside", i.e., by more developed societies (especially with regard to medical progress in combating some of the chief causes of death).
- Those early industrial countries which had a surplus of population were able to export it to a larger relative extent than is possible in the case of today's developing countries.
- High fertility is based among other things on poverty and reinforces it. Population growth and underdevelopment are mutually dependent. In contrast to the early industrial countries, most of today's developing countries for various reasons do not readily succeed in breaking out of this vicious circle by setting off a spiral of technical, social and cultural innovation.

The proportion of the population in developing countries, 78% in 1995, will increase to 86% by 2050. In absolute numbers, this means a population *increase* of 3.8 billion, more than double the total world population in 1900. In 2050, more than 8 billion humans will be living in the developing countries, although probably only a few countries will have succeeded by this time in making an adequate standard of living possible for the bulk of their population. An estimated 5 billion will continue to be poor or very poor, with adverse social, political, health and ecological consequences.

Based on the developments described, it can be seen in summing up the demographic problem in the light of realistic political and economic assumptions that the foreseeable increase in world population will contribute to the occurrence of (in part) environmentally induced conflicts. Because the problem of population increase will not solve itself through utopian developments in the poor countries, greater efforts must be undertaken toward implementing an efficient birth control policy. However, since this last is as unlikely as is a significant curbing of population growth by "progress", it must be assumed that demographic changes will occur within the range shown in Figure 2.

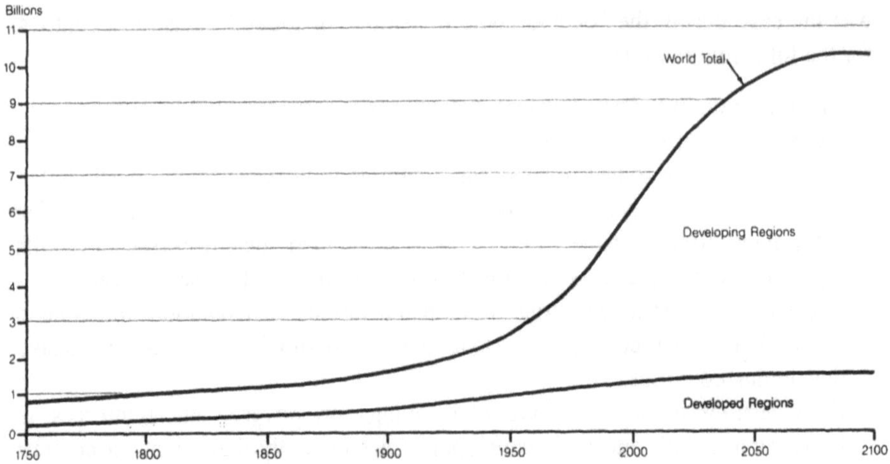

Figure 2: Population growth 1750-2100 (Source: World Resources Institute 1988: 16)

3 Population growth from the viewpoint of "Environment and Security"

Regarding the consequences of population growth, trends may be identified – aside, for the present, from environmental factors – that can potentially be significant for security. The following theses clarify this:

- *Population growth will change the present international hierarchy.* In 2025, 16 countries will have more than 100 million inhabitants: among them will be only two highly developed industrial countries (the USA and Japan) and the European Union. The regional balance of power among nations, measured by population, will shift: countries with large populations will exert more influence on processes of regional integration, move up to become significant economic partners or competitors of the industrial countries, acquire more influence in international organizations, and assert their right to have a voice in international affairs.
- *Population growth will change the international balance of security.* At present, twenty-five per cent of the world population lives in the industrial

countries, and the number is declining. Heavily populated countries – even if comparatively underdeveloped – do not have the recruiting problems of the demo-
graphically "aging" industrial countries and have the potential to mobilize substantial resources in the areas of armaments and technology.

- *Population growth will lead to increased migration pressure.* On the one hand, cross-border migration can unleash social problems in regard to national unity and cultural identity as important elements of the national self-image. On the other hand, it can potentially lead to ethnic tensions and social integration problems, which in the long-term could have foreign policy implications.
- *Population growth will make it more difficult to overcome underdevelopment.* In many countries, the political system is increasingly overburdened, mass poverty is concentrated in large cities, food security is impaired, social and ethnic conflicts fueled, anomic processes promoted and democracy and political stability endangered. Western development policy is in this context not an adequate instrument with which to achieve rapid development of the Third World. In that regard, it also cannot contribute meaningfully to curbing population growth by way of the "detour" of social development.
- *Pollution of the environment and depletion of resources will also be increased by population growth* This aspect will be examined in detail in the following, for this is a basic cause of the conflict-promoting influence of population growth.

3.1 Population growth and environmental problems

As a result of population growth and many different kinds of environmental degradation, our planet is being overused in three regards: in relation to the disposition of resources, (including fresh water and land), the renewal ability of the biosphere and the quality of the atmosphere. It is true that about three fourths of global environmental problems are caused by the industrial countries, but the developing countries are "catching up", and in the foreseeable future will themselves cause the bulk of these problems: on the one hand, because of intensified industrialization, which has an adverse impact primarily on air and water quality, and on the other hand, because of their large population. Large populations exert more pressure (the increase of identical environmental standards) on nature, landscapes and raw materials than small populations, and increased population pressure leads to settlement of regions which are especially endangered ecologically (such as earthquake areas) or are unstable (such as semi-arid regions). If this is accompanied by ecologically destructive "progress", the result is particularly problematic (see Wöhlcke 1987.)

> Depletion of non-renewable resources today is due less to population growth in the Third World than to patterns of consumption in the West. However, many of those living in the southern hemisphere will in future adopt our western lifestyle. With industria-

lization and with economic growth, per capita use of energy and raw materials is rising in Asia, Africa and Latin America, too. The consequent eco-social dilemma is inevitable, for even a modest increase in the per capita consumption in the Third World leads very quickly to a much larger global energy demand and to more emissions. In any case, every human will be able to lay claim to more energy and raw materials and to the creation of more pollution if the world population is smaller. And there will be more for the people of the South if we in the North restrain ourselves. (Münz/Ulrich 1995: 57).

Environmental problems worldwide are ultimately the result of an adverse global tendency: although the resources and ecological durability of our planet are finite, humans conduct themselves, demographically and economically, as though this were not the case. Though insight into this simple relationship is widespread, the manner in which the interests of large numbers of people are collectively organized stands in the way of satisfactory solutions. The world environmental crisis is *still* a primarily *political* problem, but the more it intensifies, the more it threatens to turn as well into a *physical* problem, which gradually eludes constructive political intervention. The adverse tendency is so hard to overcome because underdevelopment, population growth and environmental destruction form a vicious circle: high population growth leads to progressive environmental degradation (such as overgrazing and deforestation) to satisfy basic needs; this causes a permanent deterioration in living conditions and this in turn provides a motive for having as many children as possible, because in conditions of poverty they are a microsocial advantage, although macrosocially (i.e., if everyone did so) they are a disadvantage. However, not only is there environmental degradation in the developing countries as a result of underdevelopment, but there are also many kinds of environmental degradation as a result of "progress", without any satisfactory corrective environmental policy. Nevertheless, German and international development policy increasingly is adopting approaches that integrate corrective environmental policies into development concepts.

3.2 Conflicts induced wholly or in part by environmental factors

Conflicts relevant to security policy and induced wholly or in part by environmental factors are not yet very frequent, but because of the limited impact achieved so far by both international environmental policy and by foreign and security policy, it must be assumed that in future they will increase in number and intensity. In view of the fact that numerous unforeseeable factors specific to the situation always play a role in conflict generation, scenarios of this sort cannot be predicted either regionally or chronologically. Such scenarios can, however, be limited and their probability estimated:

Here the use of *divisible collective goods* becomes a problem when the amount available is exceeded by the demands of all potential users. This can be the case with non-renewable as well as with renewable resources (see Homer-Dixon 1994: 6f and 40). The total amount of the former existing on our planet is finite, even though new deposits are (still) being continually discovered. The combination of worldwide economic growth and the dramatic rise in world population accelerates the rate at which goods are becoming scarce, especially those that are not everywhere to be found, although they are everywhere in use.

A close relationship between resource scarcity and international conflicts also exists in the case of *water* (fresh water), which is becoming increasingly scarce in many regions (see Baechler et al. 1992: 8ff.). This rests on the following factors, which in each case occur in a specific mix: increasing demand because of economic development and demographic growth; growing limits on use because of pollution and toxic contamination;[4] reduction in volume of precipitation; major disturbances of regional vegetation; use of groundwater beyond its capacity for renewability; consumption of non-renewable fossil water reserves (see Baechler et al. 1993: 54ff.); salinization of groundwater and rivers as a result of the rise in sea level and of the increase in extremes of weather (in this case, of flood tides).

Many countries must "import" water (from cross-border rivers or groundwater) because their demand exceeds their own renewable resources (see Gleick 1993: 100). In some cases the dependence on "imported" river water is so great that they must rely on the good neighborliness of countries upstream from them. In the whole world there are 214 international river and lake basins (see Glatzl 1993: 309). Many of them are objects of contention, specifically in regard to water extraction, water regulation and water pollution. In the case of rivers, on which most conflicts center, those living upstream have as a rule more options than do those downstream (see Baechler et al. 1992: 34ff.), and it is therefore hard to bring them to compromise, while the very existence of the downstream countries can depend on doing so. The literature concerning international water conflicts is now abundant. In this connection, acute tensions, with serious security policy aspects, have arisen in many regions (see Elliott 1991: 26-35; Baechler 1993: 50ff.; Homer-Dixon 1994: 10ff.; op. cit. 1991: 107ff.; Glatzl 1993; Gleick 1993; Brown 1992: 52).

Likewise, there are already conflicts today over *farm- and pastureland* with security policy relevance and, as far as can be seen, they are tending to increase. The largest problem in this connection is *food security* (see Ehrlich and Ehrlich 1990: 66ff.; Korkisch 1993: 296-305). In 1798 Malthus published has famous essay "On the Principle of Population." [5] In it, he advanced the thesis that increases in population are progressive while increases in food production are only arithmetical, so that famines are inevitable; according to Malthus, they bring the number of humans into conformity with the amount of food available (see Birg 1995: 29; Neuffer 1982: 47). Malthus did not live to experience the agricultural

---

4 This resource is used not only for drinking and industrial purposes, but also as a cheap means of waste disposal.

5 Full title: "An Essay on the Principle of Population, as it Affects the Future Improvement of Society with Remarks on the Speculations of Mr. Godwin, M. Condorcet, and other Writers." Cf. Birg 1996a: 22ff.

revolution that began in the 19th century. Modern planting and breeding methods, the opening up of new farming lands, the use of artificial fertilizers and pesticides, the spread of high-yield varieties, progress in veterinary medicine, new financing and marketing possibilities – all of these made possible a substantial increase in agricultural production. In the wake of the "green revolution", they also reached the developing countries: between 1965 and 1990, food production was doubled (see Lutz 1994: 45). In the same period, however, their population also doubled, so that no per capita growth – at least, none of significance – could be registered, but at any rate, Malthus's thesis was disproved.

Although food production thus does *not* necessarily increase arithmetically, the food problem can by no means be regarded as solved for the long-term, for the following reasons:

- The land reserves still in existence are gradually being exhausted. Changing to inferior lands only brings unsatisfactory yields. At the same time, large areas of farmland are degraded or destroyed by false farming methods or by overuse (see Wöhlcke 1997a: 65ff.; Münz/Ulrich 1995: 54f.).
- The global climate change could have catastrophic impacts on agriculture in many countries in several respects: raising the average air temperature; changes in weather (including increasing the frequency of severe weather patterns); change in amount of precipitation and in the water cycle; shifts in vegetation zones; spread of deserts and semi-arid lands (see Wöhlcke 1997a: 31ff.).
- Intensification of agricultural production cannot be continued indefinitely (FAZ 12.18.1996: 16). Moreover, it requires large amounts of capital, of which not enough is available in many developing countries.
- The increasing scarcity of fresh water places limits on the exploitation of agricultural potential.
- The thesis has been proposed that the most important inventions in the area of agricultural production have already been made. The "innovation dividend" is constantly dwindling. Today it is only a matter of comparatively small advances made at great expense.[6]
- The tendency, seen during the first decades of development, for food production to grow at least as fast as the population cannot be maintained over the long-term. The tide has already turned (see Brown 1994: 9; Ehrlich and Ehrlich 1990: 68).[7]

---

6 Cf.. Brown 1994: 8. Some clear examples are also given here. The extent to which genetic engineering is the key to food security is generally questioned, because its "innovation dividend" is also probably substantially smaller than the rising demand. "With biotechnology neither providing nor promising any dramatic breakthrough in raising yields, there is little hope for restoring rapid growth in food output".
(Brown/Kane 1995; summary: http://www.worldwatch.org/pubs/ea/fh.html).

7 Since 1978, the population in 50 developing countries has been growing faster than agricultural production (cf. Münz/Ulrich 1995: 55). Concerning the fact that fish catches cannot keep pace with population growth either, cf. Ehrlich/Ehrlich 1990: 85.

- It is debatable whether, even under optimal conditions, it would be possible to produce and distribute enough food for 10 billion people – or more (see Neuffer 1982: 16 and 80f; Ehrlich and Ehrlich 1990: 20 and 67). Already today, i.e., with "only" 6 billion people, it is impossible to nourish all of them satisfactorily. Many countries are long since unable to provide basic sustenance for their population (see Korkisch 1989: 420; Ehrlich and Ehrlich 1990: 69ff.; Foundation for Development and Peace 1996: 74) and for lack of capacity to import are increasingly becoming recipients of international aid (further impairing their agricultural production). At present, about one billion people suffer from hunger and another billion are undernourished (see Ehrlich and Ehrlich 1990: 67).

  Lester Brown summarizes his assessment of the food problem as follows:

  > Our predictions lead to the assumption that environmental destruction due to population growth and the simultaneous collapse of political structures...is not only one among many possible future scenarios. In a world that goes on as if nothing were happening, this development is even highly probable. But it is not inevitable. It is possible to avoid it if the concept of security is redefined: If we realize that the most fearful threat to our future is famine, and not a foreign military attack. Such a recognition would lead to a reordering of political priorities. At their head would then be the fight against the causes of high fertility, such as illiteracy and poverty, the protection of farmland and water reserves, and massive financial investment in agriculture (Brown 1994: 11).[8]

In cases of *non-divisible collective goods*, conflicts of use also arise – especially in respect to the *atmosphere* and to what are called *sinks* (reservoirs for the absorption of pollutants) – which, however, will probably lead to scenarios relevant to security only in exceptional cases, because polluter and victim cannot be clearly distinguished and correspondingly mobilized politically. Nevertheless, there are principal polluters (industrial countries) and principal victims (developing countries). Neither adequately meet their environmental policy responsibilities At the same time, the developing countries use the environmental debt historically incurred by the industrial countries to demand economic compensation payments or exceptions from environmental regulation for themselves, which could promote a new version of the North-South conflict. This could escalate and inflame other North-South problems, if the consequences of the global environmental crisis – in part a result of continuing population growth – become more visible and tangible.

---

8 Cf. also the following: "Food security will replace military security as the principal preoccupation of national governments in the years ahead. Despite tight budgets, the resources are available to reverse the deteriorating relationship between ourselves and the natural systems and resources on which we depend" (Brown/Kane 1995: 3).

3.3 Environmentally induced migrations

It is difficult to assess the long-term security policy significance of the *indirect* effects of environmental degradation, among which mass *environmentally induced migrations* are especially to be emphasized.[9] Environmental refugees (see Wöhlcke 1992) are the losers in social and political distribution conflicts over shrinking resources. Their numbers already today have reached the hundreds of millions, and they will go on increasing in the wake of intensifying environmental problems and persisting population growth. Refugees can become the cause of new conflicts.

> Conflicts create refugees, but refugees can also create conflicts (Weiner 1995: 198).[10]

The intensification of all problems and conflicts – including ethnic disputes (see Homer-Dixon 1994: 6f.) – based on population growth, environmental degradation and scarcity of resources can lead to the erosion of the state and to the establishment of regimes with authoritarian domestic policies and aggressive foreign policies (Homer-Dixon 1994: 36; op. cit.. 1991: 77; Baechler 1993: 28 and 53; Baechler 1994/95: 45ff.). This could not only endanger regional stability, but could also strain the North-South relationship.

4 Conclusions

Not only must environmental policy be pursued more intensively worldwide; of at least equal urgency would be to curb population growth more severely, for more people produce more environmental pollution. Hopes for a rapid flattening of the population curve through a rise in the average standard of living must be assessed skeptically, since this process takes place very slowly and only where the marginalized population can be economically integrated in sufficient numbers and with adequate wages. It is often said that development is the best contraception. That may be true in theory, but in practice it is a difficult pill to obtain, because high population growth is precisely what prevents us from overcoming underdevelopment more rapidly.

Realistically we must probably assume that a significant curbing of global population growth will result neither from development nor from birth control, and that the demographic curve shown in Figure 2 must be run out within the range of variance assumed. It hardly seems possible to push through a successful policy of conserving the environment and natural resources under the demographic framework conditions to be anticipated. Probably a pessimistic view is more approp-

---

9 The general connection between population growth and migration has already be mentioned Here a particular type is emphasized.

10 Examples are found in Baechler et al. n.d.: 36ff. and Homer-Dixon 1991: 107ff.

riate: that, at best, with great efforts small, partial successes can be achieved which, however, do little to alter the overall adverse tendency.

Yet to solve global environmental problems all actors would be required really to think globally; i.e., to regard the global environmental problem as a joint international task which can only be accomplished in the framework of a comprehensive environmental partnership. In addition, we must to orient ourselves to the concept of an international joint venture society and to the need for a world environmental domestic policy, whereby the various respective interests and responsibilities must be fairly balanced.

A good brief description of the necessary steps to a world environmental domestic policy is found in the 1995 annual survey of the German Federal Republic's Scientific Advisory Council on World Environmental Change. It is, however, doubtful whether this ideal pathway will ever bring us to the goal. In other words, we must probably assume that the worldwide degradation of the environment and plundering of resources cannot be stopped despite all efforts in the area of international environmental policy, and that we will have to put up with partial successes. Already today we are unable to impose the principle of sustainability worldwide, leaving little room for hope of succeeding if and when the global population doubles (within the next 100 years). Thus conflicts are predestined, including those with security policy relevance.

References

Baechler, Günther 1993: *Konflikt und Kooperation im Lichte globaler humanökologischer Transformation*. Zürich: Forschungsstelle für Sicherheitspolitik und Konfliktanalyse.

Baechler, Günther 1994/95: "Umweltkriege als 'höchstes Stadium' der menschlichen Zivilisation?" in: Bahr, Egon/Lutz, Dieter S. (Eds.): *Unsere gemeinsame Zukunft – Globale Herausforderungen*. Baden-Baden: Nomos, 45-50.

Baechler, Günther et al. 1992: *Gewaltkonflikte, Sicherheitspolitik und Kooperation vor dem Hintergrund der weltweiten Umweltzerstörung*. Bonn: Projektstudie UNCED des Deutschen Naturschutzringes, Bund für Umwelt und Naturschutz Deutschland.

Baechler, Günther et al. 1993: *Umweltzerstörung: Krieg oder Kooperation?* Münster: Agenda.

Birg, Herwig 1994: "Weltbevölkerungswachstum, Entwicklung und Umwelt." *Aus Politik und Zeitgeschichte, Beilage zur Wochenzeitung Das Parlament*, No. B35-36/94, 21-35.

Birg, Herwig 1995: "Weltbevölkerungswachstum, Entwicklung und Umwelt – Dimensionen eines globalen Dilemmas", in: Studenteninitiative Wirtschaft und Umwelt e.V. (Ed.): *Globales Bevölkerungswachstum – Exponentiell ins Chaos?*, Münster, 25-51.

Birg, Herwig 1996a: *Die Weltbevölkerung. Dynamik und Gefahren*. München: Beck.

Birg, Herwig 1996b: *Wundersame Abnahme des Bevölkerungswachstums.* (Leserbrief). Frankfurter Allgemeine Zeitung, 17.12.96, 13.

Brown, Lester 1994: "Die Wissenschaft schreibt Fragezeichen." *Politische Ökologie*, Vol. 12, Issue 38, 8-11.

Brown, Lester/Kane, Hal 1995: *Full House: Reassessing the Earth's Population Carrying Capacity*. Washington (Worldwatch Institute): Norton; Zusammenfassung in:

http://www.worldwatch.org/pubs/ea/fh.html.

Brown, Neville 1992: "Ecology and World Security" *The World Today*, Vol. 48, Issue 3, 51-54.

Deutsche Stiftung Weltbevölkerung (Ed.) 1996: "Weltbevölkerungswachstum erfolgreich verringert.": *DSW newsletter*, No. 20.

Deutsche Stiftung Weltbevölkerung (Population Reference Bureau) (Ed.) 1996: *Weltbevölkerung 1996*. Hannover and Washington D.C

Ehrlich, Paul/Ehrlich, Anne 1990: *The Population Explosion*, London: Hutchinson.

Elliott, Michael 1991: "The Global Politics of Water." *The American Enterprise*, Vol 2, Issue 5, 26-35.

Frankfurter Allgemeine Zeitung 1996: *Riesen und Zwerge*, 18.12.96, 16.

Glatzl, Christian 1993: "Wasser – Konfliktstoff der Zukunft." *Österreichische Militärische Zeitschrift*, Vol. 31, Issue 3, 309-314.

Gleick, Peter H. 1993: "Water and Conflict. Fresh Water Resources and International Security." *International Security*, Vol. 18, Issue 1, 79-112.

Homer-Dixon, Thomas F. 1991: "On the Threshold. Environment Changes as Causes of Acute Conflict." *International Security*, Vol. 16, Issue 2, 76-116.

Homer-Dixon, Thomas F. 1994: "Environmental Scarcities and Violent Conflict. Evidence from Cases." *International Security*, Vol. 19, Issue 1, 5-40.

Klüver, Reymer 1993: *Zeitbombe Mensch. Überbevölkerung und Überlebenschance*. München: dtv.

Korkisch, Friedrich 1989: "Die demographische Explosion der dritten Welt – Konfliktpotential des 21. Jahrhunderts" *Österreichische Militärische Zeitschrift*, Vol. 27, Issue 5, 417-422.

Korkisch, Friedrich 1993: "Das Welternährungsproblem als Keim künftiger Krisen." *Österreichische Militärische Zeitschrift*, Vol. 31, Issue 4, 296-305.

Lutz, Dieter S. 1994: "Ist die 'Überbevölkerung' eine Friedens-Bedrohung?" *Politische Ökologie*, Vol. 12, Issue 38, 45.

Münz, Rainer and Ralf Ulrich 1995: "Bevölkerungswachstum: ein globales Problem", in: Opitz, Peter J. (Ed.): *Weltprobleme*. Bundeszentrale für Politische Bildung, Bonn.

Neuffer, Martin 1982: *Die Erde wächst nicht mit*. München.

Schug, Walter 1993/94: "Zum Hungern verurteilt?", in: Deutsche Gesellschaft für die Vereinten Nationen e.V./Deutsche Stiftung Weltbevölkerung (Ed.): *Weltbevölkerung und Entwicklung. Die Herausforderungen des globalen Bevölkerungswachstums*. Bonn/Hannover, 62-65.

Stiftung Entwicklung und Frieden (Ed.) 1996: *Entwicklung, Kulturen, Frieden: Visionen für eine neue Weltordnung*. Bonn.

WBGU (Wissenschaftlicher Beirat der Bundesregierung Globable Umweltveänderungen-German Advisory Council on Global Change) 1996: *World in Transition: Ways Towards Global Environmental Solutions. Annual Report 1995*. Berlin: Springer.

Weiner, Myron 1995: "Security, Stability, and International Migration", in: Lynn-Jones, Sean M./Miller, Steven E. (Eds.): *Global Dangers. Changing Dimensions of International Security*. Cambridge (Mass.)/London, 183-218.

World Resources Institute und International Institute for Environment and Development (Ed.) 1988: *World Resources 1988-1989*. In Collaboration with the United Nations Environment Programme. New York: Basic Boks.

Wöhlcke, Manfred 1987: *Umweltzerstörung in der Dritten Welt*. München: Beck.

Wöhlcke, Manfred 1992: *Umweltflüchtlinge*. München: Beck.

Wöhlcke, Manfred 1993: *Der ökologische Nord-Süd-Konflikt*. München: Beck.

Wöhlcke, Manfred 1997a: *Ökologische Sicherheit – Politisches Management von Umweltkonflikten*. Ebenhausen: Stiftung Wissenschaft und Politik.

Wöhlcke, Manfred 1997b: *Bevölkerungswachstum: Konsequenzen für die internationale Politik*. Ebenhausen: Stiftung Wissenschaft und Politik.

Part B:

Characterization and Taxonomies

Environmental Degradation in the South as a Cause of Armed Conflict

Günther Baechler

1 Introduction

Human transformation of the environment is a significant cause of the emergence, continuation and escalation of conflicts between collective actors in economically emerging and politically unstable regions. Environmentally caused armed conflicts are manifestations of abortive socio-economic and political development. At the same time, the social and political outcomes of underdevelopment that can be traced to environmental degradation and resource overexploitation have become an intra- and inter-state security problem. *Developmental and security dilemmas* thus intertwine to become a complex of problems giving rise to environmentally conditioned armed regional conflicts of differing intensity and form. But apocalyptic scenarios of environmental disasters or alarming prognoses of global environmental wars are not tenable. Environmentally caused conflicts escalate across the threshold of violence only under certain socio-political conditions.

This chapter will describe the specific conditions under which this happens. The different lines of argument will be discussed in the light of *two hypotheses* that were formulated for the Environment and Conflicts Project (ENCOP).[1] An ENCOP categorization of environmentally conditioned conflicts will be used to examine the levels of conflict and the actors or parties to 'environmental conflicts'. The syndrome approach (WBGU 1997) is used to relate conflicts to specific forms of environmental degradation.

This is followed by a hypothetical model of those environmental conflicts that have led to violence in the context of the syndromes named. The chapter ends with a general discussion of the role of the environment as a source of conflict.

1 The first part of this chapter is based on the results of the Environment and Conflicts Project (ENCOP), which ran for several years. The complete report was published in three volumes in the fall of 1996 (Baechler et al. 1996; Baechler and Spillmann, eds. 1996, Vols. II, III).

2 Environmentally caused armed conflict: A phenomenon of developing and transitional societies

Since environmentally conditioned conflicts, like other conflicts, are societal and political events, a causal analysis that starts with the type and depth of the environmental problem is necessary but not sufficient. For a complete explanation, other attributes of conflict such as actor characteristics, interests and actions, and the results of their actions, must be included. This was established through investigating more than 40 cases of environmental conflict, about half of which remained below the violence threshold and half of which crossed it.

Our first hypothesis is therefore:

Environmentally caused conflicts due to the degradation of renewable resources (water, land, forest, vegetation) are generally manifested in socio-ecological crisis regions in developing and in transitional societies if and when social fault lines can be instrumentalized in such a way that social, ethnopolitical and power struggles, and international conflicts, are triggered or intensified, sometimes leading to violence.

Empirical observations provide ample evidence for the assumption that human transformation of the environment is concerned with underlying development phenomena of a historical nature, in which countries that have poor problem-solving capacities suffer the most. Developing and transitional societies or, more precisely expressed, marginalized areas in these countries are affected by an interplay among environmental degradation, social erosion and violence that intensifies crises. Crisis areas prone to conflict are found in arid and semi-arid ecoregions, in mountain areas with highland-lowland interactions, areas with river basins sub-divided by state boundaries, zones degraded by mining and dams, in the tropical forest belt, and around expanding urban centers. Historically situated, culturally bound societal relationships to nature are subjected to upheaval and put acutely at risk in subregions of Africa, Latin America, Central and Southeast Asia and Oceania.

The argument of marginalization can be applied only conditionally to conflicts over shared water resources (or fish stocks, which ENCOP did not investigate). Inter-state conflicts between upper and lower riparians can occur in marginal regions in neighboring countries. But strategic and security issues are in the foreground for these conflicting parties; agricultural and development problems are secondary. This is especially true for regional water conflicts in the Middle East based on the historical territorial conflict between Arabs and Israelis.

Seen geographically, the areas of environmentally caused armed conflict largely correspond to those of global conflict. This means that armed environmental conflicts are a part of those armed conflicts that occur mostly in regions of developing countries. Conflicts tend to occur in the South, a pattern observed since the Second World War. Most environmentally caused armed conflict is between actors within one state (A). In a few conflicts, there is a tendency towards internal conflicts becoming internationalized (B); the major reasons for this vary. They are usually the results of migration and flight; some migrants driven by landlessness and environmental degradation do not move to more fertile ecoregions or larger cities in the same country, but cross national boundaries in the hope of finding better land or paid employment, creating potential social or ethnopolitical conflict in another country. War refugees seeking shelter in neighboring countries can also be the consequence of violent intrastate conflicts with ecological dimensions. Another element of internationalization is new nation-building, as in the Central Asian republics after the dissolution of the Soviet Union. Thus intractable intrastate conflicts, such as when regional water supplies are distributed by a centralized authority, suddenly take on an international dimension.

In contrast, inter-state conflicts (C) are carried out between sovereign states from the very beginning. Where they are environmentally conditioned, they are the outcome of ecological degradation and resource (water, air) exploitation not confined to national boundaries. International disputes arise because of asymmetrical ownership of upper and lower river areas, particularly between states that are dependent on the common use of an international river basin. These do not, however, lead to the use of military force, which usually remains a threat.

Distinguishing between the three levels – *internal, internal with inter-state aspects and inter-state* – only serves as a rough orientation. The lines of demarcation between these are fluid in every sense; most conflicts are fed by internal social contradictions and crises. Internationalized or inter-state conflicts are an expression of conflicts that expand or intentionally escalate. Causal analysis therefore needs a finer breakdown in order to categorize conflicts and link types of environmental degradation to their socio-economic consequences and the affected parties to the conflict. Based on the actors involved and the three systematic levels (A, B and C) mentioned above, we will determine seven types of potentially or currently violent environmental conflicts. Sharp divisions between these types are not always possible in reality; there are conflicts that contain elements of several types at the same time.

2.1 Center-periphery conflicts

The relationship between inhabitants of peripheral regions and the centers of developing countries, which latter include the national elite and international investors, is often precarious because of environmental transformation. In dealing with environmental problems, this precariousness is apparent in the *difference in opportunities for taking action* according to where power lies. While modernizing centers have a range of economic, environmental and energy options they can pursue because they are equipped with power, socio-economic room for maneuver is very limited for large parts of the rural population.

The catalysts of escalating center-periphery conflicts are mostly large-scale agricultural projects for export business, dams and mining projects. Capital intensive high technology and energy systems belonging to corporations oriented to global markets collide with traditional or indigenous societies that are based on a low technology, subsistence economy, or small farm economies that use little outside energy inputs. Such agro-industrial superprojects in areas that have hardly been integrated into the market economy transform the local population's society-nature relationship against its will. Mechanization and, depending on the region, collectivization or privatization of agriculture displaces traditional cultivation methods and the land use and land tenure practices associated with them. Soil and water degradation as a consequence of large-scale monocultures for export, or of dam construction or mining projects, encourage the erosion of living orders, and block economic development possibilities in the periphery, in a process accelerated by a reciprocal effect on relative overpopulation in rural areas. The inevitable consequence is increased competition for water, settlement space, arable land for self-sustenance, and employment. Social groups belonging to the periphery that are marginalized or resettled because of large-scale projects, without recourse to alternatives for earning their livings, see themselves as the losers in modernization.

2.2 Ethnopolitical conflicts

Transformation of the environment becomes a source of ethnopolitical tension and conflict if conflicting parties have diverging socio-economic interests and try to influence the outcome of an environmentally caused conflict by emphasizing superficial (race) traits. Ethnicity is instrumentalized and, depending on what phase the conflict is in, serves as an identification pattern or element in mobilization. As with the center-periphery conflict, at the core of ethnopolitical conflict is the stress of modernization. The difference lies not between a defined center and its periphery but rather along specific group traits within a multi-ethnic society. Regional population pressure combined with scarce and degraded resources also contribute to the hardening of interethnic relations.

Excessive agricultural exploitation of land, timber and available water resources leads to distribution conflicts over scarce environmental goods, often concurring with demographic dynamics. In many places in underdeveloped agrarian states, the traditional dualism between horticulture and arable farming on the one hand, and nomadic and large-scale cattle breeding on the other, is behind, if not at the center of, ethnopolitical conflict. Disputes over fertile land break up the cultural ecological niches of regionally separated societies. Intensive use of ecologically relatively stable areas and sensitive, unfavorable areas through an increasing number of rural producers has led to the loss of widespread territorial separation of ethnic groups' habitats, to be replaced by social stratification and ethnic mixing less localized in space. This process has been intensified by colonization and modernization, as well as by wars.

At the same time that central governments plan large areas of good land for monocultures, usually in cooperation with multinational agro-industrial corporations, there is less and less fertile land set aside exclusively for the use of small producers. Seasonally, arable land serves as pasture and as hunting ground; pasture is used by arable farmers. In spite of competitive interest, traditionally producing groups, whether tribal associations, clans or ethnic units, are generally not prepared to give up the rights, transmitted over centuries, that they trace back to the first appropriation of unclaimed land as a *res nullius*. Historically, land use and distribution conflicts between supporters of traditional rights and protagonists of private or state ownership of property are 'solved' by the use of deliberate measures using force. This is especially true if they come in the wake of ethno-social stratification and ethnopolitical hierarchy-building in a society. More recently, such conflicts have been masked by the concomitant negative ecological symptoms of abortive structural development and brutalized through the easy availability of modern weapons. Thus they take on a qualitatively new dimension and particular virulence. Examples can be found in numerous conflicts in Africa's Sahel belt, such as in the Nuba mountains of Sudan or in the settlement areas of the Somali, the Arbore and the Borana in southern Ethiopia.

2.3 Internal migration conflicts

Conflicts of this kind arise from the voluntary or forced migration or resettlement of people from one eco-geographical region to another within their own country. In conflicts linked to interregional migration, geographic origin is the primary criterion of the social and political relationship between the actors in the conflict; this distinguishes regionalized (internal) migration conflicts from ethnopolitical and center-periphery conflicts. Interregional migration causes members of different, not necessarily neighboring regions to confront each other. A resident population finds itself in threatening competitive situations that are to be 'fought out' with newcomers who are seeking work and land. Because of their former activities, groups that have migrated usually develop interests in using environmental resources other than those of the residents, who possess more detailed understanding of local nature-society relationships. Such migration is often structurally determined, such as when drought and soil erosion cause movement from unfavorable areas to more productive or (peri)-urban areas; migration, forced resettlement and expulsion are linked to large-scale agro-industrial projects.

Very differing interaction and behavior patterns arise. In overpopulated and degraded mountainous regions with nomadic cultures, there are significant migratory movements to irrigated areas or to population centers in the forelands with sedentary arable farming cultures. Former cattle breeders have difficulty adjusting to unfamiliar irrigated agriculture (Central Asia). Farmers also migrate from eroded highlands to more fertile lowlands settled by semi-nomads (Ethiopia). A third migratory movement arises when semi-nomads escaping drought and soil erosion seek refuge in semi-arid or sub-humid mountainous regions inhabited by arable farmers (Jebel Marra in Western Sudan). The underlying patterns to these three interregional interaction systems between highland and lowland inhabitants are comparable. Migratory movements are determined by the socio-ecological difference between living conditions in relatively favorable as opposed to unfavorable areas. Observation of this difference shows it ending in a vicious circle: conditions in the relatively unfavorable area that have slowly become inhabited because socio-economic and demographic pressure in the relatively favorable area was too high, quickly become ecologically precarious. Groups influenced by climate fluctuations or permanent climate changes therefore move out of marginal, unfavorable areas back into relatively favorable areas that are for their part already under heavy pressure. The capital of the country is often seen as a favorable area, where a tense mixture arises of city residents and newcomers in great numbers.

Regionalized conflicts usually remain limited to an area or only have a small geographical scope; they do not affect the whole country or the capital as the center of power. Violent confrontations occur either in (peri)-urban areas and are mixed with violence and gang crime, often involving former soldiers, or they occur in the disputed ecoregion itself.

Interregional migration or resettlement nevertheless leads to political confrontation over the power of members of regions that were formerly of secondary or marginalized political or economic importance. In countries where there are strong regional disparities that are not balanced by legal and democratic mechanisms (financial adjustment, structural aid funds, etc.), interregional migrations can lead to political struggles for state power if new regionalist groups are able to successfully penetrate the ruling elite or otherwise expel them from power.

2.4 Cross-border migration conflicts

If environmental migrants voluntarily cross or are expelled over national boundaries and settle in rural areas close to the border or in the cities of another country, they constitute a serious potential for social conflict, sometimes with ethnopolitical features that can be instrumentalized. Refugees of environmental degradation intensify conflict situations, particularly in those places where the economic situation is eroded, political instability or traditional strife already exist, or where lines of conflict are deepened or drawn anew by migration and population pressure. Violence accelerates this process and opens gulfs, often stemming from pre-colonial times, between formerly hostile clans, ethnic groups and nations.

Environmental flight usually means the persistent infiltration of territory in which more favorable conditions for life and survival are expected. Only exceptional situations such as extended drought lead to spontaneous mass flight. The paths of flight branch out. In many places, it is worth crossing a national boundary because foreign favorable areas are geographically closer and the people there are more similar (possibly belonging to the same ethnic group) than people in a distant capital city. In some cases, routes lead to northern industrial countries that are generally seen as favorable places due to development differentials and the spread of northern values.

Conflicts that are triggered by environmental flight from rural areas acutely threatened by degradation in developing countries fall into three categories:

- *Firstly*, problems arise from poverty and underdevelopment as such. These include the crisis of traditional agriculture and rural population growth. Most of the population in developing countries lives in rural areas. Inadequate infrastructure, unsettled tenure or diverging notions of rights, and insufficient access to credit for farmers contribute to agricultural structures of low productivity and, as a result, a low income structure. At the same time, market economics affects the dissolution of traditional structures only very selectively, so that subsistence and small farmers are further pressured by the duality of market economics and traditional production methods. Phenomena as diverse as soil erosion, heavy rains and floods, aridity and drought, salinization through irrigation, deforestation and the overgrazing of grass savannas by ever larger herds of cattle, accelerate the disintegration of traditional life orders; that is, the specific local or regional ensemble of economy, culture, neighborhood and family. After a certain point, people have no choice but to leave their homes. The erosion of cropland and pasture *due to poverty* is the kind of environmental destruction that has caused the most environmental refugees to this day. The direct reasons that lead to the decision to leave must be sought directly in the affected region.
- *Secondly*, problems arise from the processes of development, even if they are hidden and weak; these include the mechanization of agriculture, dam building, partial industrialization and urbanization. The mechanization of agriculture leads to the unemployment of countless small farmers and rural workers. These people are forced to retreat to marginal lands or to migrate to cities. While this forced movement is seen as a form of flight from the land or expulsion due to economic or political forces, diverse side effects of mechanization such as the use of fertilizers, pesticides and herbicides, and the widespread salinization of soil by irrigation systems, are what lead to the real flight resulting from environmental degradation.
- *Thirdly*, the endangering of a growing number of people who are forced by socio-economic and demographic factors to settle in marginal lands or in threatened regions, thus becoming victims of natural disasters (such as landslides), must be cited. These events are consistently seen by those affected as 'natural disasters' and not as 'social disasters', which is what they really are.

2.5 Demographically caused migration conflicts

The relative overpopulation of highly stressed or overused ecoregions provides another motive for migration, flight and resettlement. Concomitant conflicts resemble internal or cross-border migration conflicts. Population growth is closely linked to the socio-economic problems of poverty. In several countries (Rwanda, Bangladesh, Indonesia), local population pressure adds another – conflict-determining – dimension to the factors already cited. In these cases there is a demographically determined, distinct gap in the ecological and economic carrying capacity of ecoregions. This in turn has led to increased competition for scarce resources endangered by degradation. Evidence of this is reflected in increasingly smaller arable areas per capita of the population. Small cultivation areas, low crop yields and insufficient alternatives in the commercial sector force large segments of the population to migrate to the city. The link between population growth and environmental degradation is also clearly visible where population pressure on still unused areas of rainforest or mountain regions increases through migration. Another trend is the above-average rate of population growth in ecologically sensitive coastal regions. And finally, the gradual infiltration of landless people or semi-nomads and their herds into national nature protection zones or parks has considerable conflict potential.

Acute conflicts arise when migration is towards regions in which the land and/or water resources are already being heavily used by the resident population. If demographic, ecological, social and/or ethnopolitical factors accumulate, conflicts arise with a high probability of armed violence, as, for example, between Bengali environmental refugees and inhabitants of the Indian province of Assam.

Demographically caused migration conflicts are an expression of fundamental changes in the structure of societies and their international environment. Within states, the gap between center and periphery is decisive – if pressure in the periphery is too great, it is passed on in the form of migration to the city so that growing social potential for conflicts and violent crime arises in peri-urban areas. These conflicts set a dynamic force of internationalized migration in motion that – in the context of violent coups and civil wars – not infrequently assume the form of mass flight (Great Lakes region in East Africa).

2.6 International water conflicts

Conflicts in this category are between states that share a river basin. Conflicts over water use between *governmental* actors – as opposed to those between *intrastate* actors – are usually between adjoining riparian states in upper and lower river basins, or between highland and lowland inhabitants. Due to the direction in which water flows, the options for using water, and the relationships of dependency involved, are distributed very asymmetrically between the states sharing the river basin. Flowing waters are the most obvious example of the general contradiction between the natural borders of ecoregions and the historically determined political boundaries of nation states. They create a functional ecological link between regions which otherwise belong to different spheres of power.

Conflicts regarding pollution and distribution must be differentiated from each other – the first concern the qualitative degradation of resources, the second their quantitative scarcity. Pollution conflicts are disputes over indivisible public goods that usually concern regulatory limits, jurisdiction and economic costs. Since all adjoining parties are generally interested in finding a solution to the problem, these conflicts are easier to solve, using technological measures or financial compensation, than the issue of the real availability of the resources themselves. Distribution conflicts are those concerning divisible goods or zero-sum games. They touch more directly on state sovereignty and integrity than the problem of water pollution does. Water distribution is a highly sensitive issue which is treated as a threat to national security, particularly in regions that in any event suffer water scarcity seasonally or in absolute terms (in, for example, the Middle East or the Indian subcontinent). Because they can easily be instrumentalized, scarcity conflicts inevitably become interwoven with other sources of conflict in crisis areas. Both basic types can of course appear in combination with each other.

International water conflicts develop in the sensitive area between the immediate interests of the affected states in the divided resources, the distribution of power between them and the traditional pattern of confrontation and cooperation in their relations to each other. If a cooperative bilateral tradition exists in other political arenas, this has a subduing effect on water use conflicts. Precedents of successful compromising or even institutionalized mechanisms for settling disputes and cooperation reduce the danger that conflicts over water use will get out of control and escalate. In integrated regions, international agreements regarding the use of internationalized bodies of water are more frequently and quickly adopted than in non-integrated regions. The most favorable situation for reducing conflict is when intergovernmental water regimes already exist from which to proceed should new conflicts over water use arise.

Accordingly, pollution and scarcity conflicts in and between industrial states in politically integrated regions can be dealt with at the political level due to the high negotiating competence of affected states and existing regulatory mechanisms. Here, environmental conflicts can become a catalyst for cooperation if the actors participating have the will to achieve a political compromise and believe that technical solutions can be implemented. In contrast, such conflicts are a potential confrontation factor for states weak in political institutions, or in poorly integrated regions. Conflicts that cross the violence threshold must be investigated in the context of this political environment. Even disputes over the distribution of internationalized water resources in arid regions do not lead to the automatically violent resolution of conflict that some authors presume.

In the Middle East, international water conflicts are especially prone to escalation because the extreme scarcity of water is interwoven with the century's conflict between the Israelis and the Arabs. Water has always been an additional matter of dispute and an instrument in continuing the traditional conflict over boundaries, security and national identity.

As in every conflict, power relationships are a major factor in deciding the means by which a conflict between riparian states over the use of water is dealt with. In the context of high institutional involvement and cooperative tradition, the importance of power relationships is mediated by legal limits and the rules of customary behavior. The presence of a competent hegemony on the upper river basin can have a very stabilizing effect on this environment if it demonstrates willingness to cooperate in the interest of good neighborly relationships, and acts responsibly to make mutually satisfactory technical solutions possible (for example, USA–Mexico).

In a confrontational situation, the riparian position of the militarily and politically stronger state is all-important. The geographic course of a river is a power factor worthy of attention. If a country is the upper riparian and well-equipped militarily, it holds all the trump cards. It can discriminate inconsiderately against lower riparians regarding available water. If its superiority is overwhelming, cost-benefit analysis will prevent lower riparians from engaging in a violent dispute, in spite of discrimination. At the most, violent disputes will occur at the intrastate level as the outcome of migration between members of different ethnic or regional groups. If the more powerful state is the lower riparian, it can compensate politically for its less favorable ecoregional position.

To date no open wars have been proven to have been caused by water distribution issues alone. Only if a water issue coincides with an unfavorable political situation such as a historical conflict will it trigger war-like activities (in, for example, the skirmish between Israel and Syria before the Six-Day War). Asymmetrical geographic positions in the basin are then unrestrainedly exploited by the upper riparians. Scarcity originating in climate change and population pressure also intensify the situation. If dominant riparians have authoritarian regimes that enforce modernization processes against their own or neighboring populations, shared waterways will persistently be latent matters of dispute. They often provide grounds for military threats against neighboring countries or for the use of force against marginalized rural groups (see Libiszewski in Baechler et al. 1996: 117–167).

2.7 Global environmental conflicts

Conflicts rooted in neo-colonial exhaustion of natural resources are the international version of center-periphery conflicts. The special feature of such conflicts is that the conflicting parties belong to different cultures. As we have seen, the great distance between representatives of foreign interests and marginalized rural populations is not just a geographical one. This kind of conflict is characterized by uses of force that do not correspond to the scope and globality of the environmental problem. European protests against French nuclear testing were relatively moderate compared to protests made by New Zealand and other Pacific states. Bloody riots and direct confrontations between the French Army and protesting groups occurred only in French Polynesia, in close proximity to the cause itself. The same applies to the conflict between the indigenous Ogonis and Andonis and security forces connected to international oil corporations in Nigeria's delta region.

Climate change and depletion of the ozone layer demonstrate the globalization of environmental transformation, but they do not lead directly to global conflict. Global degradation promotes structural and ecological heterogeneity at the regional level. Effects benefit some and victimize others unequally. Specific statements about the socio-economic and ecological effects of climate change cannot be made based on case studies. Since the rise in sea level predicted will be a phenomenon of the intermediate and long-term future, and continuing drought in arid and semi-arid zones is not clearly attributable to human-induced climate change, statements remain speculative. Because of the dilemma of development, it is certain that the losers will be found where societal relations to nature are already precarious. If current conflicts can be traced back to global environmental phenomena, they probably mainly concern intrastate conflicts of types A1 to A4 (see Table 1). In other words, acute conflicts do not arise along the fault line between North and South, but where climate change contributes to the collapse of agricultural societies, the migration of millions of people and the implosion of political authority.

3 Conflict as the outcome of specific environmental syndromes

The previous section approached the environmental conflict syndrome by categorizing conflicts into different types according to the participating actors and their characteristics, interests and behavior patterns. For this purpose, conflict background, immediate causes, symptoms and different levels of conflict were studied and set in quasi-causal relationship to each other (see Figure 1).

This conflict typology should not create the impression that it includes all conceivable environmental conflicts. In fact, the categories refer only to certain syndromes of environmental degradation and their socio-economic and political consequences. Other syndromes and their effects, such as the impact of heavy industry in industrial states or the consequences of disasters like Chernobyl, are not included here.

In using the 'syndrome approach' of the German Advisory Council on Global Change (WBGU 1997) as a basis, it can be relatively precisely determined which combined environmental problems, under corresponding political and geographical circumstances, lead to violent conflict.

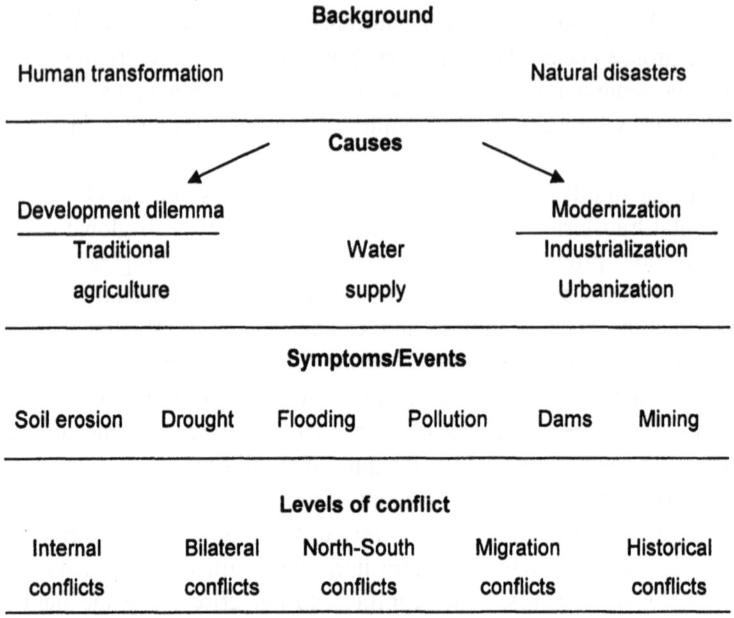

Figure 1: Environmental conflict syndrome

The German Advisory Council on Global Change assumes that there are typical patterns of interaction between civilization and environment in certain regions. It describes these as follows:

> These functional patterns (syndromes) are unfavorable and characteristic constellations of natural and civilizational trends and their respective interactions, and can be identified in many regions of the world (WBGU 1997: 112).

Syndromes reach beyond single sectors such as economics, politics or demography. But they always have a direct relation to natural resources and their human transformation. In its 1996 annual report, the Council differentiated between 16 syndromes of global change (ibid., 116) categorized into three groups: utilization, development and sink syndromes. For example, the Sahel syndrome belongs to the 'utilization' group and is characterized by the overcultivation of marginal land (see Biermann's chapter in this volume for more details on syndromes).

The following orders the conflicts studied according to syndromes. In some cases, multiple categorization is possible because two or three syndromes overlap and mutually reinforce each other (see Table 1).

Table 1: Types of environmental conflict

A. Intrastate

1 Ethnopolitical conflicts

Place	Syndrome	Intensity of violence		
		none	low	war
North Ghana	Sahel		x	
Nigeria	Sahel			x
Rwanda	Overexploitation/ Rural Exodus			x
Sudan (Jebel Marra)	Sahel			x
Sudan (North-South)	Sahel/Overexploitation			x
Tajikistan	Aral Sea/Green Revolution			x

2 Center-periphery conflicts

Place	Syndrome	Intensity of violence		
		none	low	war
Botswana	Aral Sea	x		
Bougainville	Katanga			x
Brazil (Amazon)	Sahel/Overexploitation	x		
Brazil (lower Amazon)	Overexploitation/Green Revolution		x	
Brazil (Tapajos River)	Waste Dumping	x		
Chile (Alto Bio Bio)	Aral Sea		x	
India (Narmada)	Aral Sea	x		
Nigeria (Ogoni)	Katanga		x	
Nigeria (North)	Aral Sea	x		
Papua New Guinea	Waste Dumping	x		
Philippines (Mt. Apo)	Katanga/Overexploitation	x		
Senegal (Cassamance)	Sahel/Dust Bowl			x
Thailand (Northeast)	Overexploitation/Dust Bowl	x		

3 Internal migration and displacement conflicts

Place	Syndrome	Intensity of violence		
		none	low	war
Algeria	Sahel/Rural Exodus			x
Brazil (Tocantins)	Aral Sea	x		
China (Henan)	Sahel/Rural Exodus	x		
Namibia (Ovamboland)	Overexploitation/Rural Exodus	x		

B. Intrastate with cross-border dimension

4 Cross-border migration conflicts

Place	Syndrome	Intensity of violence		
		none	low	war
Algeria (Mali, Nigeria)	Sahel		x	

5 Demographically caused conflicts

Place	Syndrome	Intensity of violence		
		none	low	war
Bangladesh (Assam)	Sahel/Rural Exodus		x	
Bangladesh (Chittagong Hill Tract)	Overexploitation/Rural Exodus			x
Indonesia (Java)	Sahel/Overexploitation	x		

C. International

6 Water/river basin conflicts

Place	Syndrome	Intensity of violence		
		none	low	war
Ganges: Bangladesh vs. India	Aral Sea/Sahel	x		
Aral Sea: Central Asia	Aral Sea/Dust Bowl	x		
Fergana Valley: Uzbekistan vs. Tajikistan	Aral Sea/Overexploitation		x	
Naryn/Toktogul: Uzbekistan vs. Kyrgyzstan	Aral Sea		x	
Mekong: China vs. Laos/Camb./Vietnam	Aral Sea	x		
Lake Chad: Nigeria vs. Cameroon	Sahel/Overexploitation	x		
Colorado/Rio Grande: USA vs. Mexico	Aral Sea/Overexploitation/Waste Dumping/ Contaminated Land	x		
Tigris-Euphrates: Turkey vs. Syria/Iraq	Aral Sea/Overexploitation/Dust Bowl	x		
Jordan: Israel vs. Syria/ Jordan/Palestine.	Aral Sea/Overexploitation/Dust Bowl	x		
Blue Nile: Egypt vs. Sudan/Ethiopia	Aral Sea/Overexploitation/Sahel	x		
Senegal: Mauritania vs. Senegal	Aral Sea/Sahel/Rural Exodus		x	
Danube (Gabcikovo): Hungary vs. Slovakia	Aral Sea	x		

7 Globalized (center-periphery) conflicts

Place	Syndrome	Intensity of violence		
		none	low	war
Nuclear tests: French Polynesia	Aral Sea/Waste Dumping/ Scorched Earth		x	

The syndrome approach shows which environmental syndromes are particularly conflict-laden and, moreover, prone to violence. These are primarily those in the 'utilization' and 'development' categories – i.e. primarily those related to eco-geographic constellations in Third World regions. The Sahel, Aral Sea and Over-exploitation syndromes are those cited most frequently. These often appear in conjunction with the Rural Exodus syndrome. The results tally with the empirical observations that led to constructing the typology of conflict. The overcultivation of marginal land (Sahel syndrome), the destructive exploitation of natural eco-systems (Overexploitation syndrome), environmental degradation caused by aban-doning traditional land use forms (Rural Exodus syndrome) and environmental damage caused by the deliberate transformation of natural areas by large projects (Aral Sea syndrome) all affect rural actors who, compared to other actors, lose access to natural, economic and political resources. They not only lose natural alternatives, but socio-economic perspectives and political options for action as well. The decision to escape a hopeless situation by using force or, conversely, to repress discriminated actors with force, is therefore part of the societal relationship to nature that can be aggregated in a short list of syndromes.

The following model of conflict puts environmental conflict syndromes into context by using hypotheses to relate individual variables to each other in a quasi-causal way.

4 A model of environmental discrimination as a source of conflict

As already shown, disparate access to natural resources is reflected in different options for action. These are manifested in international, societal, regional, ethnopolitical and power struggles over scarce and degraded resources. If there are social groups with higher economic aspirations and a large demand for resources facing limited or unclear development perspectives, degraded resources, depleted energy sources and poor competence in sustainably harnessing new energy sources, conflicts over which actors can continue to use the degraded resources, and which will be excluded, will inevitably arise. Whether organized force is used depends on what options for resolving conflict in civil society are available, the actors' ability to mobilize, their awareness of alternatives, their preferences and limitations. The discrimination of (marginalized) actors by other (dominant) actors is usually a reason for conflicts to escalate. It is not degradation as such that causes actors to mobilize into violent action, it is socio-political discrimination – inde-pendently of the single conflict type outlined above. This knowledge gives us the chance to design a general model that can, starting with discrimination, integrate the most diverse sources of conflict.

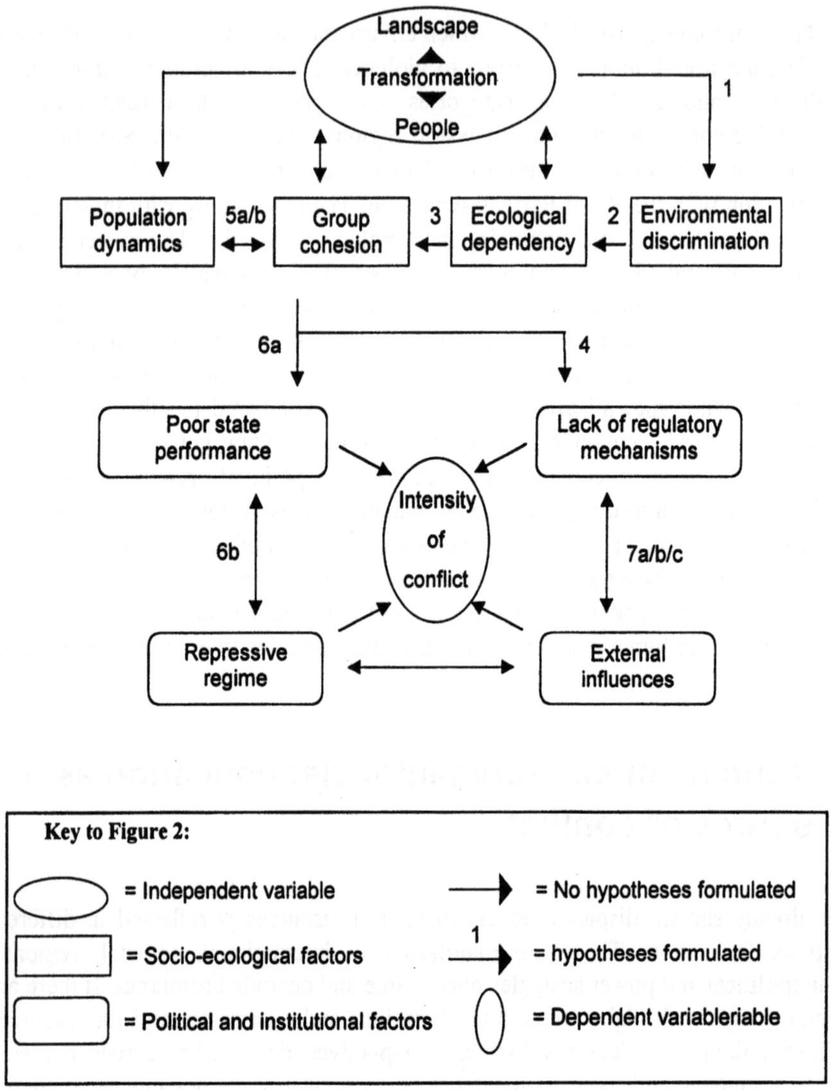

Figure 2: Model of environmental conflict

The model of environmental conflict is based on seven interconnecting hypotheses. They can be operationalized by sets of indicators. The indicators were selected so that they reflect what is to be indirectly measured. They aim at being valid and significant. However, the best indicators are useless if they are not based on data. Therefore, indicators were chosen that had already been used in analyses or are currently being evaluated for more recent work on 'sustainable development' (for example, by the UNDP, the World Resources Institute in Washington or the

World Bank). A detailed explanation of the hypotheses or choice of indicators will not be given here.[2]

4.1 Hypotheses and choice of indicators

The first hypothesis is the 'way in' to the conflict model. Starting with the phenomenon of human transformation of the environment, the heading 'environmental discrimination' is introduced. This is the condition in which different actors have clearly experienced inequity compared to other actors regarding material comfort, social security and societal participation because of their social, ethnic, linguistic, religious and regional identities. These actors have been systematically hindered or excluded from access to natural resources they depend on for sustenance and survival. Therefore, the first hypothesis is:

The greater the environmental discrimination of actors in an arena, the greater the probability that they will mobilize against the source of discrimination.

Indicators to 'measure' the dimensions of discrimination are:

- limited access to fresh water;
- high rate of child mortality under the age of five;
- per capita food production;
- low agricultural productivity per hectare;
- degree of socio-ecological vulnerability;
- low HDI[3] compared to other groups;
- low degree of participation in political and societal activities.

Not every actor who experiences discrimination will mobilize against its source. Those who are not highly dependent on natural resources or who seek substitution by alternatives might take other routes. The degree of direct dependence on nature or the landscape used as a life support system is of great significance. The second hypothesis is:

The more dependent on degraded renewable resources victims of environmental discrimination are, the more readily they can be mobilized for collective actions.

---

2 See Baechler 1997: 205–228.

3 The *Human Development Index* (HDI) is an aggregated indicator measuring human development levels. It consists of indicators for three 'significant elements of life': life expectation at birth, literacy rate and real purchasing power per person (*purchasing power parities* – PPP). The United Nations Development Programme (UNDP) developed it in 1990 with the intention of evaluating human development based on a more comprehensive socio-economic standard than per capita gross domestic product (GDP) (see Nohlen, Dieter and Nuscheler, Franz (Eds.): Handbuch der Dritten Welt, Vol. 1, 3rd edition 1993: 92–93).

Indicators of the degree of dependence are:

- availability of resources;
- high risk of soil erosion;
- low HDI as indicator of high dependency on renewable resources;
- high rural poverty;
- food imports vs. cash crop exports.

Not every discriminated actor highly dependent on natural resources will resort to violence. This step depends on various factors regarding the formation of 'strategic' groups. An actor's capacity to resolve conflicts is a function of the internal cohesion of that actor's group. The third hypothesis is:

The more cohesive a group is, the more readily discriminated actors in marginalized regions can be mobilized.

Indicators of the measure of cohesion are:

- degree of acceptance of leaders;
- degree of acceptance of the social order within the actor's group;
- number of parties within the actor's group;
- intensity of awareness of environmental degradation caused by others;
- opportunities for obtaining weapons.

The formation of groups and their cohesion has another aspect. If there are (traditional) local, practiced and accepted mechanisms for resolving conflicts, they will be an element in avoiding escalation. Therefore, the fourth hypothesis is:

The weaker regulatory mechanisms are, the more readily discriminated actors can be organized for action.

Indicators of the scope of regulatory mechanisms are:

- rural unemployment and size of informal sector;
- options to use legal instruments;
- degree of political participation;
- degree of communal self-administration;
- degree of political marginalization;
- scale of rural exodus vs. urbanization.

Demographic development is greatly significant to the escalation of environmentally conditioned conflicts. However, there is no linear relationship between a high rate of population growth and the intensity of a conflict. The fifth hypothesis takes non-linear (complex) relationships into consideration:

(a) The higher demographic pressure in a sensitive arena is, the more readily collective actors can be mobilized.

This hypothesis assumes that population pressure promotes competition for scarce resources and that actors are capable of resolving conflicts. This refers back to the third hypothesis and the ability of actors to form strategic groups. But at the same time:

(b) The higher demographic pressure in a sensitive arena is, the more difficult it is for collective actors to be mobilized.

This hypothesis also assumes that there is competition, but reflects the dispersion of actors who usually leave their rural arena for the city when scarcity leaves them no other choice. In this situation, strategic groups are not formed.

Indicators of the level of demographic pressure are:

- population growth and density;
- regional distribution (number of people per hectare of fertile land);
- urban-rural distribution and urbanization rate as indicators of land flight;
- high agricultural occupation combined with high rural unemployment.

A state's performance regarding environmental degradation and the treatment of discriminated actors plays an important role in the development of a conflict. Poor state performance can accelerate a conflict. The sixth hypothesis assumes that:

(a) The poorer a state's performance is, the more readily discriminated actors can be mobilized for armed action.

Indicators to measure state performance are:

- type of regime;
- stability of regime;
- distribution of administrative and governmental tasks;
- spending on environment, rural development and poverty reduction.

On the other hand, poor state performance, if it coincides with a repressive regime, can also be a factor prohibiting the use of force. Strong and centralized authority does not allow actors in opposition to organize themselves, accrue weapons or pursue a strategy of violence. Therefore:

(b) The more repression there is, the less likelihood there is of discriminated actors having recourse to armed conflict.

Indicators of the degree of repression are:

- number of prisoners;
- forced resettlement;
- torture and executions;
- number of people missing;
- massacres;
- political genocide.

Not only performance and repression have an effect on the intensity of force. Influences from outside the arena – which is usually intrastate, as we have seen – are also noteworthy. They can raise or lower the threshold to violence. It is a complex constellation. The seventh and last hypothesis posed is that:

(a) The lower the (apparent) threshold to violence in a wider (international) arena is, the more readily discriminated actors can be mobilized to take violent measures.

Not only state actors but also international corporations impair the life world of discriminated actors, for example by operating mines or building dams. Therefore:

(b) The clearer it is that environmental degradation is caused by international actors, the more readily discriminated actors can be mobilized to take violent measures.

Discriminated actors often find support through NGOs (for example, Greenpeace). Although NGOs take the side of the marginalized groups and animate them to protest against environmental degradation, their effect is to reduce violence since they value peaceful means for resolving conflicts. Therefore:

(c) The more prominent the role of NGOs, the less probability there is of organized force being used by discriminated actors.

Indicators of the degree of external influence are:

- presence of multinational corporations and degree of their degrading activities;
- support of corporations by elite groups in the centralized state;
- control over renewable resources;
- spillover of historical conflicts;
- lack of compensation;
- relationship between actors and NGOs.

5 Conclusions – discriminated actors in an inescapable situation

In the model it becomes clear that environmental degradation – like other sources of conflict – starts a socio-political process that is more or less prone to conflict. The model should counter the impression that serious environmental problems necessitate 'strong' answers. This kind of linearity doesn't exist. Environmental conflicts are probably more complex than many other kinds of conflict because they are strongly shaped by patterns of awareness, uncertainty in dealing with nature, competition between different levels and undefinable societal transformation processes.

In the second general hypothesis, the ENCOP team and I assume that

Environmentally caused armed conflicts arise if and only if several of the following five factors coincide:

- *Degraded resources that cannot be substituted in the foreseeable future put groups whose existence depend on the preservation of these resources in an inescapable and desperate situation. Inescapable situations are those one cannot leave rationally or intentionally.*
- *There is an absence of societal regulatory mechanisms. This is nothing other than the expression of the social and political powerlessness of state, traditional and modern civil society institutions. A political system becomes 'a might without power' if it is incapable of establishing certain social and political conditions. Societal goals such as the sustainable use of resources become unattainable.*
- *Environmental degradation is instrumentalized by state or societal actors in order to pursue group interests in such a way that the problem of resources becomes an issue of group identity.*
- *The environmentally caused conflict is in an arena where actors can organize themselves, accrue weapons and win allies either from groups facing similar problems, or from other social strata or from abroad.*
- *The environmentally caused conflict is part of an already existing conflict situation; it receives fresh impetus through subjective perception of the consequences of environmental transformation or through polarized groups' intensified competition for resources.*

Inescapable socio-economic conditions, the lack of societal regulatory mechanisms, instrumentalization, group identity, organization and armament, and the superimposition of a historical conflict, create the framework for environmentally caused armed conflict.

It is the same cultural factors and behavioral patterns that lead to environmental degradation on the one hand and armed conflict on the other: competing interests in using renewable resources, scarcity and pollution through excessive exploitation, unclear or competing legal structures and rights of tenure, and the political mobilization of collective actors who are bound up in struggles over distribution or their own defense. They are the expression of deep social changes affecting the rural population of developing countries.

In light of the fact that similar conflicts with similar causes, actors and goals are occurring at the same time in many places, the question arises whether these individual conflicts are the harbingers of larger processes of upheaval affecting agrarian structures in developing and transitional societies. If that is the case, this phenomenon and its progress can be interpreted in two ways. Environmental conflicts are either the rearguard action of increasingly marginalized rural populations in the South who have been maneuvered into capitulating to modernization and the environmentally conditioned dissolution of their life orders, and thus find themselves in an inescapable situation. Or they are the vanguard fight of future constellations of conflict that will lead to an enduring change in societal relations to nature through a fundamental reevaluation and reinvigoration of rural structures.

References

Baechler, Günther 1997: *Violence Through Environmental Discrimination. Causes, Rwanda Arena, and Conflict Model.* Dissertation. Cambridge MA, Bern (forthcoming).

Baechler, Günther et al. 1996: *Kriegsursache Umweltzerstörung. Ökologische Konflikte in der Dritten Welt und Wege ihrer friedlichen Bearbeitung.* Vol. I. Zürich: Rüegger.

Baechler, Günther and Kurt R. Spillmann (Eds.) 1996: *Kriegsursache Umweltzerstörung. Environmental Degradation as a Cause of War. Regional- und Länderstudien von Projektmitarbeitern. Regional and Country Studies of Research Fellows.* Vol. II. Zürich: Rüegger.

Baechler, Günther and Kurt R. Spillmann (Eds.) 1996: *Kriegsursache Umweltzerstörung. Environmental Degradation as a Cause of War. Länderstudien von externen Experten. Country Studies of External Experts.* Vol. III. Zürich: Rüegger.

Böge, Volker 1993: *Das Sardar-Sarovar-Projekt an der Narmada in Indien – Gegenstand ökologischen Konflikts.* ENCOP Occasional Paper 8. Zürich: Swiss Federal Institute of Technology (ETH Zürich), Swiss Peace Foundation.

Carino, Joji 1993: *Local Solutions to Global Problems: Perspectives from the Cordillera Peoples' Movement.* Bern: Manus.

Danielsson, Bengt 1993: *Tests in French Polynesia and Political Protest.* Bern: Manus.

Durth, Rainer 1993: *Kooperation bei internationalen Oberlauf-Unterlauf-Problemen – ökonomische Aspekte.* Diskussionspapier. Graduiertenkolleg für Integrationsforschung, Europakolleg Hamburg.

Ehrensperger, Albrecht 1993: *Transmigrasi – Staatlich Gelenkte Umsiedlungen in Indonesien, ihre Ursachen und Folgen in ökologischer sowie konfliktueller Hinsicht.* Bern: Manus.

Gallagher, Dennis 1994: "Environment and Migration: A Security Problem?", in: Günther Baechler (Ed.): *Umweltflüchtlinge. Das Konfliktpotential von morgen?*, 65–72. Münster: agenda.

Gantzel, Klaus-Jürgen 1987: "Tolstoi statt Clausewitz!? Überlegungen zum Verhältnis von Staat und Krieg seit 1816 mittels statistischer Beobachtungen", in: Steinweg, Reiner (Ed.): *Kriegsursachen.* Friedensanalysen Heft 21. Frankfurt/M: Suhrkamp.

Homer-Dixon, Thomas F. 1991: "On the Threshold: Environmental Change as Causes of Acute Conflict", in: *International Security*, 16/1, 76–116.

Homer-Dixon, Thomas F. 1994: "Across the Threshold: Empirical Evidence on Environmental Scarcities as Causes of Violent Conflict", in: *International Security*, 19/1, 5–40.

Kende, István 1982: *Kriege nach 1982.* Militarpolitik Dokumentation Heft 27. Frankfurt/M.

Klötzli, Stefan 1993: *Der slowakisch-ungarische Konflikt um das Staustufenprojekt Gabcikovo.* ENCOP Occasional Paper 7. Zürich: Swiss Federal Institute of Technology (ETH Zürich), Swiss Peace Foundation.

Lane, Jan-Erik and Svante Ersson 1994: *Comparative Politics. An Introduction and New Approach.* Oxford: Blackwell, Polity Press.

Libiszewski, Stephan 1992: *What is an Environmental Conflict?* ENCOP Occasional Paper 1. Zürich: Swiss Federal Institute of Technology, Swiss Peace Foundation.

Nohlen; Dieter and Nuscheler, Franz (Eds.) 1995: *Handbuch der Dritten Welt.* Vol. 1, 92–93. Bonn: Dietz.

Ogata, Sadako 1994: "Environmental Refugees. Statement by the United Nations High Commissioner for Refugees", in: Baechler, Günther (Ed.): *Environmental Refugees. A Potential of Future Conflicts*, 41–48. Münster: agenda.

Quimpo, Nathan F. 1993: *Environmental Conflict in the Philippines?* Bern: Manus.

Suhrke, Astri 1994: "Environmental Refugees and Social Conflict. A Framework for Analysis", in: Baechler, Günther (Ed.): *Umweltflüchtlinge. Das Konfliktpotential von morgen?*, 83–100. Münster: agenda.

WBGU (Wissenschaftlicher Beirat der Bundesregierung Globale Umweltveränderungen – German Advisory Council on Global Change) 1997: *World in Transition: The Research Challenge. 1996 Annual Report.* Berlin, Heidelberg, New York: Springer.

Wolf, Eric 1969: *Peasant Wars of the Twentieth Century.* New York.

Syndromes of Global Change: A Taxonomy for Peace and Conflict Research

Frank Biermann

1 Introduction

Debate in the social sciences, specifically in the environmental and security realms, has a new buzzword: 'environmental security'. As is often the case with catchphrases that are broadly debated and lend themselves to wide debate, the range of participants can lead to conceptual indeterminacy. Indeed, the ability of individual discourses to communicate with each other can rapidly be lost. It is therefore better to avoid using the term 'environmental security' and instead to juxtapose the two concepts of environment and security without suggesting any specific link. After all, it is a matter of controversy whether there is a link between environmental degradation and security at all.

I will examine a limited question in the present paper – whether degradation of the natural bases of human existence or a conflict over jointly utilized natural resources can lead to *violent international or intrastate conflict.*[1] I shall not present an answer to this question, but a method for finding the answer: The syndrome research approach, as developed since 1992 by the German Advisory Council on Global Change (WBGU) and cooperating research groups[2] with the aim to create a tool for the integrated appraisal of global environmental change.

---

1 Frank Biermann is Global Environment Assessment Fellow at the Belfer Center for Science and International Affairs, Harvard University and worked before with the Secretariat of the scientific staff at the Secretariat of the German Advisory Council on Global Change (WBGU), located in Bremerhaven, Germany. This paper should not necessarily be taken to indicate the opinion of the Council. I owe a debt of gratitude for valuable criticisms and comments to Alexander Carius, Carsten Helm, Carsten Loose, Sebastian Oberthür, Benno Pilardeaux and Udo E. Simonis.

2 These include the staff of the QUESTIONS (Qualitative Dynamics of Syndromes and Transition to Sustainability) core project of the Potsdam Institute for Climate Impact Research, the staff of the "Syndrome dynamics" projects promoted by the German Federal Research Ministry, the members of the German Advisory Council on Global Change (WBGU) and their staff, and, finally, the staff of the Bremerhaven Secretariat of the Council.

In its 1996 Annual Report titled "*World in Transition. The Research Challenge*", the Council has argued that the syndrome approach can and indeed should play an important role in the study of global change in both the social sciences and natural sciences (WBGU 1997). The present paper examines whether the syndrome approach can also be utilized in conflict research to study the linkage between environmental degradation and violent conflict.

I will first briefly present the syndrome approach (Section 2) and will go on to discuss its prospective contribution to peace and conflict research (Section 3). I will then outline two possible applications of the approach (Section 4): the hypothesis of 'environmental refugees' and the hypothesis of 'water wars'. I will argue that for both hypotheses the syndrome approach offers a fertile basis for future empirical study.

2 Syndromes of global change

Global environmental changes can no longer be analyzed in isolation from each other: To gain an overall evaluation of present processes and adverse effects, the various branches of study need to cooperate in a manner that is both cross-cutting and integrative. This realization led in the 1980s to the establishment of the International Geosphere-Biosphere Programme (IGBP), in which the natural science disciplines came together to forge a common research strategy. Later, a counterpart research network was also established for the social sciences, in the shape of the International Human Dimensions Programme of Global Environmental Change (IHDP), whose secretariat has been set up in Bonn, Germany. The new attention devoted to the social aspect of human-induced environmental crisis has led to a less restricted terminology. Now the universal term 'global change' is used instead of 'global environmental change'.

If the social sciences, from ethnology to economics, and the natural sciences, from zoology to climatology, are to work together on an integrated appraisal of global change, then the representatives of the various disciplines must first find a uniform language, a commensurate methodology and a common canon of guiding questions.

More than forty models or approaches for integrated appraisal of global change have been developed by the worldwide global change research community. A special methodology has been developed by the German Advisory Council on Global Change (WBGU, in the following "the Council") – the 'syndrome concept'. Following on from its initial formulation (WBGU 1994), interdisciplinary working groups have methodologically refined the concept and have extended it using empirical studies. In addition to the members and staff of the Council, the research teams involved in this process have included the scientists of the QUESTIONS (Qualitative Dynamics of Syndromes and Transition to Sustainability) core project of the Potsdam Institute for Climate Impact Research and of the "Syndrome dynamics" projects supported by the German Federal Research Ministry.

The scientific analysis of global change requires its reduction to a limited number of variables whose relationships can be theoretically and empirically studied. In syndrome analysis, this is done on the basis of about eighty 'trends of global change' or, as one of the medical metaphors used in syndrome analysis, *symptoms*. These include environmental problems such as 'global warming' or 'soil degradation', and also social developments such as 'market globalization' or 'mass migration'. These symptoms are in turn assigned to nine different spheres of global change corresponding essentially to the defining demarcations of traditional scientific disciplines, ranging from 'biosphere' and 'hydrosphere' to the spheres of 'social organization' and 'economy'.

The individual trends of global change interact with each other in that, for instance, some trends amplify or attenuate others. Here the syndrome approach proceeds from the assumption that within the complex of interactions among various trends in various spheres of global change, certain *patterns* can be identified: central networks of interrelations of individual developments that are closely connected to each other and whose interactions can be identified and then analyzed as a dynamic mechanism. If such partially self-reinforcing dynamic processes lead to damage to human health and the natural environment, then they can be termed *syndromes* or 'clinical profiles' of global change.

The initial assumption based on expert assessments is that there are sixteen such syndromes of global change. These then need to be studied: What course does a certain syndrome take, which trends are particularly important to the mechanism? Which regions of the world already 'suffer' today from certain syndromes? Where, by contrast, is the 'outbreak' of a syndrome impending? At what point could and should practical politics bring its instruments to bear in order to interrupt the unfolding of a certain syndrome mechanism?

Figure 1 shows the 'network of interrelations' of one of the sixteen syndromes – the Sahel Syndrome. This syndrome refers to the dynamism of poverty-related, often irreversible agricultural overexploitation of marginal land, for which the Sahel has become a metaphor. The interrelations shown here were elaborated and described in detail by the Council in its 1996 Annual Report (WBGU 1997).

How is the network of syndrome interrelations in Figure 1 to be understood? The interactions among the trends do not describe mechanistic links such as in a clockwork mechanism or in the models put forward by physicists. The arrows leading from one trend to another are indicative of hypotheses with varying degrees of substantiation in empirical research. Thus in the natural science spheres some interactions can already be viewed as quasi-laws, while for some in the social science spheres even the direction of the arrow has not yet been finally confirmed by empirical study. Therefore the networks of interrelations do not always imply *statements* concerning real-world linkages. In the social science spheres, in particular, they often still indicate *questions* concerning linkages among symptoms that the respective disciplines yet need to answer. The network of interrelations is not a mechanistic world model. Rather it may serve as a *global change research agenda* structured by means of the syndromes, with the potential to provide a platform for integrated research teams taking in all disciplines, from the social sciences to the natural sciences.

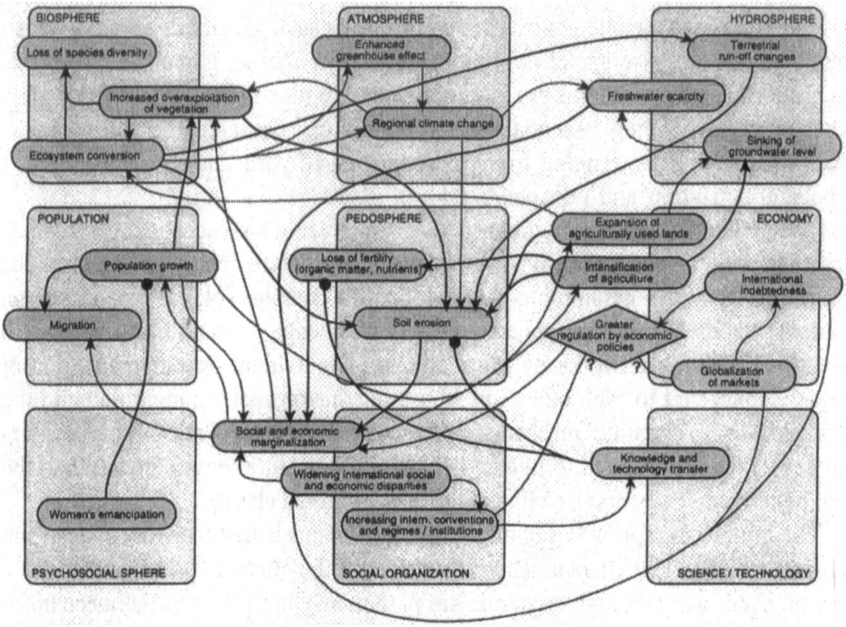

Figure 1: Syndrome-specific network of interrelations of the Sahel Syndrome Source: after WBGU 1997: 135.

3 Prospective contributions of the syndrome concept to peace and conflict research

If the syndrome approach is understood as a method for interdisciplinary analysis and assessment of global change, the next step is to consider how the social sciences could contribute to answering the questions that arise within the networks of interrelations of individual syndromes. Conversely, it is useful to consider how the syndrome approach could be utilized within the existing lines of research of the social sciences. Particularly those symptoms of global change that relate to the disciplines of international relations and peace and conflict research have until now tended to be among the 'empty spaces' on the map of syndrome analysis.

Useful links can readily be forged between 'traditional', sector-focused social science research and the integrated syndrome analysis and assessment of global change. A particularly interesting potential lies in the close cooperation among social and natural scientists, working, under the conceptual umbrella of the syndrome approach, on common lines of inquiry and initial hypotheses. This is in itself a general trend in global change research: For instance, interdisciplinary teams of scientists are already grouping around the issue of 'sea level rise', focusing on its consequences in both the ecosphere and the anthroposphere. The Land Use and Land Cover Change Programme is the first major international research program to unite organizationally the natural science community of the IGBP with its social science counterpart in the IHDP.

For such singular cooperation programs, the syndrome concept offers a conceptual umbrella under which specific interdisciplinary studies can organize. Moreover, syndrome analysis is conceptually suited to linking the study of local phenomena with the analysis of global – both ecospheric and anthropospheric – developments. For instance, what role does the globalization of markets play in the (cross-border) transfer of agricultural production methods imported from one set of local conditions into another, with the attendant, possibly severe environmental and developmental problems (Green Revolution Syndrome)?

Similarly, the syndrome approach can be productively applied to the nexus between 'environment and security'. If we apply a broad understanding of security – such as "(individual) human environmental security" (Lonergan 1997) – then the syndrome concept bundles and organizes a broad array of hypotheses concerning specific interactions, for some of which a good body of knowledge is already available. Understood thus, each syndrome – as an undesired dynamic process of change – represents a threat to the environmental security of the individual human being. Here 'preserving environmental security' becomes synonymous with 'sustainble development' or 'sustainability' (Simonis 1991), this in turn being equal to the absence of syndromes.

However, to apply a narrower concept of 'environment and security' is analytically more interesting, i.e. to concentrate on the propensity of environmental change to engender conflict. Thus, for instance, an 'increase in international (or intrastate) conflicts' is understood in syndrome analysis as a symptom of global change. For example, the construction of large dams or other large-scale projects (Aral Sea Syndrome) can lead not only to many forms of damage to the natural environment but also to international conflicts, as illustrated by the conflict between Turkey, Syria and Iraq over the water resources of the Euphrates and Tigris. This means that an increase in the 'dam construction' trend will tend to lead to an increase in the 'conflicts' trend, without this having to be empirically identifiable in every case. Not deterministic, but stochastic (quasi-)laws are assumed.

Which factors play a role in engendering conflict or in the form in which a possible conflict is resolved is a cardinal question posed within the 'syndrome analysis research network' to the expertise of peace and conflict research. Conversely, the expert knowledge of peace and conflict research serves to adjust and refine the syndrome concept. For instance, if we follow the proposition of Gleditsch (1997b) that transboundary environmental damage or international resource use conflicts in many cases do not lead to increased conflict but rather to increased cooperation, then the relevant linkages in the identified networks of interrelations would need to be adjusted accordingly. This question is currently a subject of debate in the syndrome research teams.

Above and beyond this, peace and conflict research can not only make a contribution to refining the syndrome concept by taking up the questions of the latter and integrating them in its own research program. Peace and conflict research could use the syndrome concept precisely to *answer* such questions.

For instance, the empirical study of the propensity of environmental problems or resource utilization disputes to lead to conflict requires, for operationalization, a *taxonomy* of environmental problems or resource utilization disputes. This may not be true in every case, as there are indeed contributions to the debate that proceed from only a single generic class of environmental problems as an independent variable for the explanation of 'environmental conflict' (see Sprinz in the present volume). Nonetheless, for cogent empirical studies and theoretical conclusions it would appear more appropriate to distinguish between different types of environmental problems or resource utilization disputes.

Using an inductive approach based on a large number of case studies of environmental and resource utilization conflicts, Baechler and others (1996) have identified seven types of such conflicts. The approach helps to concentrate research on these types, but has the shortcoming that the case studies were selected on the basis of the dependent variable. We learn, for instance, that 'center-periphery conflicts' – one of the seven types of environmental conflict – lead to violent conflicts; we do not, however, learn whether this applies to all such 'center-periphery conflicts', to most of them or to only a tiny minority (see Baechler in the present volume).

A different approach would be to classify according to individual environmental media, such as soil, air or water, whereby further subdivisions could be introduced (troposphere, stratosphere etc.). This, however, would implicitly sever environmental media from their social function: soil degradation caused by industrial agriculture in the USA is different from soil degradation caused by poverty-related overuse in Africa. This is why in the syndrome concept these two forms of soil degradation are assigned to different syndromes. The distinction is made essentially on the basis of their specific *social* characteristics (here the Dust Bowl Syndrome and the Sahel Syndrome).

We now have access to comprehensive empirical studies of intrastate and international conflicts in which environmental problems at least played a part in causation (see on this Baechler et al. 1996 and the papers of Baechler and Rohloff in the present volume). In a further step, this empirical data could be placed in relation to the sixteen syndromes of global change postulated by the Council, in order to identify the correlation of different syndromes with the number or intensity of 'environmental conflicts' observed in the context of these syndromes. Two such possible programs of empirical study are outlined below.

In summary, we may state that there are fruitful strands joining the syndrome research approach and the sector-focused social sciences, including peace and conflict research. In the context of the sustainability debate, syndromes are nothing other than a negative – and thus better operationalizable[3] – template of the sustainability of a society (see Reusswig 1997). Syndromes are non-sustainable, dynamic development patterns within societies, which – referring to peace and conflict research – can lead, with differing probabilities, to intrastate and international conflicts.

4 Practical applications

Two empirical complexes are cited particularly often in the 'environment and security' literature. The first of these is the – formulated in syndrome terms – symptom of migration or of so-called 'environmental refugees', which can be linked to various syndromes. 'Water wars' are the second frequently cited type of conflict, which can be caused by major water projects such as dams or river diversions with cross-border impacts. It is illustrated in the following for both types of conflict how syndrome analysis can make a specific contribution to the peace policy assessment of global change.

4.1 'Environmental refugees' (the Sahel Syndrome)

4.1.1 Fundamentals

If we posit environmental degradation as a cause of migration-related violent conflict, then two hypotheses need to be distinguished. The first question that needs to be examined is whether environmental problems do indeed lead to people leaving their home territories, and which environmental problems of which intensity can have such a consequence. The second (subsequent) hypothesis is that such intrastate or cross-border migration leads to violent conflict. The syndrome concept can contribute above all to evaluating the first hypothesis.

We can thus examine for all sixteen syndromes of global change – understood as typical patterns of adverse people-environment relations – to what extent people

---

3 A negative concept of sustainability does not prescribe in detail the development pathway of societies, but rather determines the set of non-sustainable paths. The 'crash barriers' of a development are thus explicitly identified that must in no event be transgressed, while a free 'action corridor' remains for society within the crash barriers. One advantage of such a 'crash barrier philosophy' – on which the Council's reports are based – is that it is easier to handle than a positive (necessarily global) concept of sustainable development. Another is that, as opposed to a 'scientifically derived' positive concept of sustainability, the identification of the concrete development pathway to be chosen remains with the legitimated societal actors – the governments.

have left their home territories in regions affected by the specific syndrome. A correlation between a syndrome and observable migration does not in itself provide any explanation, but does indicate a connection worthy of further examination. At the very least, the popular notion of 'environmental refugees' would be narrowed down to a specified number of situations.

This requires clarifying a number of further points. First of all, a specific syndrome needs to be identified – the 'pure' type of a specific syndrome needs to be operationalized in such a fashion that it can be applied to empirical findings so that certain regions can be diagnosed as being affected by the syndrome. This can be problematic because gaps in data mean that generally not all human-environment relations defined as essential to a syndrome can be included in the process of empirical syndrome identification. This means that the first reduction of reality – the determination of the basic type of syndrome – is amplified by a second reduction – operationalization – with the danger that the selection of limited data material distorts the analysis or even that the double reduction reduces the entire concept to triviality. Despite these difficulties, we have succeeded in determining a number of operationalizable identification rules for individual syndromes from which migration research and peace and conflict research can proceed. This shall be set out in the following for the case of the Sahel Syndrome, whose network of interrelations is shown above in Figure 1.

We encounter a further problem when relating regional syndromes to population displacement or to violent conflict: the question of the hen and the egg or cause and effect. Was the syndrome there first, with the consequence of migration, or was the migration there first, causing, with its complex of social and economic determinants, the environmental degradation? Although ultimately answers to this question can only be given by case studies, the syndrome concept tries to do justice to these relationships by ensuring that social factors are included in syndrome determination and that 'self-reinforcement loops' within syndromes are given basic theoretical consideration. For example, data on subsistence economy are included in the determination of susceptibility to the Sahel Syndrome, thus integrating social conditions from the outset. It is important to note that intrastate or cross-border conflicts have not as yet been used in the geographically explicit identification of syndromes, so that the data received can continue to be used for studying the causes of conflict without tautology.

4.1.2 Identification of the Sahel Syndrome

How can we now ascertain where precisely the Sahel Syndrome occurs, in order to examine on this basis whether in these regions an intensified migration is discernible? An analysis of the Sahel Syndrome was presented for the first time by the Council in 1994 in that year's Annual Report, and subsequently refined in various publications. As previously said, the Sahel Syndrome refers to a basic pattern of damaging people-environment relations characterized by a generally poverty-related overexploitation of marginal sites. Here we distinguish between the 'disposition' to the Sahel Syndrome and its 'intensity': The first concept refers to the susceptibility of a region to a syndrome, the latter refers to the extent to which a region is adversely affected by a syndrome.

As is common practice in the syndrome approach, both societal and natural factors were taken into consideration in disposition analysis. In the natural physical-geographical domain, the prime interest is to determine 'marginality': Which land is so marginal that overexploitation can lead rapidly to an often irreversible degradation of the soil? For this determination, the available data sets were used to integrate a series of individual factors including the angle of slope, variability of precipitation, constraints on plant growth due to aridity and temperature, soil fertility and the ease of access to surface water for irrigation purposes (WBGU 1997; Cassel-Gintz et al. 1997). Even if such a specific natural physical-geographical situation should reach critical levels,[4] this need not mean that damaging, possibly even self-reinforcing dynamics of environmental degradation are triggered. The example of Israel illustrates the extent to which societies with technically sophisticated land-use arrangements are capable even of pursuing export-oriented arable farming on marginal land.

Indicators expressing susceptibility to the Sahel Syndrome therefore need to include expressions of the social situation and of the options of people to pursue alternative courses of action. The more the people are dependent upon producing food on marginal land, and the less farmers and pastoralists have access to capital and technology-intensive, environmentally sound husbandry techniques, the more inescapable is the overexploitation of arable soil and pasture (with the possible consequence of migration). It is difficult to measure the direct dependence of people upon the utilization of their fragile land. As an initial measure, syndrome analysis uses the extent of a country's subsistence economy as an indicator, expressed as the ratio between the amount of food marketed and the amount of food estimated to be needed. This measure is complemented by the use as a second indicator of people's dependence on fuelwood, expressed as a function of per capita energy consumption and fuelwood burning as a proportion of total energy consumption (WBGU 1997).

A separate element of description is the *intensity* of the Sahel Syndrome, i.e. the extent to which the syndrome is already discernible in different regions. This measurement of intensity has in the meantime been carried out, mainly using data on the deterioration of soils and the exacerbation of poverty since the 1980s (Schellnhuber et al. 1997).

---

4 Even the determination of the natural physical-geographical disposition already integrates societal factors, as the possibility to utilize surface water is taken into consideration, which only makes sense in conjunction with a form of social organization that is 'capable' of irrigation. Without the surface water factor, Egypt in particular would be defined in the analysis as extremely water-poor, a frequent error of earlier studies (see in detail Cassel-Gintz et al. 1997).

If the various natural physical-geographical (ecospheric) and societal (anthropospheric) indicators are rated on a scale from 0 to 1 and linked,[5] the disposition and the intensity of the Sahel Syndrome can be graphically represented. Proceeding above all from the intensity analysis, it can then be examined whether in the 'red fields' – where the syndrome is particularly severe – intensified migration or violent conflict can indeed be found. While the present volume does not permit a graphic presentation in color, I shall briefly outline some main findings (graphics are given in WBGU 1997, 138–140).

4.1.3 Assessment

A first look at the regions in which the Sahel Syndrome is particularly severe already reveals a distinct correlation of these zones with those areas in which migration or violent (intrastate) conflict is identifiable (see the map in Schellnhuber et al. 1997: 32). This applies particularly to the Sahel proper and to parts of Sudan, where the Tuareg conflict is directly linked to the degradation of the natural bases of existence (Lume 1996), and also applies to the Horn of Africa (above all Somalia), Peru, Yemen and the regions of Brazil on the fringes of Amazonia, all of which are severely affected.

The analysis also shows the limits of the present state of syndrome research, particularly if we consider the imprecision attaching particularly to social indicators, which are often surveyed at the national level and say little about the situation of vulnerable marginal regions and peoples within the countries. A number of findings cannot be used to study causes of conflict as the possibility cannot be excluded that the impending and identified Sahel Syndrome is a fact a *consequence* of violent conflict. We may reasonably assume this in Angola, the Horn of Africa and Afghanistan – all regions currently severely affected by the Sahel Syndrome but whose economies are in tatters through many years of violent conflict, which has doubtlessly intensified the existential dependence of rural populations upon utilizing marginal sites. In Iraq, too, the many years of war can be assumed to have caused the high degree of vulnerability to the Sahel Syndrome, as otherwise similarly disposed states – such as Syria or Jordan – are far less affected by the Sahel Syndrome.

In sum, despite all deficiencies, the findings of syndrome analysis make it possible to give a more targeted focus to empirical research. The problems encountered in characterizing the Sahel Syndrome are questions for research, even the identification of which is already an advance. Why, for instance, do comparable degrees of intensity of the Sahel Syndrome correlate differently with migration and violent conflict, in other words: what distinguishes e.g. Mongolia from the West African Sahel, both of which are affected to a similar degree by the Sahel Syndrome but each of which address this problem differently at the political

---

[5] In syndrome research, these links are made using fuzzy logic, which permits indeterminate links (see on this Cassel-Gintz et al. 1997).

level? Here the syndrome approach permits a fresh perspective capable of generating valuable hypotheses for comparative empirical research on the nexus between environment and security.

4.2 'Water wars' (the Aral Sea Syndrome)

4.2.1 Fundamentals

A commonly encountered notion in the environment and security debate is that of 'water wars' that may be impending in various regions of the world (Gleick 1996). The hypothesis is that a growing population in conjunction with growing demands placed upon the quantity of water consumed overexploits regional stocks and may lead to international conflict, particularly where national boundaries cut across scarce natural resources. This debate generally cites the Near East, where conflicts over the waters of the Euphrates, Tigris and Jordan would seem to lend the most credibility to the proposition that cross-border resources tend to be a source of conflict (Albrecht 1996: 10).

But even if some would see the Israeli-Arab Six-Day War as the first war fought over water, the proposition that water wars will be inevitable is not well substantiated and continues to be disputed within the social sciences. Thus Gleditsch (1997b) makes the counterclaim that conflicts of interest arising over the use of cross-border water resources will tend rather to lead to increased cooperation than to escalating (violent) conflict. For instance, Hungary and the Slovak Republic have submitted their dispute over the damming of the Danube planned by the Slovak Republic to the International Court of Justice in The Hague, and the Boundary Waters Treaty concluded in 1909 between the USA and Canada is widely viewed as an exemplary case of comprehensive cooperation and in parts even of integration between two states in the utilization of common natural resources (WBGU 1998). However, it needs to be kept in mind that both of these cases come from industrialized countries. It remains questionable and is also doubted by Gleditsch (1997b) whether comparable conflicts of interest can also be resolved in such a peaceful manner in Africa, Asia and Latin America. Thus, although in this case a violent conflict between downstream and upstream riparians is not to be expected, China, the upstream riparian, continues to refuse to become a party to the Mekong Agreement (Thomas 1996).

The syndrome analysis approach offers a tool by which to empirically test the hypothesis of 'water wars'. In syndrome analysis, those damaging people-environment relations that are anthropogenically triggered by large-scale technological projects are categorized in the Aral Sea Syndrome class. This includes above all the national and international consequences of major dams, but also those of major irrigation projects, as exemplified by the Aral Sea, where they have been carried out with devastating environmental impacts. A comparable, not yet fully implemented mega-project is the 'man-made river' in Libya, where fossil groundwater shall be pumped out of the desert in order to irrigate coastal areas. This requires the immigration of a work force numbering two million in a sparsely populated country that presently has a total population of five million, which may harbor the danger of ethnically structured distribution conflicts (Albrecht 1996: 8).

Inasmuch as such major technological projects concern resources that span the territories of two states, we may assume an exacerbated propensity for international conflict or – thus the counter-hypothesis – increased cooperation. In a fashion similar to the Sahel Syndrome discussed above, the syndrome approach makes it possible here to identify those regions in which the Aral Sea Syndrome occurs, and, using this list of regions, to test the hypotheses of the propensity of such projects to spawn conflict or cooperation.

A further benefit of this approach is that it avoids a common flaw of previous studies which have selected empirical material on the basis of the dependent variable by only examining the potential ecological causation of conflicts that were resolved violently, while utilization conflicts that were resolved peacefully but were otherwise comparable have remained explicitly excluded. Syndrome analysis, by contrast, permits us to identify all relevant large-scale technological projects with cross-border impacts (within the scope defined by the network of interrelations), then to analyze issues of political resolution in a second step. In other words, the research design is open enough for an identification of the presence of the Aral Sea Syndrome to end with the (happy) outcome that such situations are only very rarely associated with violent conflict, tending rather to be resolved through cooperation.

4.2.2 Identification of the Aral Sea Syndrome

The Council's 1997 Annual Report (WBGU 1998) delivered a first identification of the worldwide incidence of the Aral Sea Syndrome. As in the Sahel Syndrome, socio-economic and natural physical-geographical indicators are integrated. Social indicators are of prime importance, as the quantity of water available to each individual or to agriculture in a certain region says little about the criticality of the situation. Saudi Arabia, for instance, shows how sea water desalination can be used to create a pleasant garden culture even in the desert, as long as the necessary financial resources are available.

In the cited Annual Report of the Council, the indicators of the Aral Sea Syndrome were already extended to include a first indicator of the 'propensity for conflict' of a certain situation. Among other factors, this takes into consideration whether two states are highly integrated (e.g. through customs unions, economic communities or military alliances) or whether – in the extreme opposite case – states already are or have recently been engaged in violent conflict. Such attributes of relations among states – thus the plausible assumption of the Council – condition, ceteris paribus, the probability of a bilateral resource utilization conflict being resolved violently or peacefully. Accordingly, existing conflicts were rated as exacerbating the syndrome. However, as propensity for conflict is thus already included in the identification of the syndrome, this analysis can no longer be utilized directly for the purposes of peace and conflict research, as the explanandum would be commingled with the explanans.

4.2.3 Assessment

Thus the current state of work on the Aral Sea Syndrome does not yet permit an empirical test of the hypothesis of water wars. This is understandable, as the syndrome concept was developed for a great diversity of issues and must be refined anew for each specific issue. The foundations of the concept have been laid from 1993 to 1997 in a series of individual steps. The aim is now to carry out these refinements in such a way that a (greater) practical applicability is gained in order to do justice to the ultimate and prime task of the entire global change research enterprise – policy advice with the aim of sustainable development.

The Aral Sea Syndrome offers a useful conceptual framework within which to tackle a series of hypotheses of conflict theory. It is plausible to assume that the dependence of one state upon the decisions of another is a pivotal variable in cross-border resource utilization conflicts. We know from the interdependence literature since Albert O. Hirschman's seminal work *National Power and the Structure of Foreign Trade* (1945) or the later *Power and Interdependence* of Keohane and Nye (1977) that such structures of interdependence can also be conceptualized as inter-state power relations. Nevertheless, the question of whether such power conflicts concerning the utilization of cross-border resources are resolved violently or cooperatively can only be examined if the dependencies are unequivocally identified.

The analysis of the Aral Sea Syndrome can contribute to this more precise determination of mutual dependencies. For instance, syndrome analysis has already been able to identify the criticality of freshwater utilization in individual states. Here the criticality index developed by the Council has unified socio-economic and natural physical-geographical circumstances (WBGU 1998). Furthermore, first drainage models have already been prepared which allow calculation of the dependence of a state upon the influx of surface water from one or several other states. If these data are placed in relation to the existing criticality of freshwater utilization in both the upstream and downstream riparians, we arrive at a picture of 'critical' constellations of interests between individual states or within regions (Biermann et al. 1998). This picture is gained without reference to existing conflicts. This can then be contrasted with empirical findings: Where are which constellations handled in a particularly cooperative manner, where do they lead to violent conflict?

5 Conclusions

The present contribution to the debate has the purpose of underscoring the potentials and problems of the syndrome approach in the perspective of peace and conflict research. A particularly difficult aspect in this connection is the high degree of complexity of the networks of interrelations in syndromes constructed to analytically capture the (damaging) basic patterns of global change. The larger the scale of a map the less detail it shows. In order to then engage in empirical study again, renewed reductions of the model are necessary that must steer a course between the Scylla of distortive reduction and the Charybdis of ultimate triviality of findings.

Moreover, the syndrome approach requires – as does the entire interdisciplinary research project on global change – a 'new thinking' that moves far beyond accustomed disciplines and methods. This is often problematic, and only too often are the models of integrated assessment no longer compatible with conventional problematiques. Thus it is as yet scarcely possible to fully utilize in the syndrome approach the extensive literature on international regimes[6] – not to mention the postmodern, ecofeminist or Marxist approaches in the social sciences.[7]

The syndrome approach is characterized by the considerable research effort needed to progress from trivial initial hypotheses to truly revealing statements on reality. Individual researchers or a team drawn from only one scientific discipline will scarcely be able on their own to build a comprehensive research design based upon the syndrome approach. However, this need not preclude individual researchers from addressing individual links within the network of syndrome interrelations. For this is the very strength of the syndrome approach: It permits the integration of individual researchers and of individual disciplines within a larger context, in which they can make their contribution to the overall picture by working along the lines of common guiding issues together with the other disciplines.

The syndromes elaborated by the Council and its cooperating research groups can form a productive framework for debate among the individual academic

---

6 In principle, elements of regime research – such as the identification of critical constellations of interests – can indeed be integrated in the syndrome approach, such as the distinction between 'generator, affected party and helper interests' made by Prittwitz (1990). However, as a rule, the integration of regime research would require a greater degree of differentiation of the trends of global change that have been analyzed until now, particularly in the social science sphere. Syndrome analysis does include the trend "expanding international institutions", but this is not yet differentiated enough to permit integration of further elements of regime research.

7 This does not apply to moderate-constructivistic approaches, such as those that analyze the influence of precisely defined ideas or of a political culture in international relations. Here the syndrome concept offers fertile points of contact, as consideration of the psychosocial sphere can be used to integrate ideas and political cultures. Nonetheless, this step yet remains to be taken with respect to international relations.

branches, from the natural sciences to the social sciences. The need for the *universitas* of sciences, which dissolved soon after the Renaissance, is more urgent today than ever (if in a different form) – the trans-disciplinary syndrome approach could build a first bridge.

This also applies to traditional peace and conflict research, and similarly to research on the linkages between environment and security. Syndrome analysis can function as a fertile heuristic tool by which, in a manner transcending the confines of separate disciplines, to relate empirical findings to each other, to test interrelations and to develop and empirically validate hypotheses. On such a basis, syndrome analysis can contribute to theory formation and can thus permit forecasts grounded in the dynamics of individual syndromes.

The syndrome approach may thus – and it is this that makes it of particular interest to the CCMS project presented in this volume – deliver a basis for developing 'early warning systems' for impending international conflicts, and may consequently offer a tool for conflict prevention. The syndrome concept will not make it possible to predict particular conflicts. What it can do, though, is to help academic inquiry and practical policy to devote more attention to certain regions or interrelations, to engage in more detailed study of these and also – thus it may be hoped – to intervene from the very outset in a way that minimizes the potential for conflict.

References

Albrecht, Ulrich 1996: "Krieg um Wasser?", in: *Prokla. Zeitschrift für kritische Sozialwissenschaft.* 26/1 (Issue 102), 5–16.

Baechler, Günther et al. 1996: *Kriegsursache Umweltzerstörung. Ökologische Konflikte in der Dritten Welt und Wege ihrer friedlichen Bearbeitung.* Vol. I. Zürich: Rüegger.

Biermann, Frank, Gerhard Petschel-Held, Christoph Rohloff 1998: Umweltzerstörung als Konfliktursache? *Theoretische Konzeptualisierung und empirische Analyse des Zusammenhangs von 'Umwelt' und 'Sicherheit'." in Zeitschrift für Internationale Beziehungen, ZIB* Vol. 5, No. 2, 273-308.

Carius, Alexander et al. 1996: *Environment and Security in an International Context. State of the Art and Perspectives.* Interim Report of the NATO CCMS Pilot Study. Berlin: BMU, Ecologic.

Cassel-Gintz, Martin A. et al. 1997: "Fuzzy Logic Based Global Assessment of the Marginality of Agricultural Land Use", in: *Climate Research*, Vol. 8, 135–150.

Gleditsch, Nils Petter 1997a: "Environmental Conflict and the Democratic Peace", in: Nils Petter Gleditsch (Ed.): *Conflict and the Environment*, 91–106. Dordrecht: Kluwer.

Gleditsch, Nils Peter 1997b: *Armed Conflict and the Environment. A Critique of the Literature.* Paper presented upon the occasion of the Open Meeting of Human Dimensions of Global Environmental Change Research Community, IIASA, Laxenburg, 12–14 June 1997.

Gleick, Peter H. 1996: "Fresh Water: A Source of Conflict or Cooperation? A Survey of Present Developments", in: Baechler, Günther and Kurt R. Spillmann (Eds.): *Kriegsursache Umweltzerstörung. Environmental Degradation as a Cause of War*, Vol. III, 1–25. Zürich: Rüegger.

Hirschman, Albert O. 1945: *National Power and the Structure of Foreign Trade*. New edition 1980. Berkeley: University of California Press.

Keohane, Robert O. and Joseph S. Nye 1977: *Power and Interdependence. World Politics in Transition*. Boston: Little, Brown.

Lonergan, Steve 1997: "Global Environmental Change and Human Security", *Changes*, Issue 5 (special issue), 1–6.

Lume, William 1997: "The Ecological Background to the Struggle Between Tuaregs and the Central Government of Niger", in: Baechler, Günther and Kurt R. Spillmann (Eds.): *Kriegsursache Umweltzerstörung. Environmental Degradation as a Cause of War*, Vol. III, 175–202. Zürich: Rüegger.

Prittwitz, Volker von 1990: *Das Katastrophenparadox. Elemente einer Theorie der Umweltpolitik*. Opladen: Leske & Budrich.

Reusswig, Fritz 1997: "Nicht-nachhaltige Entwicklungen. Zur interdisziplinären Beschreibung und Analyse von Syndromen des Globalen Wandels", in: Brand, Karl-Werner (Ed.): *Nachhaltige Entwicklung. Eine Herausforderung an die Soziologie*, 72–90. Opladen: Leske & Budrich.

Schellnhuber, Hans-Joachim et al. 1997: "Syndromes of Global Change", *GAIA*, 6/1, 19–34.

Simonis, Udo Ernst 1991: "Globale Umweltprobleme und zukunftsfähige Entwicklung", *Aus Politik und Zeitgeschichte*, Issue 10, 3–10.

Thomas, Caroline 1996: "Water: A Focus for Cooperation or Contention in a Conflict Prone Region? The Example of the Lower Mekong Basin", in: Baechler, Günther and Kurt R. Spillmann (Eds.): *Kriegsursache Umweltzerstörung. Environmental Degradation as a Cause of War*, Vol. III, 65–125. Zürich: Rüegger.

WBGU (Wissenschaftlicher Beirat der Bundesregierung Globale Umweltveränderungen – German Advisory Council on Global Change) 1994: *World in Transition: Basic Structure of Global People-Environment Interactions*. Annual Report 1993. Bonn: Economica.

WBGU 1995: *World in Transition: The Threat to Soils*. Annual Report 1994. Bonn: Economica.

WBGU 1997: *World in Transition: The Research Challenge*. Annual Report 1996. Berlin, Heidelberg, New York: Springer.

WBGU 1998: *World in Transition: Ways Towards Sustainable Management of Freshwater Resources*. Annual Report 1997. Berlin, Heidelberg, New York: Springer.

Conflict Research and Environmental Conflicts: Methodological Problems

Christoph Rohloff

1 Introduction

Since the end of the conflict between East and West, new security debates under the heading of 'environmental security' have led to the publication of many studies on the linkage between environment and conflict (for example, Brock 1991, Homer-Dixon 1994, Levy 1994; a review of the research is given in Baechler et al. 1996:10-17, Dabelko 1995). A central proposition of this literature is that the linkage exists because increasing environmental degradation can lead to armed conflict within and among unstable groups, societies and states.[1]

This proposition is controversial because, in the words of the ENCOP study carried out by the research group under the guidance of Baechler in Zurich, "environmentally induced armed conflict is symptomatic of the modernization crisis in many countries of the world..." (Baechler et al. 1996: 17). Thus the phenomenon of environmental conflict, if understood more as a societal than environmental problem, is an object of peace research and empirical conflict research. The central question is: will armed conflict in the twenty-first century be primarily environmentally induced?

Environmental conflicts have features that differentiate them from conventional types of conflict such as internal power struggles between competing elite groups or territorial conflicts over boundaries. For the conflicting parties, limited land and water resources as the *matter in dispute* become objects, namely scarce land to be divided or river water both sides want to use. On the other hand, if environmental degradation is the *source of conflict*, the transformational character of such conflicts becomes prominent (Baechler et al. 1996: 7). Environmental degradation can seldom be reversed and in most cases it is a rapidly advancing phenomenon whose effect none of the parties can escape. It is labeled a 'total societal event' (ibid.: 323). Direct and manipulative conflict resolution strategies are often

---

1 Overviews of other aspects of environmental security not discussed here are offered in Brock 1994 and by Carius and Imbusch in the present volume.

ineffective. The problem is persistent and demands alternative solutions that do not in turn harbor conflict. Thus a frequent reaction to scarcity of water or land is the migration of a population to a new region, where economic, religious or ethnic differences lead to tension with the local inhabitants. A result of advanced environmental degradation at its worst is the complete transformation or decay of a region and its geographic, climatic, regional, cultural and social structures and characteristics. Traditional and tested conflict resolution strategies, such as clientelism as a distributional strategy, no longer work since life is transformed in its entirety. Depending on what part of the environment has been damaged, the transforming aspect of degradation can have an effect on the ozone layer, river pollution, erosion, etc., at the local, regional or global level.

Baechler et al. (1996: 14) limit the phenomenon of environmental conflict to viewing environmental degradation not as monocausal and leading directly to armed conflict, but taking effect through catalysts, "social effects", in a complex web of causality. A conflict of interests in dividing scarce land and water resources only needs an ethno-political charge, for example, to put groups into a situation of potential escalation. Dokken and Græger (1995: 39) also limit environmental degradation to a 'background variable' and give it subordinate value for empirical research. They do expect that a new environmental approach to understanding conflict will provide more expressive, empirical material for security policy and general political practice (in Baechler et al. 1996: 13):

> The environmental conflict approach enables rich-case studies, which in turn can contribute to make the *environmental dimension of security* more empirical and therefore more applicable to the political community and the general public.

Compared to peace research and empirical conflict research, responsible and foresighted state security policy is characterized by other interests and intentions (Eberwein 1997: 9). The goals of preventive security policy are the early recognition of new sources of conflict and conflict dynamics, and correction through preventive and de-escalating measures. Security politicians can refer to empirical conflict research that records the reality of conflicts in statistics and analysis. However, empirical conflict research can only do this job retrospectively; its forecasting capacities are small and limited to expressing probabilities within a given context. It has a delayed reaction to new developments regarding conflict sources and structures and can only offer clarifying explanations.

This chapter will discuss the possibilities and limits of empirical conflict research. I will refer to different conflict types that could be part of an explanatory model for the phenomenon of environmental conflict. The criterion is whether and how far these types are suited to explaining those sources of conflict and conflict dynamics that seem to arise from an environmental conflict. I will argue that empirical conflict research must be better developed conceptually, not only to explain the emergence of violent conflict but also, for security policy reasons, to estimate the escalation potential of non-violent conflicts. Here a long overdue theory will be called for – an explanatory model for the escalation potential of conflicts in general and environmental conflicts in particular. The question that deserves special notice is which conflicts escalate into violence and how this can be explained vis-à-vis the peaceful resolution of environmental conflicts. The thesis proposed here is that the answer can be found through systematic investigations in a comprehensive model, whose theoretical approach and empirical validation both embrace peaceful and violent conflicts.

2 Possibilities and limits of empirical conflict research

General conflict theories can be situated at the level of individuals (aggression theories), the state (externalization theory, power competition theory) or the international state system (imperialism theory, modernization theory) (Waltz 1979; Pfetsch 1994). General, broadly based conflict theories at the system level often claim to explain all conflicts, but are often documented only by a few, historic-genetic case studies.

The high level of abstraction of such approaches together with a thin empirical base is in contrast to the more realistic, modeled approach of quantitative empirical conflict research. Starting with the early work of Sorokins (1937), Wrights (1942) and Richardsons (1950), the 'correlates of war' project by Singer and Small (1972) set the standards for this discipline. In the German-language literature, the empirical conflict research approach has been established by Kende (1971) and Gantzel (1986). In the KOSIMO conflict simulation project, Pfetsch and Billing (1994) published an aggregate database of peaceful and violent, internal and international conflicts.[2]

The goal of empirical conflict research is to study conflicts in theoretical and empirical contexts. Its intention is to prevent violence by understanding conflicts generated by identified actors. In its practice, the observation of symptoms can serve to develop prognoses for future conflict processes, early warning systems and courses of alternative action.

The strengths of this approach are obvious. Ideological, legitimating or mobilizing uses of general conflict theories are counteracted by sobering statistics on the frequency, development and results of conflict. An important outcome of empirical conflict research is its qualified confirmation of Kant's statement on the structural inhibition of democracies to engage in aggressive war (*On Perpetual Peace* 1795). Empirical study (Gantzel 1986, Billing 1991) has demonstrated that democracies, especially the Western superpowers, have been involved in wars as often as authoritarian systems were, but that democracies have maintained inter-

---

2 A completely revised, augmented and updated version of the database and the most important results is being prepared. See Pfetsch, Frank R. und Rohloff, Christoph: *Global Trends of Conflicts. Fifty Years of Peaceful and Violent Conflict.*

state peace with other democracies, the so-called OECD or democracy peace, since 1945. A second finding made by research has been the evidence that we live, since 1945, in an increasingly violent world. The number of conflicts that start each year does not rise continuously, but the number of wars that end each year has stagnated. This means that ongoing conflicts last longer (Gantzel and Schwinghammer 1995; HIIK 1996). Figure 1 shows this development based on 375 violent conflicts listed in HIIK's KOSIMO database. These wars and violent crises belong to the category of *protracted conflicts* (Zartman 1992). Such conflicts are characterized by the involvement of all social strata and actors, and the dispute over complex material and immaterial goods and values. Their resolution is difficult because of traditional historical antagonisms that often overlap ethno-political, religious and other conflict areas. No research has been done on how much the increasing duration of violent conflicts reflects their increasing complexity. It is plausible that environmental degradation is an additional escalation factor or source of failed conflict settlement.

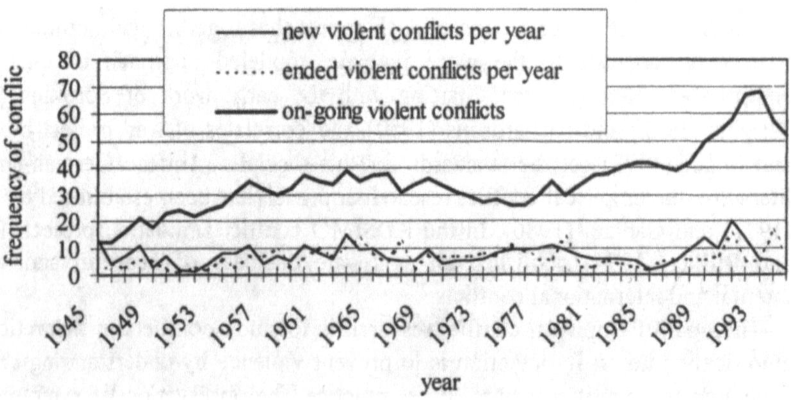

Figure 1: Development of violent conflicts

A third important research finding on matters in dispute in peaceful and violent conflicts is the fact that the number of territorial conflicts and international power conflicts since 1945 has not risen steadily; occasionally it has gone down (see Figure 3). In contrast, the number of internal power struggles, and ideological, religious and secession conflicts rose steadily until 1993. The rise in the number of resource conflicts was less drastic, whether they were resolved violently or peacefully. (Gantzel 1986; Holsti 1992; Pfetsch and Billing 1994; HIIK 1996) (see Figure 2).

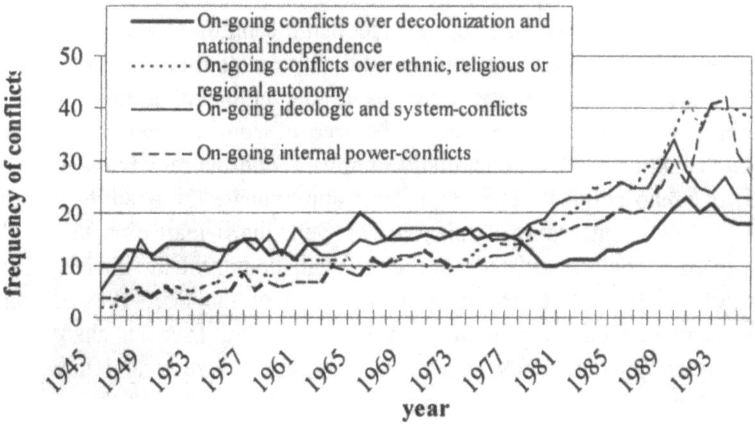

Figure 2: Development of conflicts according to matters in dispute: decolonization and national independence; ethnic, religious or regional autonomy; ideological and system conflicts; internal power conflicts

Figure 3: Development of conflicts according to matters in dispute: territorial and border conflicts; international power conflicts; conflicts over resources

But the weaknesses of quantitative empirical conflict research should not be underestimated (see the discussion in Gantzel 1996 and Pfetsch and Billing 1994). Since research took this direction in the 1960s, the definition of key terms has lacked agreement. Words such as war, peace and conflict are not uniformly defined or their meanings are deductible only from the theory being presented. The demarcation against political conflicts on a 'lesser scale' is also contested; that is, those political conflicts within democratic systems that are constitutive and contribute to a competitive, pluralist notion of dispute. Positive social, political and international developments are an outcome of conflicts of interest. So the problem is not conflict as such, but the fact that it can turn to violence at any time. In this sense, Paffenholz's essay title is valid: "It's not conflicts that must be ended, it's the wars" (1995). On the other hand, the enforced, deadly peace of an authoritarian system only appears to be free of conflict. Here, social conflict *potential* is not captured by quantitative empirical conflict research. Gurr's surveys (1995) of ethno-political conflicts at least approximate the quantity of potential conflicts. But in international relations, as well, the demarcation against 'lesser scale', internationally regulated conflict constellations remains unclear. Thus the ratification of the UN Convention on the Law of the Sea (1982) meant that numerous marine boundary conflicts were transferred from an anarchic power contest to an internationally standardized system. Whether and how far such conflicts remain the object of international power struggles, and how far the norms are accepted by all concerned, remains to be seen. This applies by analogy to trade conflicts over certain goods or trade stipulations that are supposed to be regulated by the World Trade Organization (WTO), but that are part of international power struggles and that can escalate into violence, thus remaining relevant to empirical conflict research. Another problem for this kind of research is that certain concepts used to explain conflict processes, such as cultural affiliation or the 'big man' phenomenon observed in traditional authoritarian systems, can not always be aggregated to produce comparable correlates. The danger of a culturally deterministic exaggeration of correlates is high and became apparent in many polemic articles on the Bosnian War. Quite often ethnic groups involved in the struggle were said to have "an aggressive nature all along", rather than asking what could have caused the aggression or which responsible actors were pursuing what interests. Altogether, empirical conflict research has produced very little directly comparable results – at best they were tendential – let alone interdisciplinary cooperation with the fields of social psychology or ethnology.

Empirical conflict research proceeds from the assumption that complex social relations such as violent political conflicts can be portrayed, classified and interpreted by reducing them to their characteristic variables. Researchers then maintain that they can identify common and comparable features in spite of the singularity of each conflict. In general, empirical conflict research serves the purpose of generating new and examining existing theories, while a largely quantitative procedure is used inductively to predicate the probable occurrence of observed phenomena (see Mendler and Schwegler-Rohmeis 1989).

In categorizing conflicts, Glasl (1992: 47f) differentiates between (1) matters in dispute, (2) forms of conflict and (3) characteristics of the conflicting parties. The conflict type chosen for research purposes depends on what information is being sought. Differentiating conflicts according to matters in dispute would be sensible for investigating conflict resolution and corresponding negotiation techniques. Such a taxonomy can refer to 'real' or 'artificial' matters in dispute, i.e. things that objectively exist, or the attitudes and behavior of the conflicting parties towards each other. It is more common to distinguish between conflicts of values and interests. Interest conflicts concern the availability or distribution of a good; value conflicts concern its evaluation.

Categorizing conflicts according to their form is useful for investigating actors and conflict instruments. Dahrendorf's differentiation between latent and manifest conflicts can be mentioned here (1958). Categorizing according to institutionalized and non-institutionalized conflicts is relevant for empirical research (Mack and Snyder 1957). Armed and non-violent conflicts that concern the national or sovereign consciousness and interests of a state are both understood as non-institutional, unregulated and unresolved conflicts. This difference enables a pragmatic separation of non-violent conflicts from 'non-conflicts' that are legal processes or constitutionally acceptable paths. And finally, conflicts can be categorized according to the conflicting parties' characteristics, whereby the most commonly used differentiation is between ruler, powerful elite, center and state on one side, and ruled, powerless, periphery and resistance groups on the other. Rapoport (1960) further differentiates between symmetrical and asymmetrical relationships.

Besides Glasl's three categories, a fourth basic differentiation according to sources of conflict may be posed. 'Environmental degradation' as a source of conflict is the independent variable in this way of ordering things. The peaceful or violent conflict process would be the dependent variable in need of explanation from case to case.

3 Empirical conflict research in practice

The above review of the options and limitations for empirical conflict research and the current criteria used to categorize conflict is followed by an analysis of categories that refer explicitly to environmental conflicts and to those that concern violent conflicts in general. The intention is to find out more about the power of single models to explain environmental conflict.

3.1 Typologies of environmental conflict

To answer the question of what conditions cause an environmental conflict to break out in violence, the authors of the ENCOP study (Baechler et al. 1996) suggest that these conflicts be empirically recorded and that categorization be made inductively on this basis. On three levels of observation ('intrastate', 'intrastate with transboundary dimension', 'international'), seven partially overlapping environmental conflict types can be identified. What these types of conflict have in common is that they are rooted in a certain form of environmental degradation. This category becomes the independent variable. Based on empirical investigation, the following types have been listed (see Baechler's chapter in this volume).

Intrastate:

- Center-periphery conflicts in which modernizing centers possessing more alternatives clash with traditional or backward peripheral groups;
- Ethno-political conflicts in which parties with diverging socio-economic interests mobilize, group or are instrumentalized along ethnically defined traits;
- Regional migration conflicts that arise when a population forced to migrate confronts the resident population of another area.

Intrastate with transboundary dimension:

- Internationalized migration conflicts, resembling regional migration conflicts, that emerge when environmental refugees seeking a new basis of existence must cross international boundaries;
- Demographically caused migration conflicts that are the outcome of over-population and overexploitation of land.

International:

- International conflicts over water in which the issue of distribution and access to water becomes a boundary conflict between states; and
- Global environmental conflicts that are related to global transformation phenomena such as climate change or ozone layer depletion.

Baechler et al. view environmental conflicts as societal and political events, understood not just as chains of causality that start in environmental degradation. The sources of conflict include system structures, actors' characteristics and the long-distance effect of actions (Baechler et al. 1996: 292). But methodological questions arise as to what conditions the peaceful or violent course of environmental conflict. The ENCOP methodology inductively uses a selection of about twenty violent and twenty non-violent conflicts. It does not explicitly compare similar, operationalized variables to identify the special traits of peaceful and violent conflicts. An additional comparison of environmental conflicts to conflicts with escalation potential where environmental degradation was not observed nor used as a primary variable (only as an intervening one), could enable more significant conclusions on the conditions leading to either peaceful or violent environmental conflict.

In his empirical conflict research study on "Environmental Scarcities and Violent Conflict – Evidence from Cases", Homer-Dixon (1994) explores the question of whether environmental degradation causes violent conflict, and if so, how. He identifies six different forms of environmental change that have led to violent conflicts. Using various hypotheses to investigate this possible connection, his results differ little from those of the ENCOP study – environmental degradation is seldom a direct source of violence, especially not in international relations. But if ethno-political, internal social and regional tensions are added, violent conflict becomes more probable. Homer-Dixon's method is interesting regarding our aim to determine criteria for escalation potential:

> By selecting cases that appeared 'prima facie' to show a link between environmental change and conflict, we sought to falsify the null hypothesis that environmental scarcity does not cause violent conflict. By carefully tracing the causal processes in each case, we also sought to identify how environmental scarcity operates, if and when it is a cause of conflict (1994: 7).

Unfortunately, the study does not always separate 'conflict' and 'violent conflict'. But this difference is important if we are seeking an explanation why some conflict situations escalate into violence and others do not. The situations that were peacefully settled and that lead to the null hypothesis are just as interesting for our question as the cases in which violence was observed. But they were not included in the large volume of cases that Homer-Dixon investigated. In a debate between Levy (1994) and Homer-Dixon, Levy criticizes the focus of the investigation on the independent variable 'environmental scarcity' and demands a more flexible theory. He asks whether weak political systems could be a source of conflict (see, for example, Kalevi 1996) and thus be considered as much an independent variable for violent conflicts as environmental degradation. Because both authors focused on violent forms of conflict, their debate has little influence on the question of the conditions that lead to violent conflict as might be identified by making comparisons to conflicts that were resolved without violence.

3.2 Categories of war

In an attempt to understand and explain the source and outbreak of violent conflicts and war by counting the frequency with which they occur, a number of lists of wars have been compiled since the Second World War. The following overview may suffice here. Subsequently, the limits of this approach shall be illustrated for a selected example.

Table 1: Overview of lists of wars (according to Pfetsch and Billing 1994: 58)

Authors	Year	Focus	Period	Cases
Blechmann and Kaplan	1978	US use of military force	1946-1975	215
Brecher et al.	1988	international crises, foreign policy crises	1929-1979	278
Butterworth	1976	interstate security conflicts	1945-1974	310
Coplin and Rochester	1972	Disputes	1920-1968	121
Cukwurah	1967	boundary disputes	./.	./.
Deitchman	1964	military engagements	1945-1964	46
Dingemann	1983	armed conflicts	1945-1982	71
Frei	1975	international conflicts	1960-1974	65
Gantzel/Meyer-Stamer	1986	Wars	1945-1986	159
Gantzel/Schwinghammer	1994	Wars	1945-1992	184
Gödecke et al.	1983	wars, coup d'états	1945-1983	./.
Haas	1968	Conflicts	1945-1965	108
Haas	1983	Conflicts	1945-1981	282
Haas, Butterworth, Nye	1972	Disputes	./.	./.
Holsti	1966	international conflicts	1929-1965	77
Holsti	1977	international conflicts	1919-1975	86
Holsti	1983	international conflicts	1929-1979	94
Jütte and Grosse-Jütte	1981	international conflicts	1946-1976	246
Kaplan	1981	USSR use of military force	1956-1979	19
Kende	1982	Wars	1945-1982	148
KOSIMO (Pfetsch /Billing)	1994	peaceful and violent conflicts	1945-1990	546
KOSIMO (Pfetsch/Rohloff)	*1997*	peaceful and violent conflicts	1945-1995	661
Leiss and Bloomfield	1967	local conflicts	1945-1964	52
Levine	1971	mediation in conflicts	1816-1960	388
Levy	1985	general wars	1585-1945	10
Luard	1988	principal wars	1945-1986	128
Luard	1988	frontier wars	1866-1986	88
Luard	1988	ideological civil wars	1918-1986	38
Luard	1988	colonial wars	1918-1986	27
Maoz	1982	interstate disputes	1821-1976	
Northedge and Donelan	1971	international disputes	1945-1970	50
Richardson	1960	deadly quarrels	1819-1949	./.
Ruloff	1987	Wars	1792-1983	150
Small and Singer	1982	civil and interstate wars	1816-1980	224
Siverson and Tennefoss	1982	interstate conflicts	1815-1965	256
Sorokin	1937	Wars	1100-1925	862
Wainhouse	1966	global organizations	1946-1965	21
Wood	1968	Conflicts	1898-1967	127
Wright	1965	civil and international wars	1482-1940	278
Zacher	1979	international conflicts	1945-1977	116

The 'Hamburg approach' attempts a comprehensive categorization of wars. In the following, I shall focus on this one approach, without claiming it to be representative of similar approaches by other authors. Whereas in the ENCOP study the manifold paths to violence are understood as a variable dependent on environmental degradation, the Hamburg approach concentrates on possible independent variables that lead to a dependent variable, namely war. This approach was formulated by István Kende (1971) and further developed by Gantzel (1986) and Siegelberg (1994). Assuming that violent conflict arises from the fault lines between tradition and modernity under the pressure exerted by the world's 'total capitalization', the AKUF (*Arbeitsgemeinschaft Kriegsursachen-forschung*) research group around Gantzel in Hamburg has monitored and analyzed the wars since 1945. According to István Kende (1982), war is

> a violent mass conflict that demonstrates *all* the following features:
>
> - Two or more armed forces are engaged in battles, whereby regular armed forces of the *government* (military forces, paramilitary groups, armed police units) are *on at least one side*;
> - On both sides there is a minimum of *centrally directed* organization of the war parties and battle, even if it is no more than organized armed defense or strategic-tactical, planned attacks (guerrilla operations, partisan wars, etc.);
> - Armed operations take place with a certain *continuity* and not just as occasional, spontaneous clashes. Both sides operate according to a planned strategy, regardless of how long fighting lasts and whether it occurs within an area of one or several societies (Gantzel and Schwinghammer 1994: 31; italics in the original).

This provides the basis for identifying four types of war that can appear in mixed variations. These types reflect the relationship of the conflicting parties to each other – they include the identification of dominating actors in the center and dependent groups on the periphery.

- *Anti-regime wars* in which the political dominion, the government's power, the social order or the societal system is disputed within a state;
- *Other intrastate wars* in which the government is not being disputed, but where the issues are particular ethno-political interests, limited claims to geographical areas, autonomy or separation, political hegemony, or religious domination, etc.;
- *Inter-state wars* in which the armed forces of two states oppose each other in the classical sense;
- *Decolonization wars* of the 1950s and 1960s in which a colonial power was a directly active party on one side.

The 'Hamburg approach' distinguishes four other kinds of conflict (ibid.: 44f). These regard the matters in dispute or war goals that led to a war. This catego-rization differentiates between possible sources and motives that have led to war. They are:

- *System conflicts* in which the reshaping, reforming or restoration of a system is contested;
- *Power conflicts,* and in a narrower sense, conflicts of dominion, diversionary conflicts, special interest conflicts, tribal or ethno-political conflicts and military-strategic conflicts;
- *Territorial conflicts*, wars of expansion, territorial restoration, revision or autonomy/secession;
- *Foreign power conflicts* regarding colonial power and indirect, external hegemony.

In this connection, Gantzel and Schwinghammer mention environmental conflict as a possible future kind of conflict that could be added to the above four (ibid.: footnote 44). Environmental conflict could also be a class of power conflict. This would give environmental conflict more the status of a matter in dispute than of a type of war with its own causality. Single cases of observed environmental degradation would receive object status, whereby the transforming aspects of environmental conflict described by Baechler et al. (1996) would not receive consideration. Referring back to our 'environment and security' question of how far environmental conflict can lead to armed conflict, similar methodological problems arise. Since the focus lies exclusively on violent conflicts, and those conflict situations that do not lead to war are left out, this approach does not allow conclusions to be drawn on the threshold to violence or the conditions needed to cross it. But a tenable conflict theory must first define these conditions theoretically and then examine them empirically, guided by probability predications.

The Heidelberg Institute for International Conflict Research (HIIK)[3] (founded in 1991 and thus newer than AKUF) sees its KOSIMO (conflict simulation model) conflict database as integrating existing databases and improving comparative and analytical options for empirical data collection (Billing 1992; Pfetsch and Billing 1994). Conflicts that do not lead to violence are regarded as 'latent conflicts' and 'crises'. The definition of conflict is taken partially from Ernst-Otto Czempiel (1981):

> A conflict as a generic term encompasses opposing interests (differences in position) regarding national values (independence, self-determination, boundaries and territories) of some duration and range between at least two parties (states, state groups, state organizations, organized groups) who are determined to decide these issues in their own favor (Pfetsch and Billing 1994: 15).

KOSIMO differentiates between two sorts and for each sort two intensities of conflict. Accordingly, a conflict can assume four levels of intensity, with different phases of the same conflict differing in intensity (Billing 1992: 40f):

- *Latent conflict*: Formulation of and focus on differences in position that are perceived by each side. The conflict is, however, kept 'at a low flame';
- *Crisis*: A conflict that uses non-violent, diplomatic, economic and political instruments, including non-military threats;
- *Violent crisis*: Conflicts that threaten to break out into war or that include spontaneous, sporadic, irregular or unorganized use of the military;
- *War*: The systematic, collective use of military force that produces casualties, causes destruction and continues for a certain length of time.

3 The KOSIMO database now encompasses 661 conflicts from 1945 to 1995, 375 of them violent, of which 104 were or are wars. Each conflict is coded with 30 actor and structure variables. Details on archive use, analyses, extracts and publication lists are available from HIIK (see Annex A).

Conflicts are also differentiated in four different categories according to the positive or negative use of instruments (in the sense that instruments can have escalation potential):

- Negotiations (with and without mediation);
- Authoritarian or judicial resolution;
- Threats and pressure;
- Sporadic or systematic uses of force.

Basically, conflicts can deal with seven different goods or values, whereby several goods and values can be at issue in one conflict at the same time. In the majority of the 661 observed conflicts, struggles concerned two or three different goods; in contrast, conflicts over territory, boundaries and resources are often observed as having no further matters in dispute. Internal power struggles are mostly connected to independence and autonomy conflicts. The seven goods and values that lead to conflict, according to KOSIMO's database, are:

- Territories, boundaries, marine territorial boundaries;
- Decolonization, national independence;
- Ethnic, religious or regional autonomy;
- Ideological or system conflicts;
- National power struggles;
- International, geo-strategic power struggles;
- Conflicts over resources.

Following AKUF's conflict categorization, the KOSIMO approach can address environmental conflict as a 'matter in dispute', if resources are understood as scarce, contested goods. Such conflicts could be classified as conflicts over access and distribution of resources or as territorial conflicts, with classification enlisting the object character of environmental degradation as its typical trait and not the transforming character of environmental conflict. Including non-violent and violent conflicts in a database, and thus creating a dynamic conflict model, makes it possible to trace or monitor environmental conflicts in all their phases, differentiated according to intensity. Herein lies the real, not yet fully utilized value of the KOSIMO database. In cooperation with environmental research institutes, better empirical knowledge could be gleaned from data on both environmental degradation and peaceful and violent conflicts.

4 Conclusions

From this selective overview and methodological critique of the theory and practice of empirical conflict research, including research on environmentally related conflict, the following conclusions can be drawn on the link between environment and conflict, and how these links may be studied.

There is a lack of a theoretical approach to 'environmental conflict' with a systematic structure which links the environmentally relevant aspects of conflicts and the conflictual aspects of environmental processes. This conceptual gap exists because environmental degradation as such is 'non-violent' in the sense of military force being used against people, whereas violent conflicts, in particular, have been the object of research. Little can be said about degradation catalyzing or inducing conflict in one isolated sample without comparing such a situation to peaceful conflict in which degradation also plays a role. To be exact, those cases of environmental degradation should be included that found a cooperative solution or that did not lead to conflict between those affected but rather were accepted by all sides, whether knowingly or not.

The non-trivial aspect of environmental change is its destructive *effect* on local, regional and global human reproduction structures. The non-trivial aspect of conflict is its potential to use lethal *force* (see Eberwein 1997 and in this volume). A theoretical approach that links environment and conflict should explain why and when some conflicts in which environmental degradation has been observed escalate into violence, and why others are resolved peacefully.

Part of such a model has long been an integral part of environmental research – the theory-building and empirical investigation of the constructive and destructive effects of environmental change on natural and human systems, e.g. with limit and threshold values.[4] The missing part, a theoretical approach that would identify the conditions that either lead to or prevent violence, i.e., an explanation of the non-trivial aspect of conflict, has yet to be formulated. Brock (1991: 409) attempts to show possible links between war and peace, and environmental issues, by using an eight field matrix. He differentiates between positive and negative causal, instrumental, definitory and normative links. This matrix clearly shows the possible positive links between environment and conflict that have not been given enough consideration by empirical conflict research. Billing (1992) uses the KOSIMO database, that includes violent and non-violent conflicts, to introduce an escalation and de-escalation model for *international* conflicts; he arrives at several conclusions on probabilities for political systems, issues of conflict and certain structural conditions in the international system that determine the height of the threshold to violence. Figure 4 shows the steady increase in the frequency of violent internal conflicts concurrent to the decrease in violent inter-state conflicts. This does not mean that there are fewer international conflicts than there were 50 or 30 years ago; all it means is that the proportion of inter-state conflicts using force has steadily decreased. Billing's model for violence thresholds has not yet been applied to internal conflicts.

---

4 Including the Potsdam Institute for Climate Impact Research (PIK) or the German Advisory Council on Global Change (WBGU); see also Sprinz's chapter in this book.

Figure 4: Violent internal and international conflicts from 1945 to 1995

The same question of when and why one conflict escalates, and others of the same 'type' do not, guides peace research. Peace research designed by Volker Matthies arrives at conclusions about escalation conditions using a qualitative comparison of peaceful and violent zones (see Matthies 1996: 31). In this sense, the author tried his hand at a global empirical study and partially graphic representation of peaceful and violent zones or successful and unsuccessful measures for maintaining peace (Rohloff 1996). The results underline the necessity of understanding peace, peaceful conflict resolution and peace structures not as normative or moral postulates, but in order to compare them systematically with each other and with violence. Miall (1992) makes a comprehensive attempt to unite peaceful and violent conflicts on a conceptual level; he categorizes conflicts into peaceful and violent, and resolved and unresolved. His results corroborated research using the KOSIMO database (see Pfetsch and Rohloff for details); of 81 investigated conflicts, most conflicts found peaceful resolution. These included territorial conflicts and conflicts over the access and distribution of resources.

In turning the focus from conflict research to peace research on peace-consolidation, violence prevention and mediation, violence and war should be interpreted as exceptions, not as rules and unavoidable, for instance as a necessary part of the modernization process. For empirical conflict research, this means using an unprejudiced, comprehensive concept of conflict that is not limited to its potentially destructive side.

> Peace and war research are two sides of the same coin, namely a comprehensive science of peace (Matthies 1994: 49).

Finally, such an approach would allow the phenomenon of 'environmental conflict' to lose its current appeal as a new apocalyptic scenario for the media and would release the term 'environmental security' from its narrow, violence-oriented, military interpretation.

References

Baechler, Günther et al. 1996: *Kriegsursache Umweltzerstörung. Ökologische Konflikte in der Dritten Welt und Wege ihrer friedlichen Bearbeitung*. Vol. I. Zürich: Rüegger.

Billing, Peter 1992: *Eskalation und Deeskalation internationaler Konflikte. Ein Konfliktmodell auf der Grundlage der empirischen Auswertung von 288 internationalen Konflikten seit 1945*. Frankfurt/M: Peter Lang.

Blechmann, B. M. and S. S. Kaplan 1978: *Force without War: US Armed Forces as a Political Instrument*. Washington D.C.

Bonacker, Thorsten and Peter Imbusch 1996: "Begriffe der Friedens- und Konfliktforschung: Konflikt, Gewalt, Krieg, Frieden", in: Imbusch, Peter and Ralf Zoll (Eds.) 1996: *Friedens- und Konfliktforschung. Eine Einführung mit Quellen*, 63–102. Opladen: Leske + Budrich.

Brecher, M., J. Wilkenfeld and S. Moser 1988: *Crises in the Twentieth Century*. Handbook of International Crises, Vol. 1. Handbook of Foreign Policy Crises, Vol. 2. Oxford: Pergamon Press.

Brock, Lothar 1991: "Peace through Parks: The Environment on the Peace Research Agenda", in: *Journal of Peace Research*, 28/4, 407–423.

Brock, Lothar 1994: "Ökologische Sicherheit. Zur Problematik einer naheliegenden Verknüpfung", in: Wolfgang Hein (Ed.): *Umbruch in der Weltgesellschaft*, 443–458. Hamburg: Deutsches Übersee-institut.

Butterworth, R.L. 1976: *Managing Interstate Conflict, 1945–1974. Data with Synopsis*. Pittsburgh: Pittsburgh University Press.

Coplin, W. D. and J. M. Rochester: "The Permanent Court of International Justice, the League of Nations and the United Nations: A Comparative Empirical Survey", in: *American Political Science*, Review 66, 529–550.

Cukwurah, A.1967: *The Settlement of Boundary Disputes in International Law*. Manchester: Manchester University Press.

Czempiel, Ernst-Otto 1981: *Internationale Politik*. Paderborn: UTB Verlag Ferdinand Schöningh.

Czempiel, Ernst-Otto 1997: "Alle Macht dem Frieden", in: Senghaas, Dieter (Ed.): *Frieden machen*, 31–47. Frankfurt/M: Suhrkamp.

Dabelko, Geoffrey D. and David D. Dabelko 1995: "Environmental Security: Issues of Conflict and Redefinition", in: *Environmental Change and Security Project Report*, Issue 1, 3–13. Washington, DC: The Woodrow Wilson Center.

Deitchmann, S. L. 1964: *Limited War and American Defense Policy*. Washington.

Dingemann, R. 1983: *Bewaffnete Konflikte seit 1945. Zwischenstaatliche Auseinandersetzungen, Befreiungskriege in der Dritten Welt, Bürgerkriege*. Düsseldorf: Econ.

Dokken, Karin and Nina Græger 1995: *The Concept of Environmental Security – Political Slogan or Analytical Tool. Report prepared for the Royal Ministry of Foreign Affairs*. Oslo: PRIO. (cited in Baechler et al. 1996)

Eberwein, Wolf-Dieter 1997: *Umwelt – Sicherheit – Konflikt. Eine theoretische Analyse*. WZB Discussion paper P 97-303. Berlin: Wissenschaftszentrum für Sozialforschung Berlin.

Frei, Rudolf 1975 : "Erfolgsbedingungen für Vermittlungen in internationalen Konflikten", in: *Politische Vierteljahresschrift*, 16/4, 447–490.

Gantzel, Klaus-Jürgen and J. Meyer-Stamer 1986: *Die Kriege nach dem zweiten Weltkrieg bis 1984*. München: Weltforum.

Gantzel, Klaus-Jürgen and T. Schwinghammer 1995: *Die Kriege nach dem zweiten Weltkrieg 1945 bis 1992*. Münster: Lit.

Glasl, Friedrich 1992: *Konfliktmanagement. Ein Handbuch für Führungskräfte und Berater*. Bern: Haupt.

Goedecke, P., E. Stuckmann and M. Vogt 1983: *Kriege im Frieden*. Braunschweig.

Gurr, Ted Robert 1994: "Peoples against States. Ethnopolitical Conflict and the Changing World System", in: *International Studies Quarterly*, Issue 38, 347–377.

Haas, Ernst B. 1968: *Collective Security and the Future of the International System*. Denver.

Haas, Ernst B.: "Conflict Management and International Organizations 1945–1981", in: *International Organizations*, 37/2, 189–256.

Holsti, Kalevi, J.: "Resolving International Conflicts: A Taxonomy of Behavior and Some Figures on Procedure", in: *Journal of Conflict Resolution*, 10/3, 272–276.

Holsti, Kalevi, J. 1983: *International Politics. A Framework for Analysis*. 2nd edition. Englewood Cliffs: Prentice-Hall.

Holsti, Kalevi, J. 1996: *The state, war, and the state of war*. Cambridge: Cambridge University Press.

Homer-Dixon, Thomas F. 1994: "Environmental Scarcities and Violent Conflict. Evidence from Cases", in: *International Security*, 19/1, 5–40.

Jütte, R. and A. Grosse-Jütte (Eds.) 1981: *The Future of International Organization*. London: Pinter.

Kaplan, Robert. S. 1981: *The Diplomacy of Power: Soviet Armed Forces as a Political Instrument*. Washington D.C.: Brookings Institution.

Kende, István 1982: "Über die Kriege seit 1945", in: Deutsche Gesellschaft für Friedens- und Konfliktforschung (Ed.): *DGFK-Hefte*, Nr. 16, Bonn.

Leiss, A. C. and P. Bloomfield 1967: *The Control of Local Conflict. A Design on Arms-Control and Limited War in the Developing Areas*. 4 Vols. Cambridge: Cambridge University Press.

Levine, E.P.: "Mediation in International Politics", in: *Peace Research Society papers*, No. 13, 23–43.

Levy, Marc. A. 1995: "Is the Environment a National Security Issue?", in: *International Security*, 20/2, 35–62.

Levy, J. S.: "Theory of General War", in: *World Politics*, 27/3, 344–374.

Luard, E. 1988: *Conflict and peace in the modern international system: a study of the principles of international order*. 2nd edition. Basingstoke: Macmillan.

Mack, R. W. and R. C. Snyder 1957: "The analysis of social conflict – towards an overview and synthesis", in: *Journal of Conflict Resolution*, Vol. 1, 212–248.

Maoz, Z. 1982: *Paths to Conflict International Dispute Initiation 1816–1976*. Boulder: Westview.

Matthies, Volker 1996: "Friedenserfahrungsforschung und Friedensursachenforschung", in: Matthies, V., C. Rohloff and S. Klotz: *Frieden statt Krieg. Gelungene Aktionen der Friedenserhaltung und der Friedenssicherung 1945 bis 1995*. Reihe Interdependenz Nr. 21, 7–17. Bonn: Stiftung Entwicklung und Frieden.

Matthies, Volker 1994: *Immer wieder Krieg? Wie Eindämmen? Beenden? Verhüten? Schutz und Hilfe für die Menschen*. Opladen: Leske + Budrich.

Mendler, Martin and Wolfgang Schwegler-Rohmeis 1989: "Weder Drachentöter noch Sicherheitsingenieur. Bilanz und kritische Analyse der sozialwissenschaftlichen Kriegsursachenforschung", in: *HSFK Forschungsbericht*, Nr. 3. Frankfurt/M.

Miall, Hugh 1992: *The Peacemakers. Peaceful Settlement of Disputes since 1945*. Oxford: Macmillan.

Miall, Hugh 1992: "Peaceful Settlement of post-1945 Conflicts: A comparative study" in: Rupesinghe, K. and M. Kuroda (Eds.): *Early Warning and Conflict Resolution*. New York: Macmillan.

Northedge, F. S. and M. D. Donelan 1971: *International Disputes. The Political Aspects*. London: Europa.

Paffenholz, Thania 1995: *Nicht die Konflikte müssen beendet werden, sondern die Kriege: Möglichkeiten der Transformation von innerstaatlichen Kriegen mit nicht-militärischen Mitteln*. Frankfurt/M: HSFK.

Pfetsch, Frank R. and Peter Billing 1994: *Datenhandbuch nationaler und internationaler Konflikte*. Baden-Baden: Nomos.

Pfetsch, Frank R. 1994: *Internationale Politik*. Stuttgart: Kohlhammer.

PRIO and UNEP (Eds.) 1989: *Environmental Security. Report contributing to the concept of Comprehensive Environmental Security*. Oslo: Peace Research Institute Oslo.

Rapoport, A. 1960: *Fights, games and debates*. Ann Arbor.

Richardson, L. F. 1960: *Statistics of Deadly Quarrels*. Pittsburgh: Stevens.

Rohloff, Christoph 1996: "Frieden sichtbar machen! Konzepte und Befunde zu Friedenserhaltung und Friedenssicherung", in: Matthies V., C. Rohloff and S. Klotz: *Frieden statt Krieg. Gelungene Aktionen der Friedenserhaltung und der Friedenssicherung 1945 bis 1995*. Reihe Interdependenz Nr. 21, 18–32. Bonn: Stiftung Entwicklung und Frieden.

Ropers, Norbert 1997: "Prävention und Friedenskonsolidierung als Aufgabe für gesellschaftliche Akteure", in: Senghaas, Dieter (Ed.): *Frieden machen*, 219–242. Frankfurt/M: Suhrkamp.

Ruloff, D. 1987: *Wie Kriege beginnen*. 2nd edition. München: Beck.

Senghaas, Dieter 1969: *Abschreckung und Frieden. Studien zu Kritik organisierter Friedlosigkeit*. Frankfurt/M: Europäische Verlagsanstalt.

Siegelberg, Jens 1994: *Kapitalismus und Krieg: eine Theorie des Kriegs in der Weltgesellschaft*. Münster: Lit.

Singer, J. David and Melvin Small 1972: *The wages of war. A statistical handbook*. New York: Wiley.

Siverson, R.M. and Tennefoss, M.R.: "Interstate Conflicts 1915–1965", in: *International Interactions*, 9/2, 147–178.

Small, M. and Singer, J.D. 1982: *Resort to Arms*. Beverly Hills: Sage.

Sorokin, P.A.1937: *Social and Cultural Dynamics*. New York: American Book Company.

Vogt, Wolfgang R. (Ed.) 1995: *Frieden als Zivilisierungsprojekt – Neue Herausforderungen an die Friedens- und Konfliktforschung*. Baden-Baden: Nomos.

Wainhouse, D. W. et.al. 1966: *International Peace Observation*. Baltimore: John Hopkins.

Wood, D. 1968: *Conflict in the twentieth Century*. London.

Wright, Quincy 1965: *A Study of War*. 2 Vols. 2nd edition. Chicago: Chicago University Press.

Zacher, M.W. 1979: *International Conflicts and Collective Security 1946–1977*. New York: Praeger.

Part C:

Modeling

Environmentally-Induced Conflict – Methodological Notes

Wolf-Dieter Eberwein

1 Environmental problems in the context of conflict and security[1]

The reason why this contribution is confined to "methodological notes" is simple. These notes, which form part of the vanguard of the theoretical and systematic analysis of the relationship between environment and conflict, will introduce a third item into this already exceedingly complex equation, namely security. The concept of security, which will have to be defined more precisely in this context, acts as the connecting link between environment and conflict (see Eberwein 1997a). As such it is of relevance from the point of view of international politics. There is major controversy in academic circles as to whether the environmental question is of any significance to security policy. The propositions following contend that there is such a relationship. The first two propositions assume that there are complex relationships of cause and effect between environment, security and conflict, but that these have not been sufficiently explained because they stem from the close interconnection between the ecosystem, the human system and the political system. Proposition 3 postulates that 'environmental problems' are relevant to security policy when they are cross-border in nature. Proposition 4 on the other hand rejects the assumption that environmental conflicts as such exist. Rather it is assumed that environmental problems can contribute to the outbreak of intrastate and interstate violence. In the fifth proposition it is argued that the characteristics of environmental problems differ fundamentally from classic security policy ones. The next two propositions elaborate the implications of the

---

1 I would like to thank the publishers as well as the participants in the July 1997 workshop in Berlin, where I presented the first version of this document, for their critical comments. They helped to clarify certain inaccuracies. Any remaining errors should be put down exclusively to the author. To note as well is that I refrained from updating a number of citations. I want to refer the reader to a more recent publication which includes more recent data sources (Eberwein and Chognacki 1998).

above for research strategy. In terms of research policy this means that more attention will have to be paid to prevention and early warning (Proposition 8). Politically however, the question remains to what extent security policy institutions are suited to deal with the challenges assumed to result from the interrelation between environment, security and conflict.

Proposition 1: The link between environment and conflict is plausible but not well founded theoretically.

If we make the customary distinction between renewable and non-renewable resources, the relevant problem is the (increasing) scarcity of renewable resources. Resource scarcity is affected by global climate change. Assuming that resources are becoming increasingly scarce, then the next question is where and to what extent we can reckon with a dramatic shortage of resources and whether this will ultimately lead to violent distributional conflicts. In the case of water for instance (see Falkenmark 1995) it is highly probable that we will have to cope with a systematic shortage within the next 10–20 years. At present we cannot know for certain how the world of politics will react to this. We know just as little about how the people affected will react. Incidentally, this is a general rule for the effects of global climate change on individual regions (see White 1996). Therefore, however well-founded the proposition is that global climate change will lead to major problems of resource shortages, all opinions on the subject are speculative. This is because they are based on implicit assumptions about future developments on the one hand and the political and social reactions of the people affected on the other. Such anticipation of the future will always be speculative, irrespective of how plausible it is. However, this type of speculation can be carried out in an entirely systematic manner using an experimental approach which must nevertheless be both empirically based and theoretically justified. More will be said about this below. First of all we must address the problem of the theoretical basis.

Proposition 2: As trivial as it may be to assert that the relationship between the ecosystem, the human system and the political system is complex, the consequences are significant.

The assertion that the relationships between the ecosystem and the human system on the one hand and the political system on the other are complex may be trivial in itself, but in no way is this true of the consequences. Figure 1 may provide some clarification. The ecosystem and the human system have a direct effect on one another but each of them has its specific internal dynamics (e.g. CO_2 concentration). The political system can be viewed as a separate subsystem of the human system. It is subject both to internal dynamics and to feedback effects from the human system. The first problem is that the dynamics of the political system are determined at least in part by discrete processes. Changes therefore occur step by

step. By contrast, the human system and the ecosystem tend to be determined by non-linear processes (see Homer-Dixon 1995) resulting in continual changes which can alter trajectories dramatically over a short period of time.
Secondly, there is a problem of demarcation, in the sense that it is not absolutely clear, particularly where the human and political systems are concerned, what constitutes a separate process. In contrast to 'natural' phenomena, this depends on the conceptualization of the theorist who makes these distinctions. They then have to be substantiated by empirical research. Thirdly, there are particular consequences resulting from the speed and the time horizon which characterize the three subsystems. Global or regional environmental problems emerge over years if not over decades. Political changes can occur within a very short time. This becomes relevant when the question of practical possibilities of intervention is raised: Short-term or even medium-term successes of intervention into the ecosystem, and partly also into the human system (e.g. population growth), have no effect. In the political system on the other hand, drastic changes can be brought about intentionally in the short-term. Therefore it makes no sense from a theoretical point of view to break up the presumed effective links into separate, individual cause-and-effect relationships and consider them in isolation. This means that there is a case for a strategy of integrated modeling, regardless of the analytical necessity of selectively identifying the various interdependencies . In this context an integrated modeling strategy means devising simulation models in which the various interdependencies of the three areas are reproduced in order to analyze systematically their direct and indirect effects. This approach has been adopted in the field of political science with GLOBUS, a global computer simulation model (see Eberwein 1997b). However, it is uncertain whether this global approach is suitable. Possibly a smaller, more focused model dealing with a specific region is more appropriate. Further thoughts on this subject will be formulated below. Instead we must return for the time being to the security policy aspect of environmental problems.

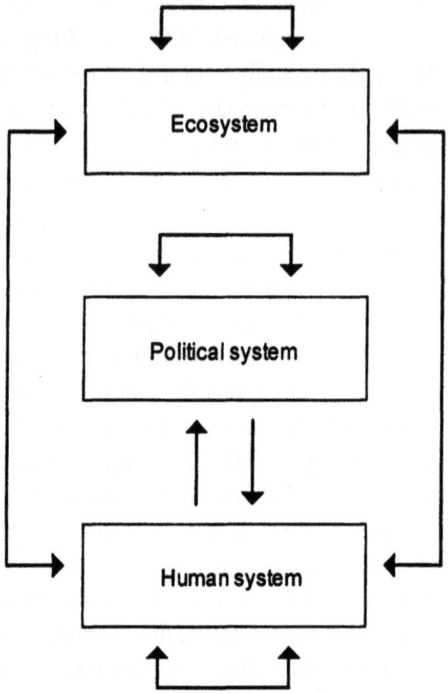

Figure 1: Ecosystem, human system and political system

Proposition 3: There are two distinct types of environmental conflicts, those that are potentially security policy relevant and those that are not.

Environmental problems are encountered at various levels. Geographically speaking, they are found at the local, regional, national and international levels. It goes without saying that a local environmental problem will be irrelevant to security policy as long as security is understood to mean international security. The relationship between environmental problems and external security is given when the former are politically relevant and directly related to the latter. Environmental problems are security policy relevant if:

a) they can lead to organized collective violence, and if
b) they affect at least two contiguous states.

Therefore, environmental problems are relevant to security policy when state security is threatened. This is a means of avoiding the need to include all the conceivable levels of security and affected parties in the definition (individuals, groups, etc.). If an all-inclusive approach were adopted, the concept of security would be reduced to a meaningless notion from the analytical point of view (see Eberwein 1997a: 4–9) because the phenomenon would become infinitely complex or empirically empty. The security policy aspect will be discussed in proposition 5. In this section we will concentrate on the relationship between environment and conflict. This raises the question of what form the postulated relationship between environmental problems and conflict may take. Two quite distinct types of conflict can be identified – intrastate and interstate conflict. Therefore every conflict which includes the political system or in which politics are likely to be involved is potentially relevant to security policy if, and this further qualifies the definition, it is likely to assume a cross-border character. This is also possible if a conflict is initially purely intrastate in nature. In the event of cross-border problems, it is only by working together that states can establish commonly binding rules to overcome environmental problems. They are key parties to conflicts, both interstate and intrastate. They are thus the central actors in the event of environmental problems, which are equated with scarcity of resources. When scarcity of resources has cross-border origins a security policy problem can arise if the shortage threatens the viability of individual states. According to Walt (1991: 212), this type of problem always comes about when there is a military threat from a third state or alternatively when, as Kolodziej (1992: 427–428) notes, the welfare and wealth of individual societies, and thus the legitimacy of the state, is threatened by material deprivation. This particular enlargement of the concept of security challenges the traditional view of security policy. This leads to this issue whether there is a specific category of conflict which can be identified unequivocally as an environmental conflict.

Proposition 4: There is no such thing as a specific environmental conflict. Environmental problems are part of the structural conditions leading to the outbreak of conflict.

This proposition makes a further qualification, disputing that unequivocally identifiable environmental conflicts exist as a separate category. This suggestion may seem exaggerated but probably it is not. According to Holsti's list (1991) of the correlates of war, there has practically never been a war that can be put down to a single cause. Apart from this, it is not certain whether the reasons given by the conflicting parties are actually those by which a conflict can be explained. The term environmental conflict suggests that this type of conflict can be reduced exclusively to environmental problems as the cause. Additionally it is frequently assumed that environmental conflicts are primarily intrastate conflicts. The reason why interstate conflicts are excluded is that 'resource wars' are rare or, as Levy points out, in international conflicts environmental factors are at best remote causes. It is also frequently argued that violent conflicts are increasingly intrastate in nature, a claim which is somewhat premature (see Eberwein 1997c). If the implicit assumption is dropped that there is an analytically, unequivocally separate category of conflicts, namely environmental conflicts, then this opens up the debate to a consideration of the analytically well-founded hypothesis that future environmentally-induced problems may contribute to the outbreak of both intra-

and interstate violence. Violence is assumed to be a specific form of conflict behavior which is based on the conscious decisions of individuals or groups. However, while violence can be intended, it must not necessarily be so. These decisions are influenced positively or negatively by structural factors. As a generic term environment corresponds to a category of structural criteria. Yet structural criteria are only prerequisites for conflict behavior itself. Violence, on the other hand, forms part of the process of escalation between two or more conflicting parties and is based on the conscious decisions of the actors to deploy these methods. In other words, structural criteria (see Bremer 1995) help to explain the outbreak of conflicts; but they cannot explain the process by which conflicts escalate to violence if it is assumed that manifest conflict behavior is determined by strategic notions of victory, i.e. using violence to win. When environmentally-induced conflicts are distributional conflicts, the state plays a vital, direct or indirect role because it establishes the institutional framework for distribution or redistribution within society. Accordingly, conflict develops primarily in the zone of tension between the human system and the political system, as illustrated in Figure 2. At the national level, such conflicts concern the legitimacy of the existing political system. This problem of the legitimacy of existing distributional and redistributional processes is also raised at the international level, even though in somewhat different terms. The main problem of conflict theory lies in the fact that the following three key phenomena have not been explained: i) the process by which intrastate and interstate conflicts escalate to violence; ii) the 'horizontal' escalation process from intrastate to interstate conflicts, and; iii) the 'horizontal' escalation process in the other direction from interstate to intrastate conflict. In theory, it can be assumed that increasing environmentally-induced scarcity of resources may lead to distributional conflicts and, moreover, that these types of conflict may lead to violence. This is not implausable because they are linked to the legitimacy of political systems. Considering that the state is involved in the process and violence can spread across borders, it is inevitable that environmentally-induced conflicts overlap with the classic security question in this specific context.

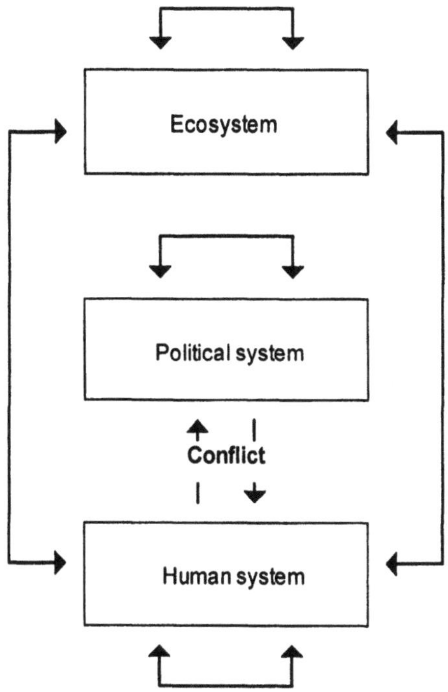

Figure 2: Environment and conflict

Proposition 5: The difficulty of classifying environmental problems in terms of security policy stems from their specific characteristics which distinguish them from classic security policy issues.

According to the traditional view of security policy there is a fundamental distinction, introduced by Haftendorn (1992: 3), between national, international (i.e. regional) and global security. In this view, there are security policy problems which can be interpreted in international terms only. It is assumed that the state alone has the authority and the duty to take the appropriate measures to ward off threats. This is in keeping with the classical idea of territorial defense which supposes that all states are actually able to survive on the basis of the principle of self defense. However, this overlooks the fact that national security in this sense is ultimately a form of relationship and that this definition implies that every state is basically threatened by the others. The explanation given for this is the anarchic structure of the international system and the absence of a central authority. An alternative argument could be that, according to this conception, one-sided or many-sided threats are always regarded as real. However, by no means is this the inevitable, logical implication. In the case of both regional and global security, which Haftendorn regards as possible security configurations, the principle of self-sufficiency no longer applies by definition. Rather, collective action is inescapable. In both cases, there is a functional and spatially definable interdependence between at least two states or in the extreme case the entire international community. In my opinion therefore, environmental problems are relevant to security policy when this type of functional and spatial interdependence exists. This relationship is represented graphically in Figure 3. In concrete terms this means that security problems observed at national level also have regional effects, and vice versa. The Middle East serves as a perfect illustration of this. A global security problem may in turn have regionally specific effects, as exemplified by the causal complex of global climate change. It can be assumed that such interdependencies between levels are an inherent feature of environmental problems. The classic example is water, when groundwater reserves are used across borders or when more than one state pollutes and makes use of the same rivers. Such environmentally-induced problems only become relevant to security policy if they lead to the confrontation between the parties, precisely because they cannot be solved unilaterally. Within the academic debate, the fact that environmental problems are frequently denied the status of a security policy issue can be traced to the fact that environmental problems lack the following key features of classic security policy problems:

- a clearly identifiable enemy, owing to the impersonal nature of the threat;
- the urgency of the threat (temporal dimension);
- the absence of a territorial link between the cause of the threat and its effect; and, finally, by extension
- the lack of a common denominator (unity of adversary, spatial location and immediacy).

Environmental problems are complex because they involve both political and economic interests and this makes collective, consensual agreements difficult to achieve.

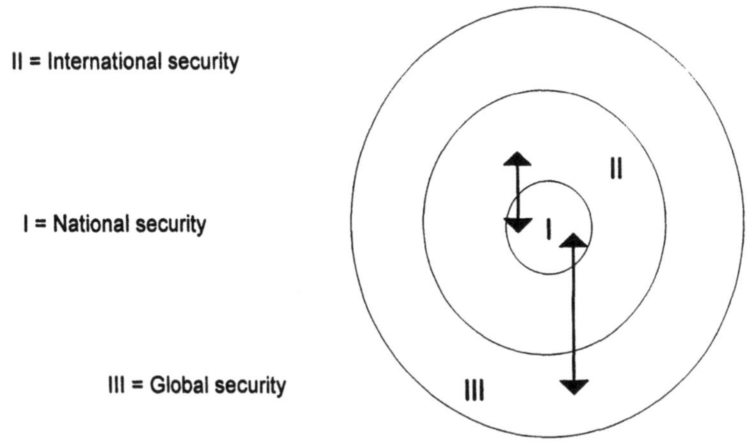

Figure 3: Levels of security in the international system

Proposition 6: Case studies are useful for analyzing the nexus between environment and conflict but they are not enough in themselves.

If, as argued above, the question of environmentally-induced conflicts has not yet been explained adequately in theoretical terms, this raises the question as to what extent research conducted until now is in a position to fill these gaps empirically by means of case studies. It goes without saying that case studies are worthwhile and necessary. The 'process tracing' method (Goodrich/Brecke 1996; Brecke 1997) may be particularly suited. This method attempts to retrace the course of conflicts which are thought to be attributable to environmentally-induced problems on the basis of general theoretical categories. However, this raises the problem that the connection between structural conditions and manifest conflict behavior (in the escalation process) would then appear to be predetermined. The main theoretical problem of escalation processes is the link between the spatial (horizontal) spread and the contemporaneous spread (intensity) of the conflict, which is only a parameter in these kinds of analysis. The insights gained by this method are difficult to generalize. It does not offer an avenue to identify the special characteristics of emergence processes systematically (see Eberwein/Saurel 1995: Part 4). Our main task therefore is to find additional, methodological tools which both allow for the possibility that violent conflicts can be traced back to a whole series of structural causes and make it possible to analyze the complex interactions among the three aforementioned systems, i.e. the ecosystem, the human system and the political system. As far as the first aspect is concerned, 'topological' methods such as those of Brecke (1997) or Duquenne (1992) seem to be useful. These mainly involve sophisticated 'classificatory' processes which not only help to identify conflict syndromes but also facilitate estimates of the probability of a particular set of factors to lead to manifest violence. Methods such as binomial regression, as used by Bremer (1992, 1993), are also valuable because they lead to statements of probability based on the presence of particular factors. On the other hand, the problem of analyzing causal linkages and their effects over the time cannot be tackled using such methods. They are all based on historical events and this means that conceivable future developments are left out of the equation. Assuming that the link between environment, security and conflict will only develop its full explosive force in the future, our analytical horizon can only be extended with the help of an experimental approach in which we elaborate systematic scenarios. These can be carried out as purely intellectual experiments. As global climate research shows, simulation models are an invaluable tool. Only with the help of these models is it possible to analyze complex, dynamic cause-effect relationships systematically. Nonetheless, this is an experimental approach, precisely because it is impossible to predict future developments. At best we can show, on the basis of assumptions about the behavior of actors and with the help of this kind of model, what the consequences might be. This is what climate researchers do with respect to the greenhouse effect. Whether CO_2 emissions really will reach the levels upon which calculations are based is uncertain. However what can be demonstrated is that global warming will reach a given level in the event of a given amount of carbon dioxide and that the corresponding consequences will follow. This argument for an experimental approach is based on the assumption that working out the consequences of complex, dynamic cause-effect relationships exceeds our mental capacities. Computers, however, can cope with this very well. With their help, we can verify complex, theoretical propositions that only the human brain can produce. However, in order to ensure that such experimental methods do not exhaust themselves in theoretical musings, it is essential that empirical research follows this track. This raises the question as to whether we yet have enough information at our disposal to form empirically-founded theories.

Proposition 7: There is a problem of data which is not confined to the question of the availability of relevant indicators alone.

In the systematic analysis of the link between environment and conflict, empirical research is confronted with a largely unresolved problem. Environmental problems are, at least in part, cross-border in nature. In this connection, Graeger refers to 'eco-geographical regions' (1996). These regions do not correspond to existing state boundaries. This means that empirical research is faced with a unit of analysis problem. Eco-geographical regions are defined by environment-specific criteria which do not coincide with the state borders forming the usual unit of research. As long as we are dealing with manifest interstate conflicts, the problem does not arise. However, it does arise immediately in the case of intrastate violence or in general, when the outbreak of conflicts has to be studied at the preliminary stage, i.e. before potential outbreaks of violence. We work on the assumption that this type of conflict, like any other, breaks out initially at a local level and only then begins to spread. Yet how can we solve this problem which covers both eco-geographical and national areas but is also divided up into smaller subordinate units? It must be kept in mind that the dynamics of conflicts are largely determined by actors and their ability to mobilize the public. If this is so, then the usual macro-quantitative strategy of recording conflicts in the form of information on events at a national level is not sufficient. This overlooks the mobilization aspect which, as Eberwein and Saurel (1995) demonstrate, has both a temporal and a spatial dimension. The extent to which these conceptual considerations can be transformed effectively into a systematic data survey scheme based on smaller spatial and, at the same time, cross-border units, is a practical research matter which shall not be discussed further here. It is possible that this strategy is too costly in relation to its results. However, the conceptual problem of spatial dimensioning does at least need to be debated in the future. As far as the availability of data potentially relevant to the theme of environment and conflict is concerned, the problem is not so much a lack of data but rather the incomplete knowledge about available data records or a failure to refer back to existing records. This can be illustrated using the following three areas with which the author is familiar.

International conflict: Several databases are available on this. A particularly useful one is the MID database of the *Correlates of War* project (Gochman/Maoz, 1984).[2] It covers all the interstate conflicts which have taken place in the world since 1816 in which one state has at least threatened another with the use of military force. It contains various stages of escalation. It is currently being complemented with a list of conflict issues. In addition there is the database of Hamburg University's AKUF (Arbeitsgemeinschaft Kriegsursachenforschung) research group, which is particularly suited to the study of intrastate conflict – not least because of its good documentation (Gantzel/Meyer-Stamer 1986; Gantzel et al. 1992).

Intrastate conflict: Apart from the aforementioned AKUF database, a mention should be made of the data on ethno-political conflicts compiled by Ted Gurr (1994), who has covered this type of conflict throughout the world since 1945. On the subject of changes of regime or attempts to change regimes, Gurr's POLITY III database[3] is bound to be of use. It also covers characteristics of the political system that may be relevant to the postulated distributional conflicts.

Disasters: Data about natural disasters are available but have to be gleaned from various sources and have not yet been compiled into a systematic database. The handbook of the German IDNDR Committee (1995) contains a compilation based on data from the Munich-based Münchner Rückversicherung reinsurance company. One possible exception is the CRED (Center for Research on the Epidemiology of Disasters) EM-DAT database (Sapir/Misson 1991).

As far as I know, empirical research on environmental conflicts has not yet taken advantage of these databases. This has a number of consequences. First of

---

2 The latest database, which goes up to 1992, can be retrieved from the Internet along with the codebook at the following address:
http://www.polsci.binghamton.edu/peace(s)/mid_data.htm

3 Internet: ftp://isere.colorado.edu/pub/dataset3/polity3/

all, a purely macro-quantitative approach using the state as the unit of analysis is problematic. Secondly, the subnational level of actors and the transnational (or regional) context are essential to the study of conflict escalation, as otherwise the spatial and temporal escalation process towards violence cannot be analyzed. Thirdly, there is a wealth of data that are suited to the purpose. However, there is some doubt as to whether the data currently available are sufficient. Finally, it is by no means certain that appropriate methodologies are used.

Proposition 8: It is conceivable for empirical analysis to pursue two complementary aims: early warning and prevention.

Up until now, arguments are based on the premise that massive conflicts will emerge in the future if the environmentally-induced scarcity of resources combined with population growth continues unabated. The call for an experimental approach implies that research policy will be aimed at prevention, i.e. longer-term prevention and early warning. This means identifying potential conflicts as early as possible. Prevention is the considerably more ambitious objective because it presupposes that science can provide plausible theoretical and even, where possible, empirically validated models of the outbreak and escalation of conflicts. This also entails the ongoing extrapolation of data on developments in regions potentially at risk. However, it is unresolved what this implies for practical policy and when, where and how preventive interventions in existing processes should be carried out. With environmental problems, there is indeed the essential difficulty that they cannot be solved in the short-term. If we accept Matthies's view, then general confusion prevails instead of intellectual lucidity and the problem is aggravated by the abstract and somewhat diffuse ideas over what preventive diplomacy consists of and what it can achieve (Matthies 1996: 22; see also Debiel 1996). This leads us to the question of early warning indicators. To date, science has not met its aspirations to yield early warning systems on a systematic basis, particularly where conflicts are concerned. Even if they were available, there would be no point in them if they did not lead to a rapid response (Grünewald 1995: 5). Until now it has always seemed as if:

> in almost all conflicts, states waited until faced with a major humanitarian crisis before taking up action of their own (Russbach/Fink 1994: 8; Matthies 1996: 25–26).

Proposition 9: If environmental problems are relevant to security policy, then it is essential that existing security policy institutions adapt to this in the long run.

The main basis of our argument has been that environmental problems exist and that they threaten the ability of states to survive, although the threat is not of a military nature at first. This threat affects welfare and wealth. The second basis is that this could give rise to distributional and redistributional conflicts which could in turn escalate into violence. If this is so, then the conclusion must be that there is a category of environmental problems which should be regarded as an integral part of security policy. Inevitably this raises the question as to whether existing security policy institutions – primarily the military and diplomacy – are suited for the task. Here is not the place to address the problem of the militarization of the environment (see for instance Käkönen 1992 and Carius/Imbusch in this volume). Suffice it to say that it seems plausible to assume that these institutions cannot cope with these kinds of issues because the armed forces are still trained primarily for national defense. Already "humanitarian operations" have overstretched the capacity of the troops involved (see MCDA 1995). Whether the diplomatic services are conceptually and technically equipped for the systematic observation and analysis of long-term developments of the type concerned is an unanswered question. However, if regional and global environmental problems aggravate existing political and social, intrastate and interstate inequalities and conflicts, then there is no doubt that both the military and the diplomatic service will have to undergo major processes of adaptation.

2 Résumé

The views expressed herein are based on the assumption that environmental problems can lead to the outbreak of conflicts and that these in turn may escalate into violence. Consequently, cross-border environmental problems are of potential relevance to security policy because they can give rise to interstate violence. In addition to this there are intrastate conflicts which are always liable to escalate horizontally, i.e. across borders. The contribution environmental problems make as a part of the structural factors which lead to the outbreak of conflicts has not been theoretically resolved. Nor have the conditions which determine the dynamics of conflict escalation been clarified. In empirical terms, the problem is to transform the spatial and temporal dimension of the conflict process into operational indicators that do justice to the circumstance that eco-geographical regions and national territories may not necessarily coincide. There is also the practical difficulty that environmental problems are either denied security policy status or are categorized over-hastily in a security policy context not organized to deal with this type of problem. A prime characteristic of the effect of environmental problems on the viability of state systems is their long-term nature, i.e. there is no urgency to resolve them now. There is also a lack of any identifiable enemy and, finally, causes and effects are decoupled spatially and temporally. Both prevention (as a long-term task) and early warning (as a short-term task) are necessary. Both imply changes in existing institutions and the development of appropriate tools. This includes developing and continuously updating relevant indicators and implementing appropriate procedures. Shortcomings, where such can be discerned, lie more in the theoretical and conceptual field than in the availability of relevant data.

References

Brecke, Peter 1997: *Using pattern recognition to identify Harbinger configurations of early warning indicators.* Paper presented to the ECPR Workshop in Bern, 27 February – 4 March 1997.

Bremer, Stuart A. 1993: "Democracy and Militarized Interstate Conflict, 1816–1965", in: *International Interactions*, 18/3, 231–249.

Bremer, Stuart A. 1995: "Advancing the Scientific Study of War", in: Bremer, S. A. and T. R. Cusack (Hrsg): *Advancing the Scientific Study of War*, 1–33. Amsterdam: Gordon and Breach.

Bremer, Stuart A. 1992: "Dangerous Dyads – Conditions Affecting the Likelihood of Interstate War, 1816–1965", in: *Journal of Conflict Resolution*, 36/2, 309–341.

Debiel, Tobias 1996: "Not und Intervention in einer Welt des Umbruchs – Zu Imperativen und Fallstricken humanitärer Einmischung", in: *Aus Politik und Zeitgeschichte*, Bd. 47, Heft 94, 29–38.

Deutsches IDNDR-Komittee für Katastrophenvorbeugung e.V. 1995: *Journalisten-Handbuch zum Katastrophenmanagement.* Bonn (Eigenpublikation).

Duquenne, Vincent 1992: *General latice analysis and design program (GLAD).* Paris: CNRS and Maison des Sciences de l'Homme.

Eberwein, Wolf-Dieter 1997a: "Umwelt – Sicherheit – Konflikt, eine theoretische Analyse", in: Discussion paper No. 303. Berlin: Wissenschaftszentrum Berlin.

Eberwein, Wolf-Dieter 1997b: "Weltmodelle", in: Albrecht, U. and H. Volger (Eds.): *Lexikon der Internationalen Politik*, 562–564. München: R. Oldenbourg.

Eberwein, Wolf-Dieter 1997c: "Replik auf Gantzel – Quantität und Qualität – Beide gehören zusammen", in: *Ethik und Sozialwissenschaften,* Oktober 1997.

Eberwein, Wolf-Dieter and Chognacki, Sven 1998: *Disasters and Violence 1946-1997. The link between the natural and the social environment.* Papers P98-304. Berlin: Wissenschaftszentrum Berlin.

Eberwein, Wolf-Dieter and Pierre Saurel 1995: *The dynamics of collective behavior: modeling mass mobilization in the GDR.* Paper presented to the Panel "Formal Theory" at the second Pan-European Conference on International Relations, 13–16 September 1995. Paris: Fondation Nationale des Sciences Politiques.

Falkenmark, Malin 1995: "The dangerous spiral: Near-future risks for water-related eco-conflicts", in: International Committee of the Red Cross 1995: *Water and War – Symposium on Water in Armed Conflicts*, 10–28. Report. Geneva.

Gantzel, Klaus Jürgen and Jörg Meyer-Stamer (Eds.) 1986: *Die Kriege nach dem Zweiten Weltkrieg bis 1984 – Daten und erste Analysen*. München: Weltforum.

Gantzel, Klaus et al. 1992: "Ein systematisches Register der kriegerischen Konflikte 1985–1992", in: *Interdependenz*. Materialien und Studien der Stiftung Entwicklung und Frieden und des Instituts für Entwicklung und Frieden, Nr. 13.

Gochman, Charles S. and Zeev Maoz 1984: "Militarized Interstate Disputes, 1816–1976: Procedures, Patterns, and Insights", in: *Journal of Conflict Resolution*, 28/4, 585–615.

Goodrich, Jill and Peter Brecke 1996: "Paths from environmental change to violent conflict", in: *Working Paper Series of the Georgia Consortium on Negotiation and Conflict Resolution*. Paper 96-8 (November). Atlanta.

Graeger, Nina 1996: "Review Essay: Environmental Security", in: *Journal of Peace Research*, 33/1, 109–16.

Grünewald, François 1995: "From prevention to rehabilitation: action before, during, and after crisis – The experience of the ICRC in retrospect", in: *International Reveiw of the Red Cross*, Issue 306, 263–281.

Gurr, Ted Robert 1994: "Peoples against states: ethnopolitical conflict and the changing world system", in: *International Studies Quarterly*, 38/3, 347 ff.

Haftendorn, Helga 1991: "The Security-Puzzle: Theory-Building and Discipline-Building in International Security", in: *International Studies Quarterly*, 35/1, 3–17.

Holsti, Ole R. 1991: *Peace and War: Armed Conflicts and International Order 1648–1989*. Cambridge: Cambridge University Press.

Homer-Dixon 1993. "Environmental Scarcities and Violent Conflict: Evidence from Cases", in: *International Security*, 19/1, 5–40.

Käkönen, Jyrki (Ed.) 1994: *Green Security or Militarized Environment*. Brookfield: Dartmouth Publishing.

Kolodziej, Edward A. 1992: "Renaissance in Security Studies? Caveat Lector!", in: *International Studies Quarterly*, 36/4, 421–438.

Levy, Marc, A. 1995: "Is the Environment a National Security Issue?", in: *International Security*, 20/2, 35–62.

Matthies, Volker 1996: "Vom reaktiven Krisenmanagement zur präventiven Konfliktbearbeitung?", in: *Aus Politik und Zeitgeschichte*, 47/94,19–28.

MCDA Reference Manual 1995: "The Use of Military and Civil Defence Assets", in: *Relief Operations 1995*. UN Department of Humanitarian Affairs, November 15. (http://www.reliefweb.int/library/mcda/refman/idex.html).

Moore, Curtis A. 1997: "Warming up to hot new evidence", in: *International Wildlife*, 27/1, 21–25.

Russbach, R. and D. Fink 1994: "Humanitarian action in current armed conflicts: opportunities and obstacles", in: *Medicine and Survival*, 1/4.

Sapir, D. G. and C. Misson 1991: "Disasters and Databases: Experiences of the CRED EM-DAT Project", in: *UNDRO News*, September/October, 13–22.

Schönwiese, Christian 1996: "Weltweite Klimaänderungen: Grundlagen, Probleme und neue Prognosen", in: *Universitas*, Jg.51, Heft 606, 999–1008.

Walt, Stephen M. 1991: "The Renaissance of Security Studies", in: *International Studies Quarterly*, 35/3, 211–239.

Modeling Environmental Conflict

Detlef F. Sprinz

1 Introduction[1]

Scientific consideration of the topic of 'environment and security' is part of the renewed discussion of the concept of security towards the beginning of the third millennium. Academic research faces a dual question: Is it scientifically fruitful to expand the concept of international security, which was originally restricted to research into the causes of wars, and to what extent is the environmental dimension of security policy suitable for practical policy-making? Answering the second question ideally presupposes a fruitful discussion of the results of scientific research, although policymakers expect answers in this field from science which is just evaluating the first generation of its research programs – and many questions are *necessarily* subject to a debate about content and methodology. This paper takes up this challenge by presenting a methodological critique of some empirical-qualitative research programs that are important for politics and political science (Section 2), and briefly compares some advantages of empirical-quantitative research methods (Section 3). Subsequently, it will be shown how environmental thresholds can be conceived of as a sufficient condition for the outbreak of violent conflict and how research on environmental thresholds can be incorporated in an empirical-quantitative research design on environmental problems and the outbreak of violent conflict (Section 4).

---

1 For comments on a previous draft, I thank Matthias Buck, Alexander Carius, Kerstin Imbusch, Sebastian Oberthür, Matthias Paustian and Ralph Piotrowski. The fruitful discussions and cooperation with Galina Churkina on threshold values in the natural and social sciences prompted the search for methods for determining environmental threshold values. This work on environmental threshold values has been supported financially by NATO's Scientific and Environmental Affairs Division under a Collaborative Research Grant.

2 Methodological critique of some empirical-qualitative projects

Two research programs with an empirical-qualitative orientation are dominating the political and scientific discussion of the effect of environmental problems on the outbreak of armed conflicts within and between states: that of Homer-Dixon, and that of Baechler and Spillmann.[2] For reasons of brevity, the criticism of the two programs can only be outlined.

In his first concept, Homer-Dixon develops a research design that lets macro-variables (such as population pressure, economic activity, and institutions) be the cause of environmental effects – which, in turn, result in unwanted social effects and potential armed conflict (see in particular Homer-Dixon 1991:53). For special problems, such as a decline in agricultural productivity or economic activity, high-resolution cause-and-effect models of path-tracing analysis are developed, which are intended to explain the origin of conflicts due to scarcity and group identity or the conflicts originating from relative deprivation. These detailed theories are often considerably too complex for systematic tests[3], and omit major theoretical aspects. The first problem results from the testing of nonequivalent causal structures, which make the empirical comparison of results very difficult or impossible; and the second problem results, for example, from an insufficiently comprehensive concept, since political intervention (e.g. environmental policy measures or policies for restricting actual or potential violence (see Section 4)) are not included systematically in the research program – although it is true that such variables are mentioned in Homer-Dixon's general set of ideas.

This original model was revised by Homer-Dixon in view of his empirical findings (Homer-Dixon 1994). However, some theoretical aspects remain unclear: sometimes, variables appear as *causes* (in path-tracing analysis) of environmental problems ('environmental scarcity'), but on other occasions, they seem to be *measures* of environmental problems. But variables can only fulfill one of these two functions, not both at the same time. On the other hand, "social and technical ingenuity", rightly emphasized by Homer-Dixon, is unfortunately not integrated into the concept.

Although Homer-Dixon does base his work on numerous detailed case studies, his selection of cases leads to a serious methodological inference problem. Only such cases were selected that involve *both* environmental problems and the outbreak of armed conflicts (Homer-Dixon 1996). This simply does not permit causal analysis[4] of inference whether environmental problems lead to armed conflicts – since the selection criteria make precisely this (correlative) connection.

---

2 A comprehensive critique of the literature on the environment and armed conflict is given in Gleditsch (1998).

3 Gleditsch (1998) speaks of "untestable models".

4 On the preconditions for causal analysis links, see Cook and Campbell (1979: 18) int. al.

Furthermore, it remains unclear which *method* of inference is chosen, especially since the uniqueness of each case is emphasized (Homer-Dixon and Percival 1996: 3). Overall, Homer-Dixon's research strategy shows a number of validity problems (see Cook and Campbell 1979: Chapter 2), which considerably restrict the use of the results for policymaking.

In addition to Homer-Dixon, Baechler and Spillmann have undertaken a large-scale study of the link between environmental problems and armed conflict as part of the Environment and Conflicts Project (ENCOP: Baechler et al. 1996; Baechler and Spillmann 1996a; Baechler and Spillmann 1996b). Unlike Homer-Dixon's explicit theoretical work, their project lacks an *ex ante* formulation of research hypotheses. On the methodological side, it remains unclear which factors must systematically accompany environmental degradation in order to account for the outbreak of armed conflict. Furthermore, exceeding environment threshold values (see Section 4) is occasionally mentioned as a precondition for armed conflicts, but this important idea is not taken up systematically and operationalized throughout the case studies. As with Homer-Dixon, a method of inference which systematically evaluates the impact of causes (e.g. environmental degradation) on effects (armed conflict) is lacking.

The research projects headed by Homer-Dixon as well as Baechler and Spillmann succeeded in directing the attention of political decision-makers on both sides of the Atlantic to the important research question of whether environmental problems lead to armed conflicts. However, due to the methodological weaknesses of the two programs, we unfortunately still do not know whether this correlation exists to a substantial degree, or not. Thus, researchers find themselves in the not uncommon, but less than desirable situation that neither the hypothesis that environmental problems lead to wars, nor the alternative hypothesis (environmental problems do not lead to armed conflict) can be refuted. Because of theoretical and methodological weaknesses, the case history remains incomplete, the diagnosis of the (non-)existence of a general cause-and-effect relationship between environmental problems and armed conflict cannot yet be made, and remedies in the form of recommendations for political action seem quite premature in view of the difficulties encountered in the previous steps.

But these research shortcomings are by no means unusual for such a new field of research as environment and security. Almost every field of research refines its theory and methods over the course of time, as part of a fruitful scientific discourse. The suggestions in the two sections to follow are intended to contribute to the advancement of the field.

3 Advantages of empirical-quantitative research approaches

Empirical-qualitative and empirical-quantitative research approaches are not necessarily derived from differing research strategies – the stages of research are largely similar in both cases (see Mitchell 1998; Sprinz 1998). However, the research design and execution of empirical-quantitative research programs are usually more rigorous.

In an empirical-quantitative research design, researchers must first define the phenomenon to be explained (dependent variables), in order to delimit the object of research (see Table 1). Next, it is advisable to develop a specification of hypotheses, which give the research design a theoretical framework, incorporate previous research results, and permit testing of alternative or complementary research hypotheses. Thus one's own research project is integrated in the broader research enterprise of the policy area under consideration. These first two stages also determine which variables are to be newly collected, or are to be derived from existing sources of data. The scaling of both dependent and independent variables is important, since this affects the selection of the statistical method for analyzing the co-variation of dependent and independent variables, as well as the extent of variation of each variable. In a fourth stage, the statistical method is selected, and the results are interpreted – with the validity of the results (fifth stage)[5] being decisively influenced by the previous decisions (the second and third stages, in particular). Finally, recommendations for further research, as well as recommendations for public policy (where applicable), are often provided.

Table 1: Research stages within an empirical-quantitative research design

Research stages
1. Selection of the phenomenon to be explained
2. Development of the theory and specification of hypotheses
3. Determination of the data sources and scaling of variables
4. Selection of the statistical method and interpretation of the results
5. Checking the validity of the results
6. Consequences for future research, and possible recommendations for public policy

Source: See Dreier 1997; Schnell et al. 1995

5 See Cook and Campbell 1979.

Empirical-quantitative research often has advantages over qualitative research, and can make a major contribution to scientific research and public policy based on it, especially in the field of 'environment and security' (see Table 2). A clear specification of the theoretical model before performing research permits conceptual synthesis, which has not yet been achieved in qualitative research in the field of environment and security. In addition, the collection of data in quantitative research compels the researcher to specify the level of analysis, i.e. the unit of data collection within a geographical area (e.g. 10 x 10 km^2) and specified period (see Gleditsch 1998). These specifications shape the interpretation of the results, and show explicitly how theoretical concepts are (often incompletely) operationalized. Furthermore, the choice of the method of inference allows the results to be obtained on the basis of a documented algorithm (rather than the subjective impression of a researcher) – they can then be tested by means of replication (including other methods of inference) and sensitivity analyses to see how well they stand up. In addition, empirical-quantitative methods can often be illustrated with modern methods of static (e.g. tables and diagrams) and dynamic (e.g. video representation of the dynamics) visualization. The choice of empirical-quantitative methods does not solve all research problems automatically, but it makes them apparent, permits independent replication, and promotes a transparency in research that is often lacking in empirical-qualitative research projects.

Table 2: Selected advantages of empirical-quantitative research

Advantages
Clear specification of the theoretical model (*ex ante*)
Specification of the level of analysis, unit of analysis, temporal and geographical scale
Explicit operationalization of theoretical concepts
Specification of method of inference; replication and sensitivity analysis
Visualization of data and findings

Source: Sprinz forthcoming (revised)

In the following section, two aspects of an empirical-quantitative research design for explaining environment-induced armed conflict are discussed, in particular the specification of a simple general model (research stage 2), and requirements upon data collection (research stage 3).

4 The study of environmental thresholds

This section suggests an empirical-quantitative research program on environment and security. This includes crucial aspects of the overall relationship among variables and the derivation of environmental thresholds as a sufficient condition for the systematic diagnosis of environmentally-induced armed conflict.

Environmental problems are normally the result of anthropogenic activities (such as the production and consumption of goods, population growth, etc.) that damage the environment. But not every impairment of the environment is associated with severe environmental degradation that may increase the probability of armed conflicts. Rather, when an environmental threshold (see below for the definition) is exceeded, this may become a precondition for *environmentally*-induced violent conflict. Only if this environmental threshold is exceeded (or several are exceeded synergetically) can environmentally-induced armed conflict be diagnosed at all. Thus it is the specification of a sufficient variable (environmental thresholds) which allows us to limit ourselves to a subset of violent conflict (see Fig. 1).[6]

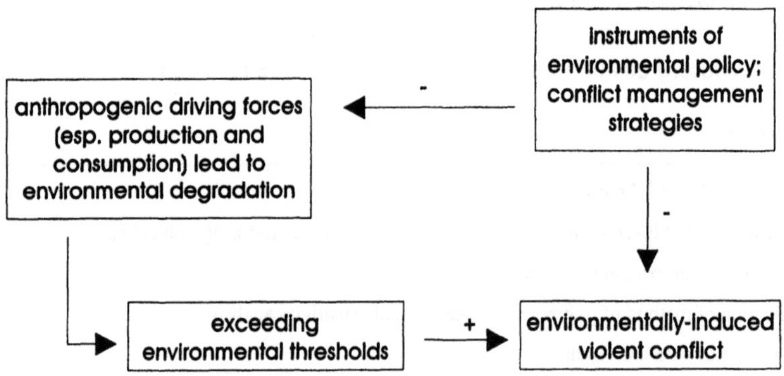

Figure 1: The relationship between environmental threshold values, armed conflict, and political intervention

This delimitation has two important implications for research. First, the scope can be narrowed considerably, especially with regard to the independent variables, since not every cause of armed conflict has an environmental component, or is causally linked to environmental problems. As mentioned above, exceeding environmental thresholds does not necessarily always lead to an outbreak of armed conflict, since governmental and non-governmental actors can employ the

6 If environmental degradation is one of many necessary conditions for the outbreak of armed conflict, studies of 'environment and security' should be included in the general research on the causes of war.

instruments of environmental policy as well as of conflict management. Second, this leads to a systematic expansion of the chain of cause and effect by including important possibilities for intervention: one the one hand, successful employment of the instruments of environmental policy (see Sprinz 1997) can lead to a modification of the anthropogenic forces of environmental degradation over time, resulting in the impact *dropping below* the environmental threshold; on the other hand, strategies of conflict management may reduce the probability of armed conflict even if environmental thresholds have been exceeded.[7] The latter seems probable under legitimate legal and social systems in democracies or if generous compensation is provided for environmental damages.

Thus, the diagnosis of whether environmental thresholds have been exceeded is central to research on environment and security. Only then can we diagnose *environment*-induced armed conflict. Otherwise, environmental problems are secondary causes – and should then be incorporated as a minor aspect in the general analysis of the causes of war. The answer to the question of whether exceeding environmental threshold values is a sufficient condition for the outbreak of armed conflicts within or between states is also of great importance for public policy, since it assists in setting political priorities.

Environmental thresholds may be described as states in which the functioning of natural systems changes fundamentally (Sprinz and Churkina 1998). For example, plants wilt at pressure values beyond minus 1.5 megapascals, since they can then no longer extract water from the soil. Determining environmental threshold values has a long tradition in the natural sciences (e.g. Parry et al. 1996) and in the field of medical research (e.g. Rosenthal et al. 1992), but it has not been much used in the social sciences. The following steps are routinely used in the USA for estimating cancer hazards in the context of medical research:

- determination whether a cause leads to an effect at all, with the help of the 'maximum tolerable dose',
- determination of a 'dose-effect function', determined by varying the dose,[8]
- determination of the magnitude of the dose actually occurring (actual exposure), and
- risk evaluation, a part of which is to transfer the results from the two previous steps into a common metric (e.g. currency units) (Rosenthal et al. 1992.).

The central relationship of the dose-effect function is illustrated in linear form in Fig. 2. The dose occurs as a cause (e.g. emission of air pollutants) that leads to unwanted effects (e.g. forest die-back). Dose-effect function A shows a constant relation between dose and response, in contrast to Function B, which deviates from

---

7 With respect to the critique presented in Section 2, this research design is both more narrowly focused and more comprehensive due to the inclusion of a response module.

8 In the field of human medical research, a safety factor of 1:100 or 1:1,000 is usually incorporated, and the dose-response function is extrapolated into this range (Rosenthal et al. 1992).

Function A from the discontinuity point (threshold value) on. From then on, Function B has a considerably sharper rise than Function A for higher values of the dose (see Fig. 2).

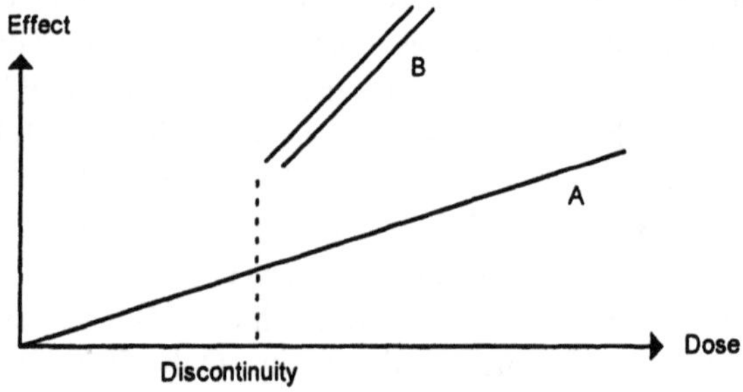

Figure 2: Threshold values and dose-effect relationships

The determination of environmental threshold values can be done in three different ways. One obvious approach is to look for the value of the dose at which any effect occurs at all (Rosenthal et al. 1992). In the case of Fig. 2, this would be reached at a value of the dose infinitesimally larger than zero. However, small effects are often (though not always) accepted in research and political practice, since absolute safeguarding of the functioning of every ecosystem or of people would be unduly expensive, and other political goals could no longer be pursued for lack of available resources. Secondly, they can be defined as not exceeding a socially 'acceptable' level of impact, such as if forests go from damage level 1 to 2 on a four-point scale (UNECE/Executive Body for the Convention on Long-Range Transboundary Air Pollution/Working Group on Effects 1991). A third way consists of systematic indication of distinct discontinuities of the dose-effect function, such as can be seen in the deviation of Function B from A from the discontinuity point on (see Fig. 2).

For research in the field of environment and security, the latter two approaches to environmental threshold values seem the most promising, since they often realistically allow a minimal dose of harmful substances. The scaling of eco-systems (e.g. into categories of vegetation zones) often models the effect side of the second research approach – in global environmental change research, corresponding transitions from one level of a classification to another are then either forecast or explained in retrospect as the result of environmental changes (e.g. changes in important regional climate parameters). On the basis of the above definition of environmental threshold values, the third type of study seems the most interesting, since it takes the predicted or observable discontinuities of the dose-effect relationship directly as its subject. The latter approach also plays a major role in research on 'surprises' (see Glantz 1997; Schneider and Turner II

1995) in the context of global environmental research, since discontinuities are ultimately surprises (about effects) in the broader sense.

Environmental thresholds are not necessarily scaled one-dimensionally, but comprise at least a number of important dimensions, which are outlined for the case of 'acid rain' in Table 3. Environmental threshold values may be absolute and abrupt (see Fig. 2), or take the form of intervals, include a temporal dimension (static or dynamic), occur as a single discontinuity or several times in the curve of the dose-effect function, and differ in their extent or their degree of reversibility by environmental policy measures. Furthermore, knowledge of each of these four aspects plays an important role, since different environmental threshold values have often evoked varyingly intensive research interest.

Table 3: Dimensions of environmental threshold values

Dimension	**Example: 'acid rain'**
Type (absolute, abrupt, intervals)	absolute and abrupt
Time (static vs. dynamic)	dynamic and long-term
Number (one vs. more than one)	one threshold value per pollutant (for a given level of deposition of other pollutants)
Degree of reversibility of overshoot (at low, medium, high costs; irreversible)	reversible (depending on the degree of optimization of international emission-reduction strategies)
Extent of knowledge about the above dimensions	high, especially about sulfur and nitrogen oxides

Source: Sprinz and Churkina 1997 (slightly revised)

The determination of the existence of environmental threshold values can be approached in a variety of ways. However, regardless of the way chosen, the geographical and temporal dimension of the dose-effect relationship considered should be specified.

In a *mathematically* abstract form, discontinuities can be expressed in the static case as

$$f(x_1) - f(x_2) > \varepsilon \quad \text{or} \quad df(x)/dx > \varepsilon$$

or in the dynamic case as

$$f(x_{t=2}) - f(x_{t=1}) > \varepsilon \quad \text{or} \quad df(t)/dt > \varepsilon,$$

i.e. the function change exceeds a specified level ε. Thus the non-arbitrary definition of ε becomes an important criterion for the diagnosis of thresholds, in particular their nature, temporal dimension, and number.[9]

9 For a more extensive discussion of discontinuities, differentiability and methods to empirically diagnose environmental thresholds, see Sprinz and Churkina 1998.

Adherents of statistical methods may take the *heteroscedasticity* (non-constant error variance) of linear regressions of dose-effect relationships as a clue to the possible existence of threshold values, since a linear regression causes constant error variances in the case of a linear functional link, while discontinuities cause a sudden increase of the error variance from the observed to the estimated function in their vicinity.[10] This method is potentially particularly suitable for routine analysis of large quantities of data; however, a minimum of *heteroscedasticity* must be specified as a criterion in order to restrict the number of meaningful threshold values.

As already discussed above, the transition between *stages in a classification* scheme on the effects side is also suitable as a criterion for determining the existence of thresholds. This method relies on the validity and reliability of the underlying measurement scale. Such scales are often relatively easy to interpret, and the factors which account for a change in the classification level are known. This latter point in turn eases the search for causal variables.

From the point of view of environmental policy, the decisive question is whether exceeding environmental thresholds is reversible. This problem has both an *ex post* and an *ex ante* dimension. In the *ex post* case, the question is whether an overshoot of an environmental threshold that has already occurred can be reversed – and if so, at what costs. In the *ex ante* case, it is the anticipation of the irreversibility of a possible overshoot of a threshold value (for example, in the case of pronounced global and regional climatic changes) that induces qualified political interventions. In both cases, the economic costs (relative to total economic activity) of reducing the dose levels back below the threshold value plays a decisive role, and thus partly determines the likelihood of the political intervention and its extent (Sprinz and Vaahtoranta 1994). In the case of technically 'impossible' reversibility, it is anticipation before environmental thresholds are reached that is of crucial importance.

In the context of determining environmentally-induced conflicts, environmental threshold values play a decisive role, since exceeding them is the sufficient condition for *environmentally*-induced armed conflicts. In this section, several empirical-quantitative approaches to determining such environmental threshold values have been shown, whose methodology can be traced explicitly, and which can be criticized systematically. At all events, the suggested approaches prevent findings about the effect of environmental hazards on armed conflicts being specific to a researcher, and therefore idiosyncratic; this also assists the derivation of conclusions for public policy.

---

[10] This method deliberately uses linear regression methods for estimating the non-linear behavior of a function. In this exceptional case, it is the violation of linear regression assumptions that is meant to serve the diagnosis of threshold values.

5 Prospects: the contribution of empirical-quantitative research

Empirical-qualitative research on the environment and security has succeeded in pushing this extremely important topic into the political limelight. Thus a potential problem has become the subject of political discussion before major dangers have occurred. Unfortunately, empirical-qualitative research to date has not succeeded in establishing whether there is a general relationship between environmental hazards and armed conflict. In this respect, politics is left in a knowledge vacuum: it cannot be demonstrated that this general relationship exists, nor that the opposite is true.

The empirical-quantitative approach proposed here concentrates on studying environmental threshold values as a sufficient condition for the outbreak of armed conflict. This would be achieved if environmental thresholds are exceeded. Otherwise, the environmental component in the causation of violent conflict is likely to be of secondary interest. Environmental thresholds can be characterized as part of an empirical-quantitative research program, utilizing several dimensions. In particular, various mathematical abstract, statistical, classificatory and economic approaches to determining environmental thresholds have been put forward. If the opportunities for environmental policy intervention and for conflict management are included, a simultaneous test of the connection between environmental degradation, violent conflict, and possible political interventions can be undertaken (see Fig. 1). We could learn why armed conflict does *not* occur in cases where, for example, environmental thresholds are exceeded, no environmental policy response is launched, but vigorous conflict management is observed. In the framework of the research design described above (see Fig. 2), findings would become possible that not only enlarge research perspectives, but also better inform public policy.

References

Baechler, Günther et al. 1996: *Kriegsursache Umweltzerstörung. Ökologische Konflikte in der Dritten Welt und Wege ihrer friedlichen Bearbeitung* [Environmental Degradation as a Cause of War. Ecological Conflicts in the Third World and Ways for their Resolution]. ENCOP Study Vol. I. Zürich: Rüegger.

Baechler, Günther and Kurt R. Spillmann (Eds.) 1996a: *Kriegsursache Umweltzerstörung: Regional- und Länderstudien von Projektmitarbeitern* [Environmental Degradation as a Cause of War. Regional and Country Studies of Research Fellows]. ENCOP Study Vol. II. Zürich: Rüegger.

Baechler, Günther and Kurt R. Spillmann (Eds.) 1996b: *Kriegsursache Umweltzerstörung: Länderstudien von externen Experten* [Environmental Degradation as a Cause of War. Country Studies of External Experts]. ENCOP Study Vol. III. Zürich: Rüegger.

Cook, Thomas D. and Donald T. Campbell 1979: *Quasi-Experimentation: Design & Analysis Issues for Field Settings*. Boston, MA: Houghton Mifflin.

Dreier, Volker 1997: *Empirische Politikforschung*. Munich: Oldenbourg.

Glantz, Michael et al. 1997: *Exploring the Concept of Climate Surprises: A Review of the Literature of the Concept of Surprise and How it Relates to Climate Change*. Boulder, CO: National Center for Atmospheric Research, Department of Energy.

Gleditsch, Nils Petter 1997: "Armed Conflict and the Environment: A Critique of the Literature", in: *Journal of Peace Research*, 35/3, 381–400.

Homer-Dixon, Thomas 1996: "Strategies for Studying Causation in Complex Ecological-Political Systems", in: *Journal of Environment and Development*, 5/2, 132–148.

Homer-Dixon, Thomas and Valerie Percival 1996: *Environmental Scarcity and Violent Conflict: Briefing Book*. Toronto: University of Toronto.

Homer-Dixon, Thomas F. 1991: "On the Threshold: Environmental Changes as Causes of Acute Conflict", in: *International Security*, 16/2, 76–116.

Homer-Dixon, Thomas F. 1994: "Environmental Scarcities and Violent Conflict", in: *International Security*, 19/1, 5–40.

Mitchell, Ronald and Thomas Bernauer 1997: "Empirical Research on International Environmental Policy: Designing Qualitative Case Studies", in: *Journal of Environment and Development*, 7/1, 4–31.

Parry, Martin L. et al. 1996: "What Is Dangerous Climate Change?", in: *Global Environmental Change*, 6/1, 1–6.

Rosenthal, Alon et al. 1992: "Legislating Acceptable Cancer Risk from Exposure to Toxic Chemicals", in: *Ecology Law Quarterly*, 19/2, 269–362.

Schneider, S. H. and B. L. Turner II 1995: "Anticipating Global Change Surprise" Hassol, S. J. and J. Katzenberger (Eds.) 1995: *Elements of Change 1994*, 130–141. Aspen, CO: Aspen Global Change Institute.

Schnell, Rainer, Paul B. Hill, and Elek Esser 1995: *Methoden der empirischen Sozialforschung*. Munich: Oldenbourg.

Sprinz, Detlef and Tapani Vaahtoranta 1994: "The Interest-Based Explanation of International Environmental Policy", in: *International Organization*, 48/1, 77–105.

Sprinz, Detlef F. 1997: "Environmental Security and Instrument Choice", in: Gleditsch, Nils Petter (Ed.): *Conflict and the Environment*, 483–502. Dordrecht: Kluwer.

Sprinz, Detlef F. forthcoming: "Empirical-Quantitative Approaches to the Study of International Environmental Policy", in: Nagel, Stuart (Ed.): *Policy Analysis Methods*. Commode, N.Y.: Nova Science Publishers.

Sprinz, Detlef F. and Galina Churkina 1997: *Environmental Security & Global Climate Change: The Role of Environmental Thresholds – A Research Proposal*. Potsdam, Missoula MT: Potsdam Institute for Climate Impact Research, School of Forestry, University of Montana.

Sprinz, Detlef F. and Galina Churkina 1998: *The Analysis of Environmental Thresholds*. Presentation to the Third Pan-European International Relations Conference and Joint Meeting of the European Standing Group for International Relations (ECPR) with the International Studies Association (ISA), Economics University of Vienna (WUW), Vienna, 16–19 September 1998, mimeo.

UNECE, Executive Body for the Convention on Long-Range Transboundary Air Pollution, Working Group on Effects (1991): *The 1990 Forest Damage Survey in Europe*. Geneva: United Nations Economic Commission for Europe.

Environmental Conflict and Sustainable Development: A Conflict Model and its Application to Climate and Energy Policy

Jürgen Scheffran

1 Introduction

In recent years, the environment has become increasingly a topic of security policy. While in the 1980s, the harmful effects of armament and war on health and the environment were at the center of the scientific and public debate (Krusewitz 1985, Westing 1990). With the end of the Cold War, it became more and more apparent that environmental degradation and resource scarcities themselves can be the objects or (additional) triggers of conflicts, which can lead to the use of force (Homer-Dixon 1991; Homer-Dixon et al. 1993; Baechler et al. 1993, 1996). So it is not surprising that a discussion of concepts of expanded or environmental security was initiated, which are meant to extend beyond narrow military concepts of security. It was criticized, among other points, that transferring a security mentality dominated by the military to the environmental sector would at the same time provide a gateway for the redefinition and relegitimization of military activities in environmental policy (Daase 1993, Brock in this volume).

In fact, most recent programs and statements by government representatives, particularly in the USA, indicate that such a process is already under way (Woodrow Wilson Center 1997). In these, environmental degradation is usually interpreted as a threat to national security, which must be countered, including by the use of suitable military means. The use of military and technological resources is also considered, as already anticipated in 1991 in the report of a panel of experts of the United Nations Office for Disarmament Affairs (see United Nations 1991, Scheffran 1992). Although such efforts may make sense under the aspect of conversion of military resources that are no longer required, it can be problematic to assign such tasks to the military itself. This is shown, for example, by the discussion of the use of 'green helmets' in environmental catastrophes, which has aroused fears in developing countries of the loss of sovereignty over 'their' natural resources.

In this paper, starting from a critical assessment of the role of models in conflict research (Section 2), an actor-oriented approach to the modeling of environmental conflicts is presented (Section 3). Based on equilibrium and stability conditions of the conflict model, the question is examined of what fundamental contribution concepts of sustainable development can make to the prevention and regulation of environmental conflicts, by influencing the equilibrium factors (Sections 4 and 5). The potential for stabilization is illustrated by the example of the climate conflict, by means of a model simulation, on the one hand (Section 6), and by the discussion of possibilities for cooperative conflict regulation through Joint Implementation, on the other (Section 7). Following a summary (Section 8), the model equations are listed in the mathematical appendix (Section 9).

2 Potentials and limits of conflict modeling

Conflict models can make a contribution to better understanding of the connection between security and environmental policy, to finding approaches to solutions to environmental conflicts, and to studying the conflict-preventing effect of concepts of sustainable development. In recent years, the model concepts of security and stability in international politics have changed considerably, along with the understanding of security. During the Cold War, security was defined mainly in the bilateral context of the East-West conflict, and models of conflict were mainly restricted to the military dimension of security policy, with the emphasis being on quantitative comparisons of armed forces (Neuneck 1995).

In the new world (dis-)order, on the other hand, numerous actors at the state, sub-state, and international levels, as well as economic, ecological, social and ethnic conflict factors must be taken into account, that are often interwoven in a complicated fashion. Conflict modeling must represent this wide variety, without drowning in the number of parameters and their interactions. A wide range of mathematical models in economics and ecology can be used to understand environmental conflicts and the role of sustainable development (Wissel 1989, Bossel 1992, Carraro and Filar 1996, Feess 1997, Krabs 1997). These include the theory of dynamic systems (differential and difference equations), stability theory, control and optimization theory, game and decision theory, statistics and the calculus of probability, as well as computer simulation.

A model is an approximation to reality, its image. In this mapping process, the complexity of reality is reduced; information is lost through the restriction to the 'essential'. The informativeness of a model is limited if factors that are not taken into account are decisive for the result. A model should be simple, so that it is comprehensible and handy to manage, but not too simple, lest its validity be limited. Mathematical models make sense mainly when the object of the problem is of manageable extent, and major system variables are quantifiable or measurable. This explains in part the major role of mathematics in the natural sciences, especially in physics, which attempts to exclude, by means of 'artificial' experimental conditions, unwanted interference effects that might falsify measurements of elementary phenomena.

It is being recognized more and more that the predictability of models and experiments in the natural sciences is limited in the vicinity of critical phase transitions. There, even 'small' changes in a few parameters can alter the qualitative behavior of the system drastically (chaos).

The more complex and indeterminate the reality is, the more difficult it becomes to find a simple and comprehensible model that is able to reflect and predict the qualitative behavior of reality. This makes modeling in the social sciences, where behavior is not determinate, difficult. The results are dependent on situations and decisions, with numerous non-objectifiable factors involved. Complete models, that cover all the relevant factors of a problem, are only possible in exceptional cases. To keep all the influencing variables under control, as in physics, is out of the question.

However, even incomplete models can provide a deeper understanding of reality, by helping us to identify and formalize the relevant system variables and their interaction, and thus sharpen our thinking (mental model). If non-modifiable parameters and control variables can be identified more easily with the help of a model, the (political) scope for action becomes clearer (action model). The goal in both cases is not to forecast reality, but to recognize connections and ways of solution that enable one to have an active influence on reality. Computer simulation allows one to test the dependence of the model's results on assumptions and variations of the parameters, like an experiment.

If the results of the model are adopted by decision-makers and actors, they can affect social and political reality, even if the model only reflects reality insufficiently. One example is the world model of the *Club of Rome* in the early seventies, whose prediction of ecological catastrophes initiated a shift in awareness of the 'limits to growth' that has contributed to ameliorating or delaying the environmental crisis (Meadows 1972; Meadows and Randers 1992). A further example is the model by the American economist William Nordhaus, who advised against avoiding global warming, since the necessary measures would be more of a burden on the economy than accepting the climatic damage (Nordhaus 1991). Despite the simplistic assumptions of the model, the U.S. government has adopted some of his conclusions, and refrained from major reductions in pollutant emissions.

Exaggerated, false, or unfulfilled expectations aroused by model developers and users have increased mistrust in complex and barely comprehensible models. One basic problem of global models is that the macroperspective chosen does not allow the individual decision-makers and actors to be identified and represented in the model. Thus these are not very suitable for modeling conflicts between actors.

Three essential roots of mathematical conflict research are the Lanchester equations of warfare, the Richardson equations of the arms race, and game theory, created by Oskar Morgenstern and John Neumann for the analysis of decisions.

Each of these approaches has been subjected to a multitude of criticisms, modifications and extensions. The Richardson equations, for example, have been criticized for treating states as structureless units that respond mechanistically, linearly and in always the same way to the size of arms potentials, with political goals and means not playing any visible role. Attempts have been made to meet such objections by incorporating nonlinearities, discrete steps in time, or decision criteria into the equations. Although this extended the applicability of the model, not all the objections can be dismissed. In a similar fashion, people have complained that game theory only applies to static decision-making situations and rationally acting decision-makers who have complete information. Game theory has since been refined substantially, and is able to describe dynamic game situations with incomplete information and even communication processes. But the assumption of 'rational choice' remains a condition that does not always apply in reality.

In the field of environment and security policy, game theory, like control and optimization theory, has found many applications, especially if the number of options for action can be limited to a few, the control variables are identifiable, and the categories of benefit and damage are economically tangible (see Carraro and Filar 1995). This is the case, for example, with the climate conflict, that is represented as a game with the options 'increase emissions' and 'decrease emissions', or in conflicts over water or fisheries, which are about the determination of optimum resource allocation. Using game-theory models, it is possible to examine the conditions and structures of interests under which the actors chose cooperative or non-cooperative behavior.

In the following, our own approach to a model of conflict is presented, which is based on interdisciplinary analysis of previous models, and tries to link decision-making models with dynamic interaction models. Its starting point is the recognition that none of the known models permits an appropriate dynamic representation of conflicts between goals and means of several actors as yet. The term 'conflict' here represents an interaction between actors who employ available means to accomplish their goals, needs, and interests. The occasion of action is a tension between the aspirations and demands of the actors and the reality perceived by them (the potential for conflict), which is an expression of their (dis-)satisfaction. The conflict potential can be reduced by adjusting reality to the aspirations, or by adjusting the aspirations to reality (as a rule, both happen). If the conflict potential is not reduced, despite repeated action, in other words, if the dissatisfaction remains or increases, the means employed can escalate all the way (including to violence). A lasting conflict leads to repeated attempts with new approaches to solving the problem, until a suitable one has been found, or the repertoire of actions is exhausted. The function of conflict-solving strategies is to mediate between the actors about the choice of means, but also of goals, in order to make these compatible, or at least to regulate them.

3 The SCX conflict model: An overview

The SCX model for conflict analysis was created in an effort to describe the dynamic interaction between actors who use available means for influencing their surroundings in order to achieve goals. The system state altered by their actions is re-evaluated by all actors at certain intervals, and occasions renewed activity. In order to pursue a goal, the actors can increase or reduce the input of means employed within the framework of given limits, and alter the direction of the input. The chosen directions of the flow of means within the given scope of action have a decisive influence on whether the actors can achieve their respective goals in the desired time, or whether they come into conflict with other actors. The model can be used to study the conditions under which an escalation of conflicts occur, and to what extent conflict escalation can be prevented or cooperation achieved by negotiations on the input of means and mutual adjustment of goals.

In formalizing this verbally expressed model relationship, the mathematical symbolism is mainly dispensed with here. The interaction of the three model variables (evaluation and goal function *S*, input of means *C*, system variables *X*) over time is examined with the help of difference equations (see appendix and Fig. 1). The concepts used are meant to structure the problem complex and make connections more visible, in order to deepen understanding of the problem even on a qualitative, conceptual level. The model has been or is being applied to various problems in the field of security policy (nuclear arms race, missile defense, disarmament, proliferation), environmental and energy policy (water conflicts, climate conflict, fisheries conflicts), and of the economy (joint implementation, wage dispute) (see Scheffran 1989, 1996a; Scheffran and Jathe 1992, 1995; Jathe 1996; Scheffran and Pickl 1997; Kaiser 1997).

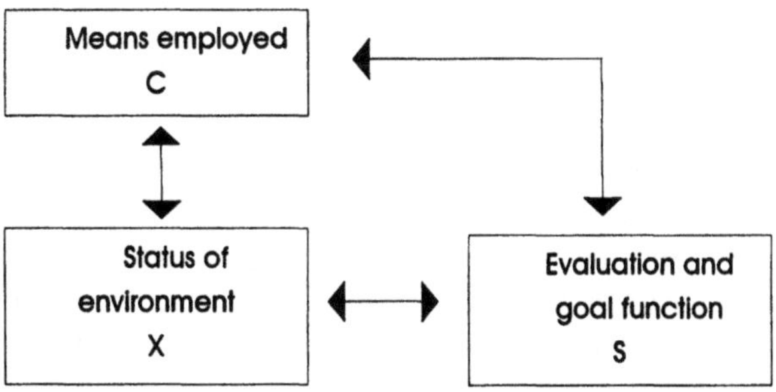

Figure 1: An actor's model variables and their interaction in the SCX model

For a conflict to be studied with the model, the three model variables must be specified concretely. Quantification of the parameters used is not necessary in all applications, and indeed is not always possible.

The *system variables X* designate all environmental parameters that are regarded by the actors as relevant to a complex of conflicts and problems considered (e.g. economic goods, or natural resources). The actors' level of knowledge and respective perceptions must be taken into account, as well. A change in the status brought about by an actor is a completed *action*, while a possible change of state in the future is an *option for action.*

The *goal and evaluation function S* of an actor corresponds to the difference between the actual or perceived value and the desired value (the goal value) for a system state *X*, that is the gap between the actual and desired status. A negative evaluation corresponds clearly to a state of dissatisfaction (value deficit), while a positive evaluation represents the degree of satisfaction (value excess). In the case $S = 0$, the desired goal has just been reached (sufficiency); here, the actual and desired state are in balance, which can occur either by adapting the actual state to the aspirations, or by adjusting the aspirations to the actual state. In the evaluation, the damage to oneself that is possible from some events (such as by military destruction, environmental destruction, loss of money or of a job) is a negative factor. There is a direct connection with the term "risk" here.

The *input of means (costs) C* comprise the instruments, efforts, and resources employed by an actor (money, energy, technology, time, information, communication, law, force), with which the current system state *X* can be influenced in the direction of the sufficiency goal ($S = 0$). The input of means may be limited upwards or downwards by the means available.

A dynamic interaction of system states, goals and means results from the fact that the means employed change the system state, which results in a re-evaluation of the situation. If the situation has worsened (if the value is therefore negative), the input of means is increased, while if it has improved sufficiently, the input of means can be reduced. Whether and how quickly a specified goal can be achieved is influenced by the effectiveness and the direction of the input of means. In the model, this is determined by the parameters *sc*, which give the increase or reduction in value per unit of means, that is, the *cost-benefit ratio* of the means employed.

If more than one actor acts accordingly, and affects environmental states that are also important for other actors, the SCX model describes a dynamic interaction of several actors, and the associated conflicts. The corresponding mathematical equations are summarized in the Appendix. In this case, the coupling parameters sc_{ij} of the interaction indicate how the means employed by actor *i* affect the evaluation function of another actor *j*, with the individual effects of all action options of actor *j* being weighted with his preference. In the limiting case, only a single option for action is pursued. As a rule, however, each actor has considerable freedom of action in choosing his preferences: with them, he can decide what he will employ his existing means for.

The interaction parameters sc_{ij} are of decisive importance for the actors' conflict behavior. They can be combined into a two-dimensional *conflict matrix* that denotes the effect of each actor on another actor. If they are positive, the actions of one actor are to the advantage of another actor; if they are negative, they are evaluated as disadvantageous. There is a *win-win* situation if both coefficients are positive between two actors, in other words the actions of the two are mutually beneficial (common security). The number of possible conflict constellations and their complexity is usually reduced by the fact that actors are mutually indifferent $(sc_{ij}=0)$ or allocate one another to groups of 'friends' $(sc_{ij}>0)$ and 'opponents' $(sc_{ij}<0)$. From considerations of stability theory, it can be derived that for two actors, the ratio

$$z = \frac{sc_{11} \cdot sc_{22}}{sc_{12} \cdot sc_{21}}$$

is decisive for the stability of the interaction. For $z > 1$ the interaction is stable, i.e. the actors can achieve their goals within the limits on the means, despite mutual impairment, since their positive effects for themselves dominate. But for $z < 1$ the dynamics become unstable, and escalate towards negative values and large inputs of means. The actors move away from their goals, despite increased efforts.

The question arises of how the actors involved in the conflict can achieve their respective goals jointly with a minimum input of means. If each actor acts only in accordance with his own end-means rationality, only the strongest may be able to achieve their goals by large inputs of means, while weaker actors with fewer means are outdone. In negotiations, on the other hand, all the parties to the conflict may be able to coordinate their preferred actions and their interaction cooperatively in such a way that they all obtain greater benefits than from purely self-interested behavior.

In order to simulate the development of the conflict dynamics over time, the goal functions of the actors are defined depending on the system state, and initial conditions are specified for the means and the system parameters used. It turns out that the SCX model can represent potential conflict dynamics of astonishing complexity (Hinze 1995, Jathe 1996). In particular, it is seen that for certain parameter combinations and initial conditions, bifurcations and chaotic behavior occur (Jathe and Scheffran 1994). In this case, the actors respond so rapidly and intensively to their evaluation of the environment that they overshoot their respective goals. There is maximum uncertainty, and everything seems possible for small variations of the model. In order to avoid unstable, risky or chaotic conflict constellations, possibilities of control can be investigated in the framework of the SCX model, with the help of game and control theory (see Jathe/Krabs/Scheffran 1997). Two questions in particular can be distinguished here, concerning cooperative stabilization of conflicts:

- For given goal and evaluation functions of the actors, how can the magnitude and direction of the input of means be controlled in such a way that all actors achieve their goals, and a cooperative conflict constellation with positive interactions and mutual benefit arises *(means-based stabilization)*?

- How can the objectives and evaluations themselves be influenced so that mutually beneficial stabilization (towards cooperation) results *(goal-based stabilization)?*

Examples of applications

Three examples from varying fields may illustrate the general concepts involved.

Military security policy: Let the system variables X here be the number of weapons of a certain category deployed (e.g. the number of tanks or missiles), means C be the government arms expenditures per year spent on the procurement of additional weapons (for disarmament, the costs of eliminating the weapons must be taken into consideration). The evaluation function S (security) is determined by the potential damage in case of war, as a function of various scenarios for the employment of the warring parties' offensive and defensive weapons. Possible goals of the decision-makers might be to avoid a degree of damage that is unacceptable to them (secured survival) or to inflict a minimum of damage on the enemy (minimum deterrence). The interaction parameters sc_{ij} correspond to the degree of mutual perception of military threat, which can stimulate an arms race.

Economics: Let the system variable X be the quantity of goods consumed (e.g. food), means C be the expenditures on this, and S the benefit, which corresponds to the gain in the quality of life for the consumer (e.g. calorie or vitamin content, or taste of the food). The benefit is supposed to exceed a threshold regarded as sufficient (the minimum requirement). If the demand for goods exceeds the supply produced or affordable, such as when too many competitors occur or the prices are high, scarcities may result which may be interpreted by economically weaker consumers as an impairment of the quality of their lives by competitors.

Environmental policy: Let system variable X be the consumption of a natural resource (e.g. energy) by a country, and means C be the necessary resource costs for this at a given resource price. Let S be the increase in prosperity (e.g. gross national product) generated by this resource flow, which is supposed to exceed a certain growth figure. The resource use, or over-use beyond a sustainable level, creates undesirable side effects (such as emissions, environmental damage, climate fluctuations) that not only harm the country itself (and thus reduce its own prosperity effect), but also do damage to other countries across international borders *($sc_{ji} < 0$)*. The mutual damage due to the resource use can lead to environmental degradation and economic destabilization, and promote the use of violence.

4 Modeling environmental and resource conflicts

Environmental conflicts in the broader sense comprise conflicts that are carried out over or by the use of natural resources, or are triggered or substantially intensified by the damage to natural resources. By contrast to conflicts over exhaustible resources (minerals, fossil fuels, territory), there are also conflicts over the degradation of renewable resources. Examples of renewable resources are agricultural products, fish stocks, favorable climatic conditions, water, soil and air. They are regenerated in metabolic cycles, depending on the functionality and stability of the ecosystems.

For a representation in the framework of the SCX model, let X be the consumption of a natural resource , C the expenditure on its use, and S its value for the prosperity and quality of life of an actor. The model is in an equilibrium condition if both the evaluation of the situation and the input of means by the actors involved is constant, that is, all changes are zero. The boundary condition for stability given in the preceding section ($z = 1$) states that the actors should act so that the positive effects of resource use at least balance the negative risk effects, since otherwise a mutual deterioration of the quality of life, and an escalation of conflict would result.

The model equilibrium corresponds to the requirement that for each actor, there shall be a balance between the desired growth in prosperity and the growth in prosperity actually produced by the resource use. Human wishes, aspirations and needs must be brought into accord with the natural and social realities. To do this, it is assumed that each actor receives a certain share of the total resource consumption, and that the resource use is linked to positive prosperity effects and negative risk effects for the actors. The equilibrium condition of the model for each actor is therefore determined by five influential factors: the total amount of resource consumption, the resource shares claimed by each actor, the resource's prosperity and risk effects, and the level of aspirations for prosperity (see Fig. 2 and Appendix). If the equilibrium condition is violated, a potential conflict situation exists. The following basic types of conflict are to be distinguished, corresponding to deviations of the factors from the equilibrium of the model:

- *Scarcity conflicts* arise because actors are suffering shortages due to an insufficient supply of resources, and employ means of conflict to overcome this shortage, which increases the scarcity among other actors. There is a shortage in particular if a minimum threshold of needs that is important for the actor's own identity is not being met. If existence is in danger below this threshold, the only alternative to self-surrender is the struggle for existence. Possible consequences include increased willingness to use force, or the loss of governmental order through pauperization, which can lead to outbreaks of violence against those who are actually or are believed to be responsible for the misery.

- *Availability, distribution and fairness conflicts* are waged over the share that the actors receive of a resource. The deviation from the average that would result in a formally 'fair' distribution per head of the population is an indicator of the distribution injustice. The potential for conflict grows with the asymmetry of resource distribution. A familiar example is the North-South conflict, since in this case, one fifth of humankind lays claim to four fifths of the wealth, and thus, of course, of the natural resources. Other examples are water and fisheries conflicts, in as far as the allocation of usage quotas is concerned.
- *Conflicts from the risks of resource use:* Here, the risk effect (self-induced or externally induced) of the resource use triggers conflicts. In addition, social conditions that have a large potential for conflict can be created by unsuitable resource usage. Examples are conflicts due to environmental degradation, nuclear power accidents, overworking of resources (oil, uranium), climate-caused catastrophes, and dams.
- *Conflict between humanity and nature:* This concerns the incompatibility between human aspirations and the natural conditions for the existence of life. It is true that nature cannot be considered a conflict actor in the strict sense, since it does not act purposefully, according to the general view, but it can 'respond' to anthropogenic disturbances with substantial harm to people and society. We may include the consequences that occur from overexploitation of a natural resource, or from overloading the sinks in excess of a sustainable or compatible level. Important here is the recognition of environmental threshold values at which grave consequences occur, for example if the health stress becomes unacceptable, nature's regenerative capacity is exceeded, or a collapse of resource production occurs (on the role of environmental threshold values, see Sprinz 1996, 1997). Some examples are conflicts resulting from overfishing, the destruction of rain forests, or the loss of biodiversity.
- *Conflicts over goals and means:* By contrast to the other types of conflict , where conflicts occur due to certain system conditions, these involve a dispute over which system conditions should be chosen. The object of the conflict is the decision as to which goals actors should pursue, how a current situation is evaluated, and what means should be employed in what way to affect a system. Getting certain models of behavior (sustainable development) accepted, or resistance to the use of military force fall under this category.

In real conflict situations, more than one of these types of conflict may exist at the same time, or influence one another. For example, in order to avoid a scarcity conflict, a distribution conflict may be carried out, which in turn increases the willingness to take risks or exceed the limits of the burden on nature. Whether and how conflicts are carried out depends int. al. on the availability of suitable means of conflict, and on the ability of the actors to assess the situation. If affected parties resist, an escalation of conflict can result, in which the actors harm one another up to the use of force, or the exhaustion of one side (an example: the Ogoni in Nigeria).

5 Sustainable development: A contribution to conflict prevention and regulation

'Sustainable development' attempts to embed the development of the world created by humankind (sociosphere) permanently in the limited framework of the natural environment (ecosphere). The Brundtland Report formulates the goal thus:

> Sustainable development is development that meets the needs of the present without compromising the ability of future generations to meet their own needs (WCED 1987).

Human needs are thus made the measure for sustainable development, in other words a largely subjective parameter based on individual scales of values. If sustainable development is indeed put into practice, many environment and resource conflicts could be avoided.

The considerations in the preceding section show that when there is a deviation from the equilibrium conditions, conflict situations can arise, especially if the aspirations for prosperity exceed the increase in prosperity actually resulting from the resource use. In the following, we try to show how, by influencing the factors occurring in the SCX model's equilibrium condition, in the context of sustainable development a balance of aspirations and available resource usage (of the desired and actual state), and thus a reduction of the conflict potential, can be created (see Appendix and Fig. 2). The five concepts listed correspond to the five equilibrium factors.

- *Eco-compatible use of natural resources (consistency):* The total consumption of resources must be compatible with the natural material and energy flows, i.e. it must not exceed a sustainable level. Limits to carrying capacity and stress bearing capacity are set by the finiteness of non-renewable resources, by the limited regenerative capacity of renewable resources, and by the limited ability of nature to absorb wastes. Maintaining regeneration can be done by creating nature reserves, restricting rates of usage and harvesting, improving regenerative capacity, cultivation and conservation of endangered species. This makes it possible to defuse the conflict between humankind and nature.
- *Improved effectiveness of resource use (efficiency increase):* How effectively a resource can be employed to satisfy needs depends on the efficiency of its extraction, conversion, utilization, and regeneration (that is, its positive prosperity effect). Resource utilization is made more efficient by savings, structural changes, technical means, and ingenuity, so that needs can be satisfied with fewer resources (decoupling resource consumption and prosperity/quality of life). Such measures can soften scarcity conflicts and the humanity-nature conflict, and counter a conflict escalation.
- *Reduction of risks (damage prevention):* If the resource consumption involves unacceptable hazards and damage (risk effects) to one's own needs or those of other actors (e.g. from accidents, anthropogenic natural catastrophes, environ-

mental toxins, or radioactivity), either the cause of the damage must be eliminated, or its consequences limited to an acceptable level (e.g. by treaties, refugee aid, disaster reduction, or conflict management). Military forces can be used in certain circumstances after the event to limit the risk (combating floods or forest fires, crisis management), but are themselves a major source of risk and conflict. So the early control and reduction of means of violence (disarmament, renunciation of force) is important. The goal should be to avoid risk-based conflicts and an escalation of conflict by prevention.

- *Influencing the social distribution conditions of resource use (distributive justice):* What share of the resource cake the actors can use depends on the social distribution processes, and thus on the structures of power and interests. The goal of a policy aiming for fairness is to restrict the use or threat of the use of instruments of power to enforce individual interests, and to improve the conditions for a fairer adjustment by more democracy and rule of law and a system of social security. This would counteract availability and distribution conflicts.
- *Altering the goal of needs (sufficiency):* The structure of needs (i.e. the evaluation and goals function*)* itself is altered and adjusted to the prevailing natural conditions. This can happen by what exists being regarded as 'sufficient', so that no further efforts to satisfy needs have to be made. Or else, new values and goals of needs can be sought for instead, which require different or fewer natural resources (alteration of the way of life). This need not be perceived as renunciation or loss, but can in fact lead to a higher quality of life (see the discussion of role models in BUND/Misereor 1996). However, existential minimum needs that are considered indispensable must not be affected, since otherwise scarcity conflicts might result.

Many approaches to sustainable development limit themselves to individual strategies. Often, they only concern measures that can most readily be handled with a scientific-technical approach: increasing efficiency, eco-compatible resource management, or subsequent disaster relief (Fritz/Huber/Levi 1995). The redistribution and sufficiency strategies, that act more in the social sphere, are often ignored because of societal interests and power structures. Integrated concepts of sustainable development should try to do justice to all five strategies, and take into account the elemental connections between the five factors that are expressed in the equilibrium condition of the SCX model (Scheffran 1996b). The general circumstances are illustrated here by the example of energy and climate conflicts.

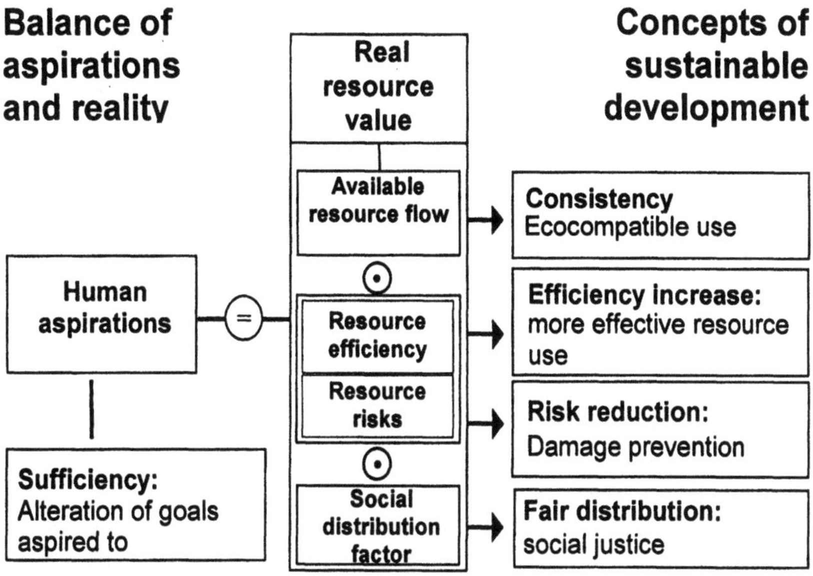

Figure 2: Sustainable development and the balance of aspirations and reality: Five strategies of conflict prevention in the context of the SCX model

6 Modeling energy and climate conflicts

The production and utilization of energy is a prerequisite for economic development; but it also alters the social and natural environment at the same time, with sometimes substantial risks and conflicts. Wars have been fought over fossil energy resources such as coal, petroleum, or natural gas. The Gulf War, in which oil was both an object of conflict (access to oil) and means of conflict (burning oil wells), is the most recent example. The utilization of energy can affect elemental vital interests of parties involved, and thus form a basis for violent conflicts. Some examples of this are the social and ecological consequences of dam-building projects, deforestation to gain firewood, radioactive wastes, and the transnational spread of radioactive pollutants after an accident in a nuclear reactor, or climate changes as a consequence of the burning of fossil fuels, which may lead to an intensification of the 'North-South conflict' and an increase in regional environmental conflicts. The spread of technologies relevant to nuclear weaponry (proliferation) through the use of nuclear power remains a problem for future security policies, as well. Overall, the conflict potential of energy supplies will probably grow in importance in the coming decades, due to the increasing energy demand with sinking reserves, the North-South gap in the energy sector, the geopolitical conflict potential of oil dependence, and growing ecological risks (more on the subject of energy conflicts is found in Bender et al. 1996).

There is also a substantial conflict potential associated with the predicted climate change as a result of the anthropogenic greenhouse effect, caused in part by fossil fuel use. *Climate conflicts* can occur over the evidence of climate change (e.g. over the uncertainty of a scientific prediction, or the interpretation of observed phenomena), or over the responsibility for climate changes and the suitable strategy for avoiding them (who is the main polluter, who must reduce emissions by how much). If the climate change occurs, there will be additional climate conflicts: risk conflicts directly triggered by climatic consequences, conflicts over prevention of damage from and protection from climatic consequence, conflicts from the asymmetry of those responsible for and suffering from effects. Numerous *actors* (individual, groups in society, groups of states, international organization) are involved in the climate conflict, pursuing sometimes contradictory *goals and interests* (prosperity and development, competitiveness, emission reductions and protection against climate change, environmental compatibility, justice) in varying coalitions. The *means* employed vary from repression, force, and disaster relief, via technological instruments (emission reduction, increased efficiency, energy savings, renewable sources of energy) and economic instruments (permit conditions, taxes and other levies, emissions trading), to cooperative measures (international treaties, financial and technology transfers, 'joint implementation'). (For more one the climate conflict, see Loske 1996, Brauch 1996.)

If suitable measures are not taken promptly, the difference between the CO_2 emissions to be expected if current trends continue, and the reduced levels needed for stabilization of the climate will continue to grow. The need for action increases as time passes. The question is whether the trends can be adjusted to the requirements, or whether the potential for conflict will be realized in catastrophes and conflicts. If existing asymmetries are reinforced, which would make the South the main relative loser from climate change, while the North is the main cause and possesses more means for handling conflict, the possible intensification of the North-South conflict would be especially problematic (Meyer-Abich 1994).

Although the complexity of the climate conflict, with its many system variables, interactions, actors, goals, and possibilities for action, cannot be represented completely in a model, some essential relationships can be illustrated. In order to depict the interaction of the energy system with the economy, environment and conflicts, and to analyze the influences of means and goals, within the framework of the SCX model, the following *coupling parameters* are significant:

- *Energy and population:* per-capita energy consumption,
- *Energy and economy:* prosperity effect (gross national product) per unit of energy,

- *Energy and environment:* changes in environmental parameters (e.g. atmospheric concentrations of pollutants, global temperature, displacement of climatic zones) through energy use,
- *Energy and risk/conflict:* damage, risks, and conflicts associated with the use of energy (e.g. costs of damage from climate change, risks from nuclear accidents and proliferation, potentials for violence and conflict).

How some of these parameters can affect the conflict dynamics in the framework of the SCX model is exemplified here with a simplified simulation. For two actors, called here 'North' (industrialized countries: Actor 1) and 'South' (developing countries: Actor 2), who are striving for a certain level of prosperity (their goal) by means of energy production, the development of energy consumption over 50 years is simulated. The evaluation and goal function is taken to be the distance from a desired gross national product per head of the population ($20,000 per person), which is about 25 times the present GNP per capita in the South, but only slightly over the present average for the North. Possible damage (risk) from climate change is to affect GNP negatively. In order to achieve their GNP goal, the actors invest money in energy production, which increases prosperity, but also the potential environmental damage. The choice of the other model parameters (population growth; emissions, costs, prosperity and risk effects per unit of energy; initial conditions for energy consumption, GNP, CO_2 content of the atmosphere) represent a standardized reference case, meant to correspond to the initial year 1990 (details are to be found in Scheffran and Jathe 1995, Hinze 1995).

As one would expect, in the numerical simulation, the attempt by the developing countries to reach the level of prosperity of the industrialized countries, despite a doubling of the population, leads to a dramatic increase in energy consumption and GNP to more than 50 times the present level. After about ten years, the South outstrips the North's GNP, but the North achieves the goal of GNP per capita sooner. The CO_2 content of the atmosphere increases about fivefold, which is associated with high risks of damage (a few percent of GNP).

In order to avoid this catastrophe scenario, in which a pessimistic case was assumed for some model parameters, a few parameters are now altered.

- The rate of growth of the population declines linearly, until a constant population is achieved in 50 years.
- The desired GNP-per-capita goal is reduced to $10,000 in the North, and $5,000 in the South.
- The CO_2 output per unit of energy is halved (through low-emission technologies).
- The gain in prosperity per unit of energy consumed is doubled (through energy conservation and more efficiency).
- The costs per unit of energy are increased (Factor 4).

Under these conditions, energy consumption in the South only grows about fourfold, while it declines by half in the North. The CO_2 content of the atmosphere increases to a slightly higher level (9 units, compared to 7.5 units today). The lower prosperity goals are achieved, despite drastic increases in energy prices, with substantially less environmental damage (order of magnitude one tenth of a percent of GNP).

From this simplified model simulation, which reflects an incorporation of the climate problem into the framework of the SCX model, it is apparent how stabilization of the system can be achieved by suitable variation of parameters. This corresponds to influencing the equilibrium factors named in Sections 4 and 5 (energy consumption, resource distribution, prosperity and risk effect, aspirations for prosperity) in the direction of sustainable development. More extensive conclusions, especially a predictive power, cannot be derived from this. In particular, the simplified approach does not show how the South is to provide the capital needed for its development, what measures can lower energy-related emissions and increase prosperity gains, and what economic effects the increases in energy prices have. For this, the structure of the model must be refined and its assumptions further specified; additional dimensions of the climate conflict must be incorporated.

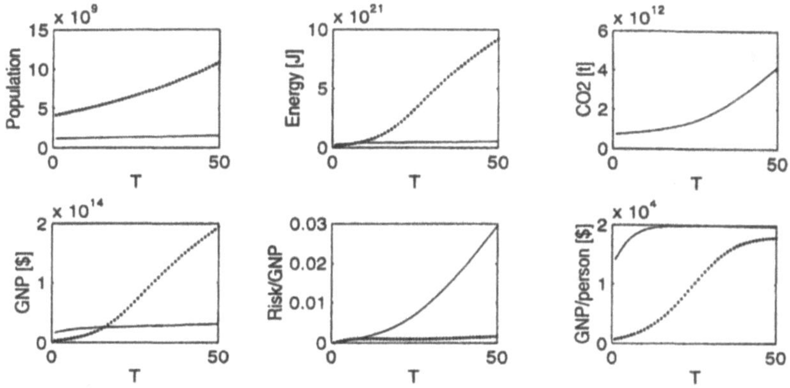

Case 1: Catastrophe scenario if the 1990 situation continues, for high prosperity goals

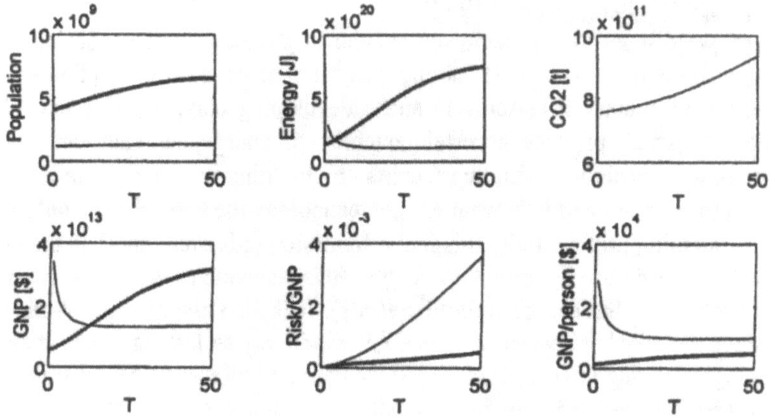

Case 2: Stabilization scenario for smaller population growth, lower prosperity goals, fewer energy emissions, more prosperity efficiency, and higher energy prices

Figure 3: Simulation of simplified SCX-model dynamics for population, energy consumption, CO_2 content of the atmosphere, gross national product (GNP), risk/GNP and GNP/capita for two actors over 50 years (North: solid line, South: dotted line) (Hinze 1995)

7 Regulating the climate conflict cooperatively: The example of Joint Implementation

In the Framework Convention on Climate Change, the industrialized countries declare their willingness to reduce their emissions of climate-affecting greenhouse gases. The developing countries, on the other hand, state that their energy demand is growing, which is associated with increased releases of greenhouse gases. The continuing boom in China's economy, in particular, could contribute to destabilization of the global climate, if suitable emission-restricting measures are not taken in good time (Loske 1993, Zhong Xiang 1996). One proposal for implementing such measures as rapidly as possible under current conditions is the concept of Joint Implementation (JI). By this is meant the cooperative attempt to reduce the global costs of emission reductions by having countries with high avoidance costs (industrialized countries) invest in countries with low avoidance costs, and be credited with the emissions thus saved (on the potential and limits of the JI concept, see Rösch and Bräuer 1997).

Here, we discuss in the framework of the SCX model, under what conditions emission reductions and cost savings can be achieved by JI. For this, an industrialized country IC (Actor 1) and a developing country DC (Actor 2) are considered, which produce a certain quantity of energy per year, and provide further power capacity by funding (such as by building new power stations). The question to be examined is in what energy technology the two actors should invest, with IC pursuing preferentially the goal of emission reduction, and DC the goal of increasing prosperity by growth in GNP (this asymmetry of goals is applied deliberately here, but can be modified easily). It is also assumed that both actors each have the choice between an old and a new energy technology, with the newer technologies having lower specific emissions per unit of energy, but also involving higher specific costs. IC is to have the opportunity of investing part of its expenditures in DC's more modern energy technologies, and the rest in its own more modern technology, while DC uses part of its expenditures for its own modern technology, and the rest for the old kind of power generation.

With the help of the SCX model, it can be shown that both actors must not only consider their own expenditures, but also those of the other party, in order to achieve their goals. In particular, the interaction parameters sc_{ij} can be represented as functions of the specific emissions and costs of the energy technologies and their prosperity effects, and of the preferences of the two actors for the respective technological options. If IC attempts to maximize its own benefit, in most cases it would have to invest all its funds in its own power stations converted to the new technology, since the newly built power stations in DC result in a relative emission increase. Correspondingly, in order to maximize its own benefit, DC will continue to invest in old technologies, as long as their cost-based prosperity effect is greater than that of the new technology. In this case, no interaction, and thus no cooperation, occurs. A deviation from this *Nash equilibrium* under constant technological conditions is only possible if both parties coordinate their preferences cooperatively in such a way that their costs are reduced, or even minimized. This can occur, for example, by IC supplying funds and technology to DC in return for greater investment by DC in emission-reducing technologies. The money saved by cooperation can be redistributed from a fund in order to control the dynamics of the SCX model, so that instabilities are avoided, and the conditions of sustainable development are met (Jathe/Krabs/Scheffran 1997, Scheffran and Pickl 1997).

8 Summary and conclusions

Based on a conflict model that maps the dynamic interaction of goals and means of several actors, a classification of environmental conflicts was undertaken, and the question of what contribution sustainable development can make to conflict prevention or conflict resolution was studied. A connection was established between the five factors of the model equilibrium state (resource consumption and distribution, prosperity and risk effect, aspiration goals) and five concepts of sustainable development (consistency, justice, efficiency increase, risk reduction, sufficiency). Using a simplified simulation of the climate conflict between North and South, the influence of relevant system parameters of the energy sector (declining population growth, reduced economic growth, lower CO_2 output, higher prosperity effect, higher energy costs) for the stabilization of the system were elucidated. The interdependence of industrialized and developing countries, and the possibilities of cooperative control of trends was studied for the example of 'Joint Implementation'.

The advantage of modeling is a clear structuring of the complex set of problems in the framework of a mental model that permits a deeper understanding of the relationships, and a search for solutions to the problems. It becomes possible to recognize and discuss the (in-)compatibility of goals and means, and to show the conditions for cooperation. Recommendations for political action that can be derived from this are limited to fundamental rules for avoiding environmental and energy conflicts, and the promotion of cooperation with the help of integrated concepts of sustainable development. Predictive power, in the sense of forecasting the future, is not sought.

The model's limitations are the insufficient quantifiability of relevant variables and parameters, such as when specifying the values, goals and means of the actors involved, or the response intensities that determine their changes over time. In some cases, only estimates and plausibility considerations are possible here, which makes validation of the model's results in reality difficult or impossible. Essential aspects of the problem must sometimes be ignored in order to make mathematical representation and simulation easier, for example, if only one evaluation function and only one category of means is selected for each actor, or the model parameters are kept constant over a long period of time. In reality, both the environmental conditions and the respective interest and power structures can in part change very rapidly and unpredictably. Some of these aspects can be taken into account in an expanded model, at the cost of increased complexity and decreased transparency of the model.

9 Mathematical Appendix

The interaction of the basic variables of the SCX model (system variables X, state-dependent goal and evaluation function $S(X)$ and means C) are grouped in the box below, with the variables having an index i for the actors and an index k for the system variables, and changes being designated with Δ. Expansions of the model to more than one goal and evaluation function and more than one type of means for each actor are possible. External influences on X, S, and C that are not due to actions of the actors can also be taken into account (for further explanations, see Scheffran 1989, 1996a).

The interaction parameters can be represented by

$$sc_{ij} = \sum_k f_{kj} \cdot s_{kij} / c_{kj}$$

with k designating various options for generation and use of resource X by actor j, with the respective resource price c_{kj} and the preference f_{kj}. Here, $s_{kij} = s^w_{kij} - s^r_{kij}$ is the net benefit effect of the resource usage, with positively acting *prosperity effects* s^w_{kij} and negatively acting *risk effects* s^r_{ij} being distinguished. The dynamics is in a state of equilibrium for

$$S_i^* = S_i^P = \sum_j f_j^x (s^w_{ij} - s^r_{ij}) X .$$

This represents the requirement that for each actor i there be a balance between the increase in prosperity sought, S_i^*, and the increase in prosperity actually produced from the use of the resource, S_i^P. It is assumed that each actor j receives a share $X_j = f_j^x X$ of the total resource consumption X that generates a prosperity effect s^w_{ij} and a risk effect s^r_{ij} among other actors. The five factors S_i^*, f_j^x, s^w_{ij}, s^r_{ij} and X correspond to the five concepts of sustainable development (see Section 5).

The fundamental equations of the SCX model for more than one actor and system variable

$$\Delta S_i = \sum_{j=1}^{n} sc_{ij} C_j$$

$$\Delta Ci = - \kappa_i (C_i - C_i^0)(C_i^* - C_i)(S_i + \tau_i \Delta S_i),$$

$$\Delta X_{ki} = \frac{f_{ki}}{Cki} C_i$$

Where:

Δ: Change of the variables in a discrete time *(e.g.* $\Delta X(t) = X(t+1) - X(t)$*)*
$i,j = 1,\dots,n$: Index for n actors
$k = 1,\dots,m$: Index for m system variables X_{ki}
$S_i = S_i^P(X) - S_i^*(X)$: State-dependent goal and evaluation function of the actor i
$S_i^P(X)$: Actual or perceived value for the system state X (actual value)
$S_i^*(X)$: Desired value for state X (goal value, desired value)
C_i : Input of means by actor i
sc_{ij} : Cost-benefit effects of means C_j on the goal functions S_i (interaction parameters)
C_i^0, C_i^* : Lower and upper limits to the input of means
κ_i : Reaction strength of means for actor i
τ_i : Memory parameter for actor i
f_{ki} : Distributive preference of actor i for the option ΔX_{ki}
c_{ki} : Partial costs (price) of a unit of X_{ki}

For more than one actor $i = 1,\dots,n$ and system variable X_{ki} $(k = 1,\dots,m)$, the model variables can be represented by vectors:

$$\vec{C} = (C_1,\dots,C_n),\ \vec{S} = (S_1,\dots,S_n),\ \vec{X_i} = (X_{1i},\dots,X_{mi}),\ \vec{sc_i} = (sc_{i1},\dots, sc_{in})$$

References

Baechler, Günther et al 1993: *Umweltzerstörung: Krieg oder Kooperation*? Münster: agenda.

Baechler, Günther et al. 1996: *Kriegsursache Umweltzerstörung – Ökologische Konflikte in der Dritten Welt und Wege ihrer friedlichen Bearbeitung* [Environmental Degradation as a Cause of War. Ecological Conflicts in the Third World and Ways for their Resolution]. ENCOP Study Vol. 1. Zürich: Rüegger.

Bender, Wolfgang et al. 1996: *Energiekonflikte – Kann die Menschheit das Energieproblem friedlich lösen?* Dossier 22, Wissenschaft und Frieden, Heft 2.

Bossel, Hartmut 1992: *Modellbildung und Simulation.* Braunschweig: Vieweg.

Brauch, Hans-Günter (Ed.) 1996: *Klimapolitik.* Heidelberg: Springer.

Brock, Lothar 1994: "Friedensforschung im Zeichen immer neuer Kriege", in: *AFB-Texte 1/94* (see also: Documentation in Frankfurter Rundschau of 9.5.95).

BUND, Misereor (Eds.) 1996: *Zukunftsfähiges Deutschland – Ein Beitrag zu einer global nachhaltigen Entwicklung.* Studie des Wuppertal-Instituts für Klima, Umwelt, Energie. Basel: Birkhäuser [a 25-page summary in English is available from the Wuppertal Institute].

Carraro, Carlo and Jerzy A. Filar (Eds.) 1996: *Control and Game-Theoretic Models of the Environment.* Boston: Birkhäuser.

Daase, Christopher 1993: "Ökologische Sicherheit: Konzept oder Leerformel?", in: Meyer, Berthold and Christian Wellmann (Eds.) 1993: *Umweltzerstörung: Kriegsfolge und Kriegsursache*, 21–52. Frankfurt/a.M.: edition suhrkamp.

Feess, Eberhard 1997: *Mikroökonomie – Eine spieltheoretische und anwendungsorientiere Einführung.* Marburg: Metropolis.

Fritz, P., J. Huber and H.W. Levi 1995: *Nachhaltigkeit in natur- und sozialwissenschaftlicher Perspektive.* Stuttgart: Wissenschaftliche Verlagsgesellschaft.

Hauchler, Ingomar (Ed.) 1995: *Globale Trends 1996, Stiftung Entwicklung und Frieden.* Frankfurt: Fischer.

Hinze, Peter 1995: *Computersimulation des SCX-Modells.* Seminararbeit WS 95/96, Fachbereich Physik der TH Darmstadt (Department of Physics of Darmstadt Technical University).

Homer-Dixon, Ted 1991: "On the Threshold – Environmental Changes as Causes of Acute Conflict", in: *International Security*, 16/2, 76–116.

Homer-Dixon, Ted, Jefrey Boutwell and George Rathjens 1993: "Umwelt-Konflikte", in: *Spektrum der Wissenschaft*, April 1993, 36–44.

Jathe, Markus and Jürgen Scheffran 1994: "The Transition to Chaos in the SCX Model of International Security", in: G. Johannsen (Ed.): *Proceedings of the IFAC Conference "Integrated Systems Engineering".*

Jathe, Markus, Werner Krabs, and Jürgen Scheffran 1997: "Control and Game-Theoretical Treatment of a Cost-Security Model for Disarmament", in: *Mathematical Methods in the Applied Sciences*,Vol. 20, 653–666.

Jathe, Markus 1996: *Methoden der nichtlinearen Dynamik und Kontrolltheorie zur Untersuchung eines Konfliktmodells.* Doctoral dissertation at the Department of Mathematics of Darmstadt Technical University. Aachen: Mainz.

Kaiser, Christian 1997: *Entwicklung eines Models zyklischen Wirtschaftwachstums.* Diplomarbeit, Fachbereich Mathematik der TH Darmstadt.

Krusewitz, Knut 1985: *Umweltkrieg – Militär, Ökologie und Gesellschaft.* Königstein: Athenäum.

Loske, Reinhard 1993: "Chinas Marsch in die Industrialisierung: Gefahr für das Weltklima?", in: *Blätter für deutsche und internationale Politik*, Heft 12, 1460–1472.

Loske, Reinhard 1996: *Klimapolitik: Im Spannungsfeld von Kurzzeitinteressen und Langzeiterfordernissen*. Marburg: Metropolis.

Meadows, Donella et al. 1972: *The Limits to Growth*. New York: Universe Books.

Meadows, Donella, Dennis Meadows and Jorgen Randers 1992: *Beyond the Limits*. London: Earthscan.

Meyer-Abich, Klaus 1994: "Im gemeinsamen Boot? Gewinner und Verlierer beim Klimawandel", in: Sachs, Wolfgang 1994: *Der Planet als Patient*, 194–210. Berlin: Birkhäuser.

Neuneck, Götz 1995: *Die mathematische Modellierung von konventioneller Stabilität und Abrüstung*. Baden-Baden: Nomos.

Nordhaus, W. D. 1991: "A Sketch of the Economics of the Greenhouse Effect", in: *The American Economic Review*, 81/2, 146–150.

Rösch, Roland and Wolfgang Bräuer 1997: *Möglichkeiten und Grenzen von Joint Implementation im Bereich fossiler Kraftwerke am Beispiel der VR China*. Dok. No. 97–03. Zentrum für Europäische Wirtschaftsforschung, Mannheim.

Scheffran, Jürgen 1989: *Strategic Defense, Disarmament, and Stability – Modelling Arms Race Phenomena with Security and Costs under Political and Technical Uncertainties*. Doctoral dissertation at the Department of Physics of Marburg University. IAFA publication series, Vol. 9.

Scheffran, Jürgen 1992: "Panzer gegen die ökologische Krise?", in: *Spektrum der Wissenschaft*, Heft 10, 128–132.

Scheffran, Jürgen and Markus Jathe 1992: "Stability and Optimal Security-Cost Ratios in the SCX Model of the Armament Dynamics", in: Sulway et al (Eds.): *Supplementary Ways for Increasing International Stability*. IFAC Workshop Series.

Scheffran, Jürgen and Markus Jathe 1995: *Modelling the Impact of the Greenhouse Effect on International Stability*. Presentation to the Conference on Supplementary Ways for Increasing International Stability, Vienna, 29.9–1.10.1995.

Scheffran, Jürgen 1996a: "Modelling Environmental Conflicts and International Stability", in: Huber, Reiner K. and Rudolf Avenhaus (Eds.): *Models for Security Policy in the Post-Cold War Era*, 201–220. Baden-Baden: Nomos.

Scheffran, Jürgen 1996b: "Leben bewahren gegen Wachstum, Macht, Gewalt – Zur Verknüpfung von Frieden und nachhaltiger Entwicklung", in: *Wissenschaft und Frieden*, Heft 3, 5–9.

Scheffran, Jürgen and Stefan Pickl 1997: *Control and Game-Theoretic Assessment of Climate Change – Options for Joint Implementation*. Presentation to the International Conference on Transition to Advanced Market Institutions and Economies, Warsaw, 18–12 June 1997.

Sprinz, Detlef F. 1996: "Klimapolitik und Umweltsicherheit: Eine interdisziplinäre Konzeption", in: Hans Günter Brauch (Ed.): *Klimapolitik. Naturwissenschaftliche Grundlagen, internationale Regimebildung und Konflikte, ökonomische Analysen sowie nationale Problemerkennung und Politikumsetzung*, 141–150. Berlin: Springer.

Sprinz, Detlef F. 1997: "Environmental Security and Intrumental Choice", in: Gleditsch, Nils Petter (Ed.) 1997: *Conflict and the Evironment*, 483–502. Dordrecht: Kluwer.

WCED (World Commission on Environment and Development) 1987: *Our Common Future* ('The Brundtland Report'). Oxford, New York: Oxford University Press.

Westing, Arthur H. (Ed.) 1990: *Environmental Hazards of War*. Oslo: PRIO.

Wissel, Christian 1989: *Theoretische Ökologie*. Heidelberg: Springer.

Woodrow Wilson Center 1997: *Environmental Change and Security Project Report*. Issue 3.

United Nations 1991: "Potential Uses of Military-Related Resources for Protection of the Environment", in: *Disarmament Study Series*, Issue 25, 1991.
Zhong Xiang, Zhang 1996: "Energy, Carbon Dioxide Emissions, Carbon Taxes and the Chinese Economy", in: *Intereconomics*, July/August, 197–208.

Part D:

Foreign and Security Policy Approaches

The Foreign Policy Dimension of Environmentally-Induced Conflicts

Bernd Wulffen

1 Status of the debate

In recent years, particularly since the end of the East-West divide, science has shown increased interest in the subject of 'environment and security'. Political scientists have taken up the question of the extent to which environmental degradation and the over-exploitation of natural resources can be or become the cause of international conflicts or even wars.

Not only the world of science has taken up this multi-faceted topic. It is also the object of increasing interest in foreign policy and security policy. The former British Foreign Minister, Malcom Rifkind, recognized that environmental factors are the reason for many disputes between countries. Here the environment is of fundamental relevance:

> To try to promote the security and prosperity of one's country without considering the environment is like building a house on sand. No prosperity will last if it is not built upon sustainable foundations. No peace will be secure unless it is grounded in equitable sharing of scarce resources or offers a sustainable future for all concerned (Rifkind 1997).

The US Vice President, Al Gore, pointed out that the USA:

> [has] gone beyond the definition of strategic interests dating from the time of the Cold War. Environmental problems such as global climate change, the depletion of the ozone layer and air pollution – aggravated by the growing world population – know no borders and endanger the health, prosperity and workplaces of all Americans (FAZ 1997).

The German Foreign Minister, Dr. Kinkel, stated on the occasion of the start-up of the Secretariat of the Framework Convention on Climate Change in September 1996:

> Environmental threats do not stop at national borders. The protection of the climate and water bodies are international tasks which cannot be handled by individual nations on their own. Environmental policy is increasingly becoming world domestic policy. The North and South, East and West must finally understand that they are one community on

> the spaceship that is the Earth and that the blue shell that protects us is getting more and more fragile.

The NATO Committee on the Challenges of Modern Society (CCMS) is currently looking into the connection between environmental degradation and security in a Pilot Study entitled "Environment and Security in an International Context" (see Lietzmann in this volume). Since then, within the framework of this pilot study a number of scientific investigations, seminars and workshops have been held. There is widespread agreement that environmental degradation and resource scarcity considerably increase the potential for international conflict and thereby the risk of armed conflicts (Dabelko and Dabelko 1995; Schellnhuber and Sprinz 1995; Carius et al. 1997).

At the same time international policy-makers have been called upon to devote more attention to the questions raised above. The US Government has complied with this demand by commissioning one of their own departments in the US Department of Defense (Office for the Undersecretary of Defense for Environmental Security) to study these problems, amongst others. Moreover, in April 1997, the US State Department presented the first in a series if annual reports entitled "Environmental Diplomacy". This is intended to provide an outline of the endeavors being undertaken in the field of environmental protection and to map out future tasks. Here the USA announced, inter alia, that it is planning to set up regional offices in twelve developing countries in order to address environmental problems together with the local governments (Claussen 1995: 40-43). The following conclusions can be drawn from the discussion to date:

- The discussion commenced in the USA and continues to be held in the USA for the most part. Yet the scientific and political debate in the USA is concentrating increasingly on questions relating to international environmental problems as the cause of conflicts and their potential to threaten national security policy and/or US security interests in other parts of the world.
- In contrast to this, the USA have to date not assumed a leading role either in international climate policy or in other major areas of the "Rio process". Neither have they ratified either the Convention on Biological Diversity nor the Desertfication Convention. In his speech before the UN General Assembly in New York on 16 June 1997, US President Bill Clinton therefore had to admit that the USA had to considerably improve in the field of environmental protection ("we have to do better").

  This makes it more remarkable that the USA take the subject of security and environment so seriously. This may be due to the fact that those environmental crises shown empirically to be of relevance to security, occur outside of the USA in principally developing countries. Yet the concrete security interests of the USA are involved, e.g. in connection with environmental refugees from Mexico, with the problem of water distribution in the Middle East or with burning oil wells in the Gulf. This security argument could be used to incline reluctant members of the US Congress to agree to the overdue ratification of the environmental conventions (Griffiths 1997: 15ff.).

- Apart from scientists and representatives of environmental policy, members of the military are showing increasing interest in this subject. It is welcomed that military quarters in numerous countries are addressing the problem of environmental protection. In Germany alone, environmental damage from former US and Soviet military presence estimates in billions of USD. This is particularly acute in the case of the new federal states (formerly East Germany) where environmental pollution dates back before reunification.

  The concern with environmental protection may be connected with the fact that the armed forces, and above all NATO, have been looking for new areas of activity since the end of the East-West conflict. In most NATO countries environmental protection enjoys high priority amongst the population.
- In the Rio process, particularly in Agenda 21, crisis prevention, including the prevention or mitigation of environmental disasters, did not play a role, although the Brundtland Report stressed this aspect as early as 1987.

  The fact that to date the Rio process has not dealt with security aspects of environmental degradation may be connected with the fact that the UNCED Conference concentrated on questions relating to the environment and development. Questions of security policy or crisis prevention were not emphasized as a direct interest to the Rio process.

  However, the question will have to be looked into as to whether the Rio process and Agenda 21 do not provide the international community with a set of instruments for crisis prevention through the channels of environment, development and foreign policy.
- Armed conflicts which occur solely as a result of environmental degradation or resource scarcity are nowhere documented. However, it is by no means denied that environmental degradation and the over-exploitation of resources may be *one* of several causes of armed conflict.

The following examples are intended to illustrate this:

- At the beginning of the 1970s the construction of a joint hydroelectric power plant by Brazil and Paraguay on the upper reaches of the Paraná River led to strong protests by Argentina, which is situated further downstream. These protests culminated in an open threat of war by Admiral Rojas, who was sympathetic to the ruling Argentinean military junta at the time. Subsequently relations between Brazil and Argentina significantly deteriorated. However, this did not result in a war between the two neighbors. Instead, Argentina decided to follow the example of Brazil and built its own hydroelectric power plant a few hundred kilometers downstream on the Paraná, also together with Paraguay. Today the relations between the countries through which the Paraná flows can be regarded as exemplary.
- Today water resources and their distribution play a particularly prominent role in many parts of the world. In the Middle East and south-western Asia they have

become a significant economic and political determinant. The countries through which the River Jordan flows, for example, all depend on the common use of this stretch of water. Likewise the Euphrates and the Tigris are vital to south-western Asia. Turkey, Syria and Iraq are de facto bound together by a common interest in these two rivers. The intention of Turkey, the country located farthest upstream, to build numerous dams on the upper reaches of the Euphrates and Tigris and to use more water from these two rivers for irrigation purposes in East Turkey, is a highly explosive political issue and could lead to severe conflicts with its neighbors.

- The situation is similar for the countries through which the Central Asian rivers Amu Darya and Syr Darya pass; these flow from the North-West Himalaya region through five states into the Aral Sea. The geographical conditions also unites Kyrgyzstan, Tajikistan, Uzbekistan, Kazakhstan and Turkmenistan in a kind of enforced alliance. Incidents in the past have shown that over-exploitation of water resources of these rivers increased the risk for conflict.
- However, the example of the Aral Sea illustrates yet another danger which may result from environmental degradation : the danger of *mass migration and environmental refugees.*

  As a result of the reduction of the amount of water flowing into the Aral Sea by more than three quarters, the size of this inland sea has decreased by more than one third. The shrinking of the Aral Sea has deprived local fishermen, farmers, traders, etc. who formerly lived along its shores for their livelihood. If they are to survive, they will have to move to another region with their families. This results in a refugee movement caused by environmental degradation.
- Refugees as a consequence of soil degradation are becoming increasingly frequent today. They can be found in Latin America (e.g. in the North of Mexico), in West Africa (Sahel Zone) or in India (Bangladesh). It is an odd fact that the so-called "Football War" between El Salvador and Honduras was partly caused by Salvadorian farmers, who occupied new farmland on Honduran territory after the soil on their own land had been exhausted, and thus contributed significantly to the tensions which finally exploded into violence on the occasion of a football match.
- Of much greater consequence, are the conflicts in the area of the Great Lakes in Africa. Here tensions had been building up for many years and resulted into violent conflict due to increasing soil degradation in Burundi and the increasing search for new farmland combined with the well-known ethnic tensions.

2 The conflict potential of environmental degradation

The principal conclusion that can be drawn from present examples is that environmental degradation and over-exploitation of natural resources may well constitute a cause of conflicts. As a rule, the examples also show that not only these two factors alone set off an armed conflict. Only when they are combined with other causes, e.g. political, social and ethnic tensions, can an armed conflict result from environmental problems (Carius et al. 1997).

This complex interconnection may be one explanation of why the topic of environment and security has only in the case a few exceptions, failed to receive as much attention in Europe than as in the USA. Political, social and ethnic crises have pushed environmental problems into the background and often turned them into a 'quantité négligeable'.

This might be a further explanation as to why the Rio process and Agenda 21 have also failed to tackle the connection between environmental protection and security.

However, today we should ask ourselves whether this view can still be accepted for the long-term future, especially against the backdrop of probable climate changes, increases in soil degradation and frequent environmental disasters, and in view of the projected increase of the global population.

For the future we have to expect that the conflict potential might significantly increase as a result of environmental degradation and the exploitation of resources in view of the changed coordinates. Then the connection between environment and security, which is today regarded by some as a merely theoretical one, might become a politically explosive topic and is indeed a question of survival.

This question can be broken down into the following areas of consideration:

- *Climatic changes:* We must expect emissions of greenhouse gases to show a marked increase over the coming decades in spite of present efforts to reduce them. Between the years 2020 and 2030, a culmination point with regard to emissions might be reached as a result of industrialization in the developing countries. This would have severe consequences not only for the group of forty 'small island states' threatened by the forecast rise in the sea level, but also for numerous other countries who may face changes in their overall climatic conditions.
- *Soil degradation:* Today we already assume that approx. 40% of the world's arable land has been degraded as a result of aridity, degeneration into steppe, and erosion. More than a billion people are affected by this. Combined with the projected changes in the climate, the situation may become significantly aggravated and lead to major social and political tension.
- *Natural disasters:* According to estimates made by the major insurance companies, the damage resulting from natural disasters is put at just under $100

USD billion per annum. The damage caused by the most recent floods in the Oder region in the Czech Republic, Poland and Germany alone has been estimated at 10 billion DM. Unfortunately we have to assume that the trend towards major environmental disasters is likely to increase, especially due to the probable changes in the climate.

- *Population growth:* If forecasts of an approximate doubling of the global population over the next hundred years are accurate, the demand for processed and drinking water, arable land and natural resources will increase substantially. Already within the next thirty years, it is estimated that water resources will be so scarce that we will have to talk of a stress situation for around one third of the world's population (ECOSOC; WBGU 1997: 81 ff.).
  However, significant population growth is also likely to result in other stress situations and to increasing social tensions with regard to other natural resources (e.g. in further over-use of soils, considerable migration movements and growing poverty amongst a large part of the global population). In general, we may say that it is extremely probable that the combination of environmental degradation and population growth will impact the potential for armed conflict at the local and regional levels.

3 Challenges for foreign policy

1. If, as the above examples show, environmental degradation and resource consumption are able to lead to an increase in the international conflict potential, then looking into the connections between environment and security is not only justified but essential.

The appeal to foreign policy makers to join in the discussion on the connections between environment and security is justified and necessary. One of the most important aims of foreign policy is to ensure the long-term peaceful coexistence of nations. It can therefore not be a matter of indifference to foreign-policy makers when factors become evident which make disruptions in the peaceful coexistence of nations seem likely or at least no longer impossible, even if this is only in the long-term. Therefore foreign policy must take seriously the risk factors recognized as probable, namely environment and resource degradation and population growth and to make an assessment of the consequences and to take timely preventive action.

2. It is interesting to note that individual states or groups of states have already reacted in a very concrete way to recognized risk factors in connection with increasing environmental degradation. At the European Union (EU) Government conference in June 1997, the principle of "sustainable development" was included in both the Treaty of European Union and in the EC Treaty. Other examples include China passing their own Agenda 21 in 1994 and Bolivia creating a Ministry for Sustainable Development. The countries of Central America have also

set up a "Council for Sustainable Development". National environmental plans and sustainability strategies in the form of Agenda 21 have since been initiated in two thirds of all industrialized countries and more than 30 developing countries (Jänicke, Carius and Jörgens 1997). At the United Nations General Assembly session in 1993, German Foreign Secretary Kinkel successfully proposed an initiative for the creation of a United Nations early-warning capacity in the event of environmental disasters.

3. However, we have to ask ourselves whether or not this is adequate for the prevention of environmentally- induced conflicts or whether we need a much more extensive, multilaterally supported set of instruments to address the dangers described over time.

It is nevertheless worth noting that in recent years an extensive set of regulations for the protection of the environment has been created at both a national and international level, which could also serve as a basis for further preventive measures with regard to security policy. We could consider extending this and strengthening its institutions, perhaps also adding security policy components. This would then make it unnecessary to create a completely new set of instruments to prevent environmental crises as a source of armed conflicts.

The process set in motion in Vienna twelve years ago to protect the ozone layer could provide us with these instruments. Indeed, the International Convention on the Protection of the Ozone Layer together with the Montreal Protocol (1987) are an important set of landmarks in the fight against the degradation of the environment and are examples to be followed.

It is also true that the process of sustainable development initiated with Agenda 21 at the UNCED conference in 1992 marks an important new beginning. However, the question arises as to whether we could make greater use of the process started in Rio in 1992 for the aims of security policy. Would it not be a major step forward, also with regard to prevention in terms of security policy, if we managed to consistently implement the model of sustainable development both nationally and internationally? Would not environmental crises *eo ipso* be made impossible from the outset by means of a sensible environment and resource policy?

This idea does indeed have something in its favor. After all, we have already succeeded in getting environment and resource protection put on the international agenda through the Rio process, in ensuring that environment, development and foreign policy have to deal with this and take it into account. In the Commission on Sustainable Development (CSD) we have a platform for the political discussion of topics connected with the protection of the environment and natural resources. This discussion must be intensified in its *risk prevention dimension.*

4. Above all, one should think about utilizing the conventions signed and ratified in the Rio process for the objectives of international security policy. It might be a good idea to take advantage of the Desertification Convention, signed in Paris in 1994 and since ratified by over 100 states, to reduce potential tensions resulting from soil degradation or problems with the supply of water.

The objective (Article 2) of the Convention is:

> to combat desertification and mitigate the effects of drought in countries experiencing serious drought and/or desertification, particularly in Africa, through effective action at all levels, supported by international cooperation and partnership arrangements, in the framework of an integrated approach which is consistent with Agenda 21, with a view to contributing to the achievement of sustainable development in affected areas.

The Desertification Convention's advantages is that it offers regional programs and allows for the coordination of donor activities. It also contains a mechanism for the resolution of conflicts that provides at least the beginnings of a mechanism which may be useful for peace-maintaining measures. Moreover, it is also important that environment, development and foreign policy make use of these ideas to prevent conflicts.

However, we should also not ignore the fact that the Rio process and the conventions resulting from it have their weaknesses. The Rio follow-up conference in New York in June 1997 (Rio+5) showed that unfortunately it has to date not proved possible to defuse conflicts of interest. The developing countries regard the UNCED follow-up as an instrument for their development (i.e. as part of a catching up process) and accordingly lay claim to substantial additional transfers in financial resources. On the other hand, the industrialized countries see it as an instrument for the protection of the environment and resources, which is the responsibility of each individual country but nevertheless does not justify any claims. This disagreement also explains the slowness of the Rio process and the constant search for formulae capable of resolving this insoluble conflict.

A further weakness of the Rio process is that it has not proved possible to underpin it institutionally with an efficient organization furnished with comprehensive functions within the framework of the United Nations system.

Above all, it is repeatedly evident that the awareness of environmental and resource degradation and the consequences of this is still very underdeveloped worldwide. It was no mere coincidence that the 3rd Conference of the Parties to the Convention on Biological Diversity, convened in Buenos Aires in November 1996, called upon the Special Session of the United Nations to do more to encourage the growth of international environmental awareness. Many people fail to realize that the degradation of the environment is an insidious and sometimes initially invisible process that requires immediate action due to its disastrous consequences.

5. In order to make the instruments offered by the Rio process useful for maintaining peace, it is essential to overcome the fundamental differences of opinion between developing and industrialized countries. This could be initiated, for example, in the form of an intensified foreign policy dialogue between North and South. The annual United Nations General Assembly or the United Nations Economic and Social Summit (ECOSOC) could serve as a platform for this.

The idea of creating an 'Environmental Security Council' has also been proposed. The starting point for this was the idea of strengthening environmental protection in the United Nations. However, creating an 'Environmental Security Council' played no role in the June 1997 Special Session of the UN. On the other hand, German Chancellor Kohl, together with the Brazilian President, the Vice President of South Africa and the Prime Minister of Singapore, took the initiative to call for a United Nations environmental umbrella organization. This new organization is to be based on existing organizations, a strengthened UNEP being one of its pillars.

If it proves possible to strengthen the Rio process in terms of its content and institutions and to overcome the fundamental differences of opinion, it is quite conceivable that greater use could be made of the conventions passed in the Rio process for the prevention of environmentally induced conflicts.

However, as long as the Rio process remains paralyzed by the profound disagreement between developing countries and industrialized countries, the option of independent forums for the prevention of environmentally induced crises should be maintained. The following arguments speak in favor of this:

- The Rio process is by its very nature oriented towards environment and development policy. UNCED was a conference on environment and development. Considerations of security policy at best played a fringe role in this.
- 'Environment and security' stands at the interface of environmental, development, security and foreign policy. It would place an additional burden on the already difficult Rio process if that were now extended to include far-reaching aspects of security and foreign policy.
- The intensive treatment of the subject of security policy would presuppose the extension of Agenda 21. This may indeed be quite welcome, especially as seen from today's point of view it must be assumed that such problems are likely to become more severe in nature. It may also make sense to add a chapter dealing with "disaster prevention" to Agenda 21.

  Nevertheless, in the present circumstances and allowing for the aforementioned this should be put aside, at least for the time being. Requests for additions unfortunately often turn out to be a *'Pandora's Box'*. They may easily lead to a fundamental questioning of Agenda 21, which could well do further harm to the Rio process.
- The Rio process does not embrace all the instruments and mechanisms for environmental protection. The aforementioned Vienna Convention on the Protection of the Ozone Layer or, for example, the Basel Convention on the Control of Transboundary Movements of Hazardous Wastes were precursors to the Rio process or run parallel to it.

  Nonetheless, it would not be expedient to deal with 'environment and security' completely independently of the Rio process. This trend seems to crop up from time to time. A successful outcome of the Rio process would have a positive effect on the international security situation. This affects above all developments in the next century. The reverse is also true, namely that an impetus from the field of security and foreign policy can also have a positive effect on the Rio process. It can give it an additional raison d'être.

- The Desertification Convention offers an approach combining regional programs and the coordination of donor activities. This at least provides the beginnings of a mechanism which can certainly be useful for peace-maintaining measures. It is important for environment, development and foreign policy to make use of such approaches in order to prevent conflicts.

4 Summary

Due to the probable future developments such as climatic changes, water scarcity, increasing soil degradation, growth of world population, etc., it is rational and expedient to look into the problems of 'environment and security'. Here a differentiated approach embracing foreign policy is appropriate.

The Rio process on the basis of the concept of sustainable development contains elements for defusing the projected conflict potential. It should be included in the proposed solutions for crisis prevention. However, it is important in this context also to see the difficulties involved in this process.

The inclusion of the security policy dimension into the Rio process should not be considered at present. Independent forums for the discussion of security and foreign policy aspects with regard to the prevention of environmentally induced crises are a better option. Here, the security policy component can have positive effects on overall demands for sustainable development.

Foreign policy could act as a catalyst within the framework of the discussion on environment and security. In order to give visible status to environmental protection in foreign relations what is required is coordinated integration in sectoral policies together with representation which could be concentrated in foreign policy.

References

Carius, Alexander et al. 1997: "NATO CCMS Pilot Study: Environment and Security in an International Context", in: *Environmental Change and Security Project Report*, Issue 3, 55–65. Washington, DC: The Woodrow Wilson Center.

Claussen, Eileen 1995: "Environment and Security: The Challenge of Integration", in: *Environmental Change and Security Project Report*, Issue 1, 40–43. Washington, DC: The Woodrow Wilson Center.

Dabelko, Geoffrey D. and David D. Dabelko 1995: "Environmental Security: Issues of Conflict and Redefinition", in: *Environmental Change and Security Project Report*, Issue 1, 3–13. Washington, DC: The Woodrow Wilson Center.

ECOSOC Doc. E / CN. 17 / 1997 / 2 / Add. 17, 2.

FAZ (Frankfurter Allgemeine Zeitung) 1997: "'Umwelt-Diplomatie' und eine grüne Außenpolitik", in: *Frankfurter Allgemeine Zeitung*, 29.4.1997.

Griffiths, Franklyn 1997: "Environment in the U.S. Security Debate: The Case of the Missing Arctic Waters", in: *Environmental Change and Security Project Report*, Issue 3, 15–28. Washington, DC: The Woodrow Wilson Center.

Jänicke, Martin, Alexander Carius and Helge Jörgens 1997: *Nationale Umweltpläne in ausgewählten Industrieländern*. Berlin: Springer.

Schellnhuber, Hans Joachim and Detlef F. Sprinz 1995: "Umweltkrisen und internationale Sicherheit", in: Karl Kaiser and Hanns W. Maull (Eds.): *Deutschlands neue Außenpolitik Band 2: Herausforderungen*, 239–260. München: Oldenbourg.

Rifkind, Malcolm 1997: *Foreign Policy and the Environment: No Country is an Island – Secretary of State's Speech on the Environment*. 14 January 1997.

WBGU, Wissenschaftlicher Beirat der Bundesregierung Globale Umweltveränderungen (German Advisory Council on Global Change) 1997: *Welt im Wandel: Wege zu einem nachhaltigen Umgang mit Wasser* (also available in English, entitled "World in Transition: Ways Towards Sustainable Management of Freshwater Resources). Berlin: Springer.

Environmental Threats and International Security – The Action Potential for NATO

Volker R. Quante

1 New dimensions in security policy

Since the end of the Cold War a new security policy has been the subject of increasing debate. Some of the fundamental questions in the context of the wider conception of security created in this connection are: What does *security* mean after the end of the Cold War? Security for whom? Security against whom? Where are the dangers, the risks or the threats today (Summer 1994: 3)?

At the time of the Cold War there was only one existential threat to the countries of the West: a massive Soviet military attack. This divided the world into two camps and entailed the risk that any conflict on the periphery of these camps could escalate into an international conflict (Brill 1995: 2).

Since then the determinants of security policy have undergone a fundamental change: security policy faces the dual challenge of changed legitimization – especially with regard to military prevention – and adaptation to new security risks.

The system of mutual deterrence has become almost obsolete, but new challenges and risks are taking the place of the old confrontation between the two power blocs (Frankenberger 1995: 12). There has to date been only limited success in subsuming these as particular security problems in a uniform problem pattern, as was possible prior to 1990 with the terms *power/system conflict* and *bloc confrontation* (Mutz 1997: 54). They differ from the one-dimensional threat of the Cold War by virtue of their variety; today we are dealing with a broad spectrum of instabilities and security risks. They are multi-faceted, multi-dimensional, come from many different directions and are difficult to predict and assess (Rühe 1995). In addition to the new potential threats summarized by the term *extended security interests* (including organized crime, international terrorism, illegal cross-border waste trading, etc.) we are also dealing with indirect security problems (e.g. more than 17,500 were killed in 27 armed conflicts in 1996 alone while the number of refugees is estimated in hundreds of thousands) (SIPRI 1997: 25-30).

The end of the East-West conflict marked a very special watershed for all areas of security policy. Within a short time all those concepts which had hitherto characterized the debate, such as *bloc theories, strategic concepts* and *military balances of power* have become meaningless (Maier 1993: 9). Security policy is therefore currently engaged in a debate on new theoretical-conceptual clarification – a process which has not yet been completed.

Security is generally understood to mean the absence of and protection from dangers and threats. In contrast to *internal security*, *external security* is defined in terms of the absence of intervention from outside and the precautionary measures taken against these. The type of intervention from outside does not necessarily have to be strictly limited: The former German Defense Minister Hans Apel recognized some time ago that the classic definition of security no longer corresponds to reality.

> In our times security has long since assumed dimensions which can no longer be solely defined in terms of the stricter concept of military security. Today security is the product of *domestic policy, economic, social and military factors.* (Wetter 1984: 186)

Seen against the backdrop of the immense political changes that have taken place in recent years, this almost visionary statement from the early 1980s is proving to be all too true.

The change in the situation in terms of security policy is reflected particularly vividly in the scientific field in the debate on a new extended concept of security. There is a consensus amongst the proponents of different viewpoints with regard to the fact that in many parts of the world the concept of security in its ideas on threats today concentrates less on military threats. In contrast to this, the concept of crisis potential and the inevitably related question as to how to solve this problem relate directly to an increasingly diffuse concept of security, which affects both the *classic concepts of sovereign-territorial* and thus *primarily military security* as well as *domestic and trans-social dimensions* of other areas of security that affect each other, such as *ecological, economic* and *social security* (Haedrich / Ruf 1996: 7).

The broadening of the concept of security doubtless does justice to current developments in foreign and security policy precisely because it is not only military threats that endanger the existence and stability of countries but also the survival and well-being of the people living there. Also with regard to the world-wide increase in interdependency the extension of the dimensions of security is proving necessary since mutual political and economic vulnerability is growing and future dangers, such as threats to the environment, are gaining prominence in the awareness of those responsible for security policy.

Since the eighties, environmental threats and attempts to tackle these have become topics on a new security agenda and determine in particular the nature of international relations. It is essential to deepen the process of integration between individual countries, groups of countries and security organizations in equal measure if the escalation of local, regional and global environmental problems is to be prevented.

NATO too took up the conceptual challenges for security policy at an early stage. In the *New NATO Strategic Concept,* the heads of state and government of the NATO member states stated in Rome already on 8 November 1991:

> It is now possible to draw all the consequences from the fact that security and stability have *political, economic, social, and environmental elements* as well as the *indispensable defence dimension.*

Thus NATO adopted the so-called *extended security concept*, which abandons the strict understanding of the classic definition of international security and whose first outlines began to be discernible for the first time in the Conference on Security and Cooperation in Europe (CSCE) Helsinki Final Act.

Other institutions of security policy also reacted to the changed challenges. Within the Organization for Security and Cooperation in Europe (OSCE) the debate developed on a "common and comprehensive security model". At the OSCE Summit in Lisbon in 1996 a number of non-military challenges for security policy were shown. In addition to ethnic tension, migration, terrorism, organized violence in addition to drugs and arms trading, environmental problems also proved to be a new topic for Europe's security policy.

Whereas almost all security institutions are aware of the new challenges, at least on the level of political statements and official declarations, possible contributions by individual security institutions to the prevention of environmentally-induced conflicts are not sufficiently considered. What the United Nations, for example, are still planning, namely to anchor a world-wide fundamental consensus for an environmental partnership in the United Nations Charter, NATO already has successfully included in its statutes for its area of responsibility.

2 International environmental policy – No reason to hope for more security

The efforts of international environmental policy to bring about a consensus find themselves in a permanent dilemma. Precisely at the Rio follow-up conference in June 1997 (Earth Summit + 5 in New York) this became increasingly evident: global environmental pollution and degradation endanger traditionally understood national sovereignty in the form of numerous phenomena. This danger could be countered by international agreements. Yet this would also necessitate the renunciation of sovereignty to a quite considerable extent, which would clearly be every bit as unsettling as the consequences for the security of a country which might result from a conflict partially caused by ecological factors.

In view of this situation the risk exists that classic thinking on security might induce individual states to deliberately undermine existing environmental agreements or to implement them slowly or not at all. This may result in a more or less hidden risk spiral similar to the arms spirals of past decades.

Against this backdrop and that of a worsening North-South conflict – without losing sight of the East-West conflict, which should not be neglected from an environmental point of view either – a fundamental debate has developed on the subject of how far national sovereignty may be permitted to exist at all in the context of free control of certain resources of supranational importance worldwide if the global objectives of environmental protection are to be achieved. National sovereignty in this connection is to be understood as self-determination, political independence, territorial integrity, domestic competence. Control of resources of supranational importance is taken to mean for example cross-border river systems and dam projects, the use of fresh-water reserves, the exploitation of migrating schools of fish, the desertification of agricultural land as a result of excessive use, rainforest clearcutting, carbon dioxide emission and the climate problem. The Earth Summit + 5 conference in New York made it clear that on this question too the opinion-forming processes have been completed in very few countries only, indeed even leading industrial countries such as the United States have not yet developed any stringent political attitude, nor do they wish to.

3 There are no purely ecological conflicts

Strictly speaking there is no such thing as a 'purely' ecological conflict, i.e. a conflict exclusively about natural resources or as a result of environmental degradation. In a complex way, ethnic differences interact with distribution conflicts, poor overall socio-economic conditions, and problems of environmental pollution/degradation. In combination they have the potential to lead to the outbreak of violent conflict.

Although all conflicts have their own specific characteristics, in order to be able to take any preventive action it is essential to concentrate on filtering out common factors and interconnections. In comparison to a military threat, environmental threats (in this paper this term is intended to cover both environmental pollution/degradation and danger to the environment) display a number of special features. These can be summarized as follows:

- Due to their complex structures it is not possible to identify any clear conflict partner to whom the causes of environmentally-induced conflicts can be clearly attributed.
- An environmental threat may exist for longer periods between individual states, groups of states and alliances, or within states without other conflict components (military, economic, ethnic) and without turning into a violent conflict.
- Environmentally-induced conflicts pass through different stages of conflict, which do not necessarily display escalation dynamics.

- As yet there are no environmental alliances over and above economic and military alliances.
- The 'defense possibilities' against environmental threats at a national/ governmental and international level are limited because there are no organized and active protection organizations for this task and there is only a very restricted 'deterrent' in the form of sanctions.
- Even if a threat to the environment is successfully combated or de-escalated, real environmental degradation is the initial concrete result.
- Although military organizations have possibilities for ending a conflict or preventing it via deterrence, they themselves do not have the specific capacities for removing the causes of the environmental threat.

A comparison of the consequences of military threat and environmental threat reveals a fundamental difference between the two. Merely *potential degradation* in the case of a military threat to the sovereignty of states and international security stands against *real* and *gradually accumulating environmental degradation*. The threat to security and sovereignty involved in concrete environmental degradation is thus a stage more real than any military threat. Furthermore, its effects do not become felt in the future only, but very much in the present.

Just how closely the different causes of conflicts can sometimes be linked is illustrated by the following breakdown. It represents an attempt to record the generalized most important reasons for the outbreak of ethnic conflicts and their connection with environmental threats and name examples of their occurrence:

- the idea of a nation-state as an identity-promoting, but often exaggerated demarcation factor (e.g. Serbia);
- fear of over-alienation or of losing specific cultural or religious values (e.g. the Baltic States);
- the constant growth in world population and the related problem of over-population in certain regions of the world, combined with an increasing scarcity of natural resources (e.g. Burundi, Rwanda, as well as Bangladesh, India and China);
- arbitrarily drawn borders with no regard for natural cultural areas or ethnic and religious allegiance (very frequent in Africa, but also in Bosnia-Herzegovina), often also without regard for existing eco-regions (Byers 1991: 65 ff.; Baechler 1993: 15);
- unevenly distributed economic resources (e.g. Middle East);
- the fight for resources which are no longer available in sufficient quantities, such as water and food (e.g. Turkey, Syria, Iraq);
- power struggles resulting from outdated but still existing clan structures (e.g. the warlords in Somalia) (Rühle and Höppner 1997: 18).

This list shows that conflicts are by no means a problem which is specific to developing countries. The common denominators in the causes for a conflict also indicate that a higher level of development is no guarantee of peaceful coexistence

– precisely when considering environmental threats as a catalyst for conflict. In the conflicts of the future, therefore, causes, parties to conflicts and lines of conflict will vary to a considerable degree. Under such conditions the consensus principle generally holds at a political level, yet this often remains a fiction when it comes to implementing effective measures to prevent and solve conflicts on the ground (regionally and locally).

As there is thus no such thing as a 'purely' ecological conflict, there is no need to develop new regulatory mechanisms which are especially geared to the 'environmental conflict'. Nevertheless, it is noticeable that, above all in the area of environmental policy, there are still no established or proven doctrines or concepts to contain environmental threats harboring conflict risks at an international level. Practicable solutions are nowhere near being found.

4 The danger of environmentally-induced conflicts is increasing

In 1996 the Swedish Peace Research Institute SIPRI registered 27 wars, three fewer than in 1995. As stated in the annual report published on June 26, 1997 (SIPRI 1997), all these wars, with one single exception (the conflict between India and Pakistan), were civil wars and violent intra-state conflicts. Although the number of crises has continuously decreased, the duration of constantly latent conflicts has not. The African continent, where military conflicts are increasingly spiraling out of control, was identified as the greatest trouble spot. In addition to ethnic and religious causes, unsolved environmental problems and an increasing scarcity of natural resources have also contributed to political instability to the point of armed conflicts in this region, above all in Sudan and the region around the Great Lakes with Burundi, Rwanda and the Democratic Republic of Congo. The principal line of conflict in the area of the environment is at present still the North-South axis. This will increase in importance in the future, as has been shown by the Environment and Conflicts Project (ENCOP) in a number of case studies (Baechler et al. 1993; Baechler et al. 1996).

The Arbeitsgemeinschaft Kriegsursachenforschung (AKUF) research grouping has analyzed that in the years between 1945 and today the number of new conflicts has increased faster that the number of those resolved (see also Rohloff for a detailed analysis in this volume). In addition to this, most conflicts liable to escalate in the near future will display an economic-ecological-social structure (Volmer 1996: 18). This was also the conclusion arrived at by the participants in the "Conflict and the Environment" workshop held by the Division for Science and Environmental Affairs of NATO in Bolkelsö, Norway, in June 1996 (Dabelko 1996: 222).

5 A global system of collective security: The United Nations

Now that the East-West conflict has been overcome, the United Nations has regained its freedom to act, but without being able to effectively fulfill its tasks of maintaining and restoring peace.

The powers of the Security Council as a central sanctioning body of the UN are wide-ranging. For example, the Security Council may intervene not only in the case of an act of aggression and breach of peace, but even if there is only a threat to peace. The fact that in their causality threats to the environment together with other causes can become a danger to peace, particularly at a regional level, is undisputed. Timely reaction renders the use of force superfluous, especially as Article 2 para 4 of the UN Charter forbids its use and the threat of its use in international relations.

The Security Council also has the power to decide against which countries (regional approach) or against which country (local approach) measures are to be taken, whilst also being able to decide between non-military and military sanctions. In its "Agenda for Peace", which introduces the new concepts of peacebuilding and peacemaking operations, the UN has created useful instruments of security policy for resolving conflicts which were not originally included in its Charter. Compared to the UN Charter, which according to Article 2 para 5 (general principle of support) and Article 25 of the Charter (binding effect of Security Council resolutions) obliges Member States to participate in sanctions imposed by the Security Council, the principles for 'peacekeeping operations' show some differences. They do not stress primarily the unilateral support of the victim against the aggressor, but the search for a lasting peaceful solution to a dispute. Further principles for implementing peacekeeping and peacemaking measures are:

- All parties to the conflict must give their consent.
- The participation of the states called upon is not binding as with sanctions for collective security, but is left to their free discretion in each individual case.
- The use of force is only permitted for the purpose of self-defense.

6 Possibilities and limits of cooperative security: The Organization for Security and Cooperation in Europe (OSCE)

During the East-West conflict the CSCE process served both as a forum for limited co-operation in various areas, including security, and for the non-violent political-ideological confrontation between the two alliances *NATO* and the *Warsaw Pact.* In the area of security policy the main achievement of the CSCE was the agreement of step-by-step confidence and security-building measures. This resulted in an increase in military transparency and laid the ground for disarmament agreements.

In the aftermath of 1989, the CSCE/OSCE took on new tasks as a cross-European institution for cooperative security, thereby gaining a new legitimation. In spite of being renamed OSCE, it is still not an inter-state international organization with a charter of foundation. However, the OSCE now has a complex institutional structure, ranging from the plenary session of heads of state and government and foreign ministers to the establishment of permanent diplomatic missions.

In the early phase of conflicts, the OSCE takes preventive action using a number of diplomatic instruments. In this context, observer missions perform tasks which may range from collecting information and assessing the status on the ground to acting as mediator. The observers can collect the complaints of the parties involved and make contacts between the actors, or monitor adherence to recognized principles – for example during elections – as a neutral body.

From the point of view of security policy amongst the permanent OSCE bodies above all the Forum for Security Cooperation and the Conflict Prevention Center should be mentioned. This gives the OSCE, as a regional security organization according to Chapter VIII of the UN Charter, certain advantages over NATO. To date, NATO does not have any instruments at its disposal for preventing conflicts and for bringing disputes to a peaceful conclusion. As Bosnia-Herzegovina shows, NATO is in a position to prevent armed conflicts and to force a cease-fire by the presence of armed forces. Yet NATO's abilities are at present inadequate to influence the processes of peaceful reconstruction, economic recovery and the enforcement of human rights. The possibilities of the OSCE to exert influence in the form of missions in crisis areas, early warning via the High Commissioner on National Minorities or the Convention on Conciliation and Arbitration thus constitute important instruments for handling conflicts, which are based on unanimous, politically binding decisions.

As an issuer of mandates in the event of conflicts to institutions furnished with corresponding military resources such as NATO, the OSCE holds an important position in the field of cooperative security. But here too, the OSCE was not able to fulfill the hopes placed in it at the beginning of the 1990s, namely that it would become a guarantor of security in Europe.

7 Is there an alternative to NATO as an organization for collective security?

Since the end of the East-West conflict all the security organizations working in Europe have been in need of fundamental reforms. Although there is still certainly room for improvement, NATO has been the most consistent in tackling reforms and has made most progress in their implementation. Through its principle of consensus it has been particularly successful in achieving a balance of national influence. From its very beginnings, NATO was conceived not just as a military alliance, but as a community of common values amongst its members. The preamble and Article 2 of its founding treaty specifically include, inter alia, the principles of democracy, individual freedom and the constitutional state as well as the promotion of stability and the well-being of its members.

The main objective of NATO anchored in its Treaty of Foundation – the safeguarding of the freedom and security of all its Member States by deterrence and defense – remains unchanged. A further task which the Alliance is to pursue is the foundation of a stable security environment on the basis of democratic institutions and the peaceful resolution of conflicts. Here political instruments are increasingly gaining importance as compared to military solutions. It is for this reason that NATO offers consultations (but not military support) to any partner in the Euro-Atlantic Partnership Council (EAPC) or in the Extended Partnership for Peace (EPfP) which sees its territorial integrity, political independence or security threatened. Thanks to these additional forums for conflict resolution NATO is able to offer other countries similar opportunities to make the perception of threat the subject of official negotiations. This makes it possible to build up an atmosphere of trust, which is in turn the prerequisite for peaceful co-existence, and to gradually develop into a new form of collective security for Europe (Reiter 1997: 377).

The outlines for this are already to be found in *"The Alliance's New Strategic Concept"* of 1991. Its broadly based approach to security policy is based on *dialog*, *cooperation*, *collective defense* and *crisis management* and *conflict prevention*. The Strategic Concept thus makes it legally possible, by virtue of the broad definition of the various aspects of security, to contribute to solving conflicts resulting from a wide range of causes – the concept mentions "political, economic, social and environmental elements". One principal focus here will doubtlessly have to be on the prevention of armed conflicts.

NATO will take center seat in a future Euro-Atlantic security order and thus play a key role in efforts to reorganize the European security architecture. However, if NATO assumes responsibility for collective security regionally and globally, it is doubly bound to the mandates of the two organizations to which the international community has assigned this task, but for which it has not made the necessary funds available – the UN and the OSCE.

Today NATO is the only security organization with the necessary political and military instruments to prevent conflict, to provide mediation and to put an end to violence. It has flexible leadership structures facilitating precise political monitoring based on world-wide telecommunications capacities and satellite information-gathering systems; it has multinational and national reaction forces, which are equipped and trained for crisis management; it has state-of-the-art computer and analyzing capacities for verification, for operation control and as an early-warning system to support political decision-making.

The collective experience of the Alliance in the field of security over almost five decades has resulted in new dimensions of crisis management, the principal focus of which is jointly assuming responsibility for crises at an early stage. This requires fewer capacities and efforts, risks and costs are lower and it is possible to take preventive action to avoid escalation. Rapid policy coordination and joint actions remain the main points in this context. This is also the purpose of the annual sessions of the Political Committee of NATO with UN experts to coordinate between NATO and the UN in the case of NATO-led peace-supporting measures on behalf and under the mandate of the UN.

NATO is currently pursuing a dual approach to crisis management:

- No distinction is made between old and new threats; the existing (and expanding) collective community of NATO is sufficient to deal with both.
- The collective responsibility for future threats will be assumed pragmatically on the basis of ad-hoc decisions.

Both solution strategies demand a high degree of cooperation with other international organizations. NATO as a Euro-Atlantic security alliance can boast a number of successes in its fifty-year history. For many decades it has been successful in preventing conflicts at an early stage precisely by bringing about a balance of national interests and influences (for example Turkey – Greece). It has shown itself to be flexible and capable of political action, and is prepared as a matter of principle to engage in dialog also with countries which are not members of its forums. In particular, it was NATO which in 1992 developed the concept of the *European security architecture*, in which the different institutions should mutually reinforce each other – the concept of *interlocking institutions*, which is intended to include and integrate the OSCE, NATO, the EU, the WEU and the Council of Europe. In the Madrid Declaration on Euro-Atlantic Security and Cooperation on 8 July 1997, NATO once again emphasized this idea. The declaration states:

> A new Europe is emerging, a Europe of greater integration and cooperation. An inclusive European security architecture is evolving to which we are contributing, along with other European organizations. Our Alliance will continue to be a driving force in this process.

However, NATO cannot handle crises and conflicts alone, as

- most crises and conflicts exist outside of the NATO area and thus do not fall under the classic Alliance defense according to Article V of the NATO Treaty;
- other security institutions have to be involved, as generally the participation of countries in their area is necessary in crisis management in order to increase or simply create credibility of the measures in the regions affected;
- it must coordinate its military operations in terms of timing and content with the supporting civilian measures, for which as a rule other organizations (e.g. the OSCE, but also the numerous non-governmental organizations) bare responsibility.

This means that the decision-making processes in NATO, unlike its residual core task, are closely linked to the formation of political will outside its area of influence.

In order to remain capable of political action in this field, NATO developed its concept of Combined Joint Task Forces (CJTF). This concept makes it possible to assemble forces from different parts of the armed forces of various NATO states under flexible command structures, under a variety of different mandates and with different objectives. Forces from non-NATO states can also be integrated in the concept and even operations under the leadership of the Western European Union are conceivable. This means that NATO is in a position to adequately meet a wide variety of threats of different levels of intensity and to cover the entire spectrum of humanitarian aid or preventive deployment of armed forces, through peace-maintaining, peacemaking or peace-enforcing measures down to war in the form of the armed defense of its Alliance area (see Figure 1).

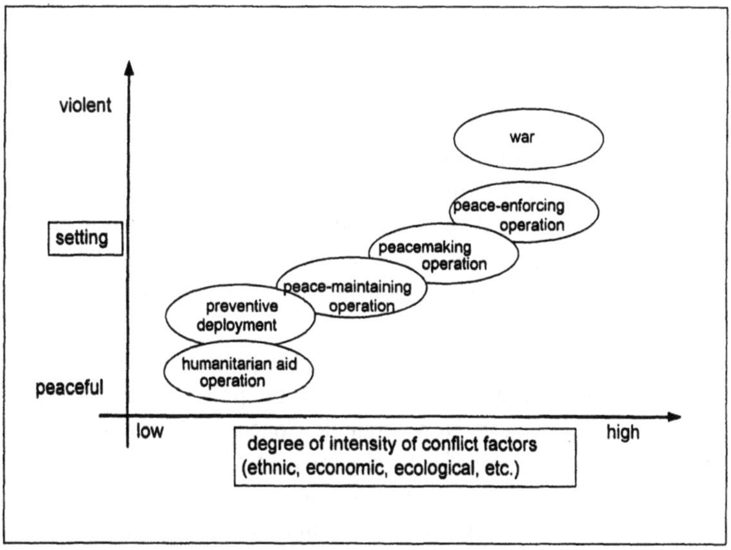

Figure 1: CJTF Concept: Spectrum of reaction possibilities

NATO thus has at its disposal a historically developed package of instruments for the preventive and reactive management of crises and conflicts, which can and must be used in interaction with the aforementioned international security policy institutions.

8 Summary

Security policy cannot bring about a state of non-conflict (Baudissin 1969). The future system of international security must emerge in a multi-faceted process, which has to be constantly adjusted in accordance with the requirements of long-term objectives and well thought-out criteria in all fields of security. The challenges to be faced by international security are today more serious and more complex than previously assumed and result from a consideration of the conflict categories existing today. The management of environmentally-induced crises and conflicts is one of the resulting task areas.

Most conflicts since the end of the Cold War differ from the type of conflicts on which the creation of the principle of collective security and the formulation of the military section of Chapter VII of the UN Charter were based: the wars of conquest and annihilation in the first half of our century. This is why the development of adequate and practicable procedures for conflict prevention stand at the center of conceptual efforts (Biermann 1995: 24).

This challenge has to be faced by all security organizations, not only the United Nations and the OSCE, but also NATO. This is not a question of competition or rivalry, but of the sensible splitting up of work to be done between the individual security institutions, where each institution is able to use its own particular strengths. The OSCE has advantages in the area of the protection of minorities, democratic institutions and human rights as well as in general observer and mediator activity. In view of the extremely low budget of the OSCE amounting to approx. US$ 30 million in 1996, an increase of expenditure in the aforementioned areas of preventive action of the OSCE seems efficient and promises a relatively favorable increase in security (Troebst 1997: 162).

This is correct for the global approach, but as a consequence raises the question with regard to the levels below this as to how conflict management can be further regionalized and transferred to the responsibility of regional security organizations outside Europe and North America.

As recent political reality has shown, NATO is practically alone in Europe in being able to fill the current vacuum in security policy action with its crisis management as described in its New Strategic Concept. Military means are often the prerequisite for making peaceful development possible.

The usefulness of military support in acute crises is finding increasing acceptance. Thus the United Nations High Commissioner for Refugees (UNHCR) in his report "The State of the World's Refugees" stated in 1995:

> As experience in countries such as Iraq, Somalia and former Yugoslavia has demonstrated, the armed forces possess an abundance of precisely those resources which are normally in shortest supply when a disaster strikes: fuel, communications facilities, building equipment, medicines, large stockpiles of off-the-shelf supplies and highly trained manpower. In addition, the military is noted for its 'can-do' mentality, hierarchical discipline and organizational skills, attributes which are extremely useful in the turmoil of an emergency.

In order to further increase the effectiveness of such crisis management the following should be laid down in an 'Agenda for Environment and Security' (e.g. in a suitable UN framework or – regionally – in the OSCE):

- identify regions where environmentally-induced conflicts are most likely to occur,
- identify when and how military and non-military solutions for the prevention of impending conflicts should come into effect.

The ongoing NATO Pilot Study on "Environment and Security in an International Context" could do a lot to prepare the ground for this.

In the transition phase until the implementation of such an institutionalized approach to conflict prevention and crisis management the actors in international security policy remain obliged to pursue a policy of pragmatism and moderation.

References

Baechler, Günther 1993: "Konflikt und Kooperation im Lichte globaler humanökologischer Transformation". *ENCOP Occasional Paper*, No. 5, Zürich / Bern.

Baechler, Günther 1996: *A Model of Environmentally Caused Violent Conflicts as a Contribution to EW*, working paper 1997.

Baechler, Günther et al. 1993: *Umweltzerstörung: Krieg oder Kooperation?* Münster: agenda.

Baudissin, Wolf 1969: Soldat für den Frieden, München

Biermann, Wolfgang 1995: "'Alte' UN-Peacekeeping-Prinzipien und 'neue' Konflikte", in: *Sicherheit und Frieden*, 1/95.

Brill, Heinz 1995: "Neue Dimensionen der Sicherheitspolitik", in: *Verteidigungsanalysen*, Ausgabe 18, Waldbröl: Amt für Studien und Übungen der Bundeswehr.

Byers, Bruce 1991: "Ecoregions, State Sovereignty and Conflict", in: *Bulletin of Peace Proposals*, Vol. 22 (1), 65-76.

Dabelko, Geoffrey D. 1996: "Conflict and the Environment", in: The Woodrow Wilson Center (Ed.): *Environmental Change and Security Project Report 1997*. Washington DC, 220–223.

Frankenberger, Klaus-Dieter 1995: "Entfesselte Welt" Frankfurter Allgemeine Zeitung, 66/1997.

Groß, Jürgen 1996: "NATO am Scheideweg?". *Wissenschaftliches Forum für Internationale Sicherheit*. Hamburg: Edition Temmen.

Haedrich, Martina and Ruf, Werner (Eds.) 1996: *Globale Krisen und europäische Verantwortung – Visionen für das 21. Jahrhundert*. Baden-Baden: Nomos.

Jaberg, Sabine 1997: "Unvermeidbare Gewalt? Chancen und Grenzen präventiver Friedenssicherung", in: *Friedensgutachten 1997*. Münster: LIT.

Maier, Gerhart 1993: *Sicherheitspolitik*. Bonn.

Mutz, Reinhard 1997: "Die NATO auf dem Weg zu einem hegemonialen Machtkartell", in: *Friedensgutachten 1997*. Münster: LIT.

Neuhold, Hanspeter 1997: "Optionen österreichischer Sicherheitspolitik", in: *Österreichische Militärzeitschrift*. 4/97.

Reiter, Erich 1997: "Die NATO-Reform entscheidet die eurasische Sicherheitsordnung", in: *Österreichische Militärzeitschrift*. 4/97.

Rohloff, Christoph 1996: "Frieden sichtbar machen! Konzeptionelle Überlegungen und empirische Befunde", in: Matthies, Volker (Ed.): *Frieden statt Krieg. Gelungene Aktionen der Friedenserhaltung und der Friedenssicherung 1945 – 1995*, Bonn.

Rühe, Volker 1995: *Amerika und Europa – gemeinsame Herausforderungen und gemeinsame Antworten.* Lecture at Georgetown University (held in English). Washington DC.

Rühle, Joachim; Höppner, Eberhard 1997: "Gefahr durch ethnische Konflikte", in: *Europäische Sicherheit*.1/97.

SIPRI *Yearbook 1997*, New York: Oxford University Press.

Sommer, Theo 1994: "Keiner will den Weltgendarmen spielen", in: *Die Zeit*, 25/1997.

Troebst, Stefan 1997: "Dicke Bretter, schwache Bohrer", in: Dieter Seghaas (Ed.): *Frieden machen.* Frankfurt am Main: Suhrkamp, 147–165.

Volmer, Ludger 1996: "Gemeinsame Sicherheit durch ökologisch-solidarische Strukturpolitik", in: *Vierteljahresschrift für Sicherheit und Frieden*. Baden-Baden: Nomos.

Wetter, Ernst 1994: *Militärische Zitate*. Frauenfeld: Huber & Co.

Part E:

Environment and Development Policy Approaches

Preventing Environmentally-Induced Conflicts Through Development Policy and International Environmental Policy

Sebastian Oberthür

1 Introduction[1]

Since the end of the 1980s, the political and scientific debate on the links between environment and security has been gaining pace. From the outset, one concern of this debate has been to identify the potentials available in certain policy domains to solve the corresponding problems. Both the security and foreign policy side on the one hand and the environment and development policy side on the other have stressed their specific capabilities and instruments (Mathews 1989; Myers 1989; Butts 1994 and 1996). On the environmental policy side, in particular, this has given rise to the fear of being counterproductively co-opted by the security policy domain (e.g. Daase 1992; Deudney 1992). In the past debate, which has largely revolved around the concept of 'environmental security', environmental policy protagonists have usually aimed at broadly upgrading environmental policy through linking it to the urgency implied by the concept of security (e.g. Mathews 1989; Myers 1989).

It is ultimately the indeterminacy and openness to many (also military) interpretations of the guiding concept of 'environmental security' that has fueled the fear of being co-opted by the security policy side. The various parties involved in the debate have imbued the term with very disparate contents. The present collection of essays casts the conceptual ballast of 'environmental security' overboard, preferring to frame the debate in more concrete terms: As a starting point, it is assumed that environmental problems can fundamentally contribute to the emergence of severe, violent conflict, and that this is one – if not the most important – link between environmental and security issues. On the basis of this understanding, it becomes possible to discuss the roles of environmental and security policy in concrete contexts.

---

1 I wish to particularly thank the editors and Kerstin Imbusch for comments on earlier versions of this paper.

One restriction to be noted here is that present research findings indicate that environmental problems can only be a source of conflict (at all) through their interplay with other factors. Thus, in the environment-security nexus as elsewhere, political strategies aimed at preventing violent conflicts can address quite different levels and factors (see Carius et al. 1997). However, in the environmental policy or sustainable development perspective chosen here, those solutions are of prime interest that, while focused on crisis prevention, tackle the underlying environmental problems. Furthermore, there are clear indications that environmental problems indeed harbor a *considerable* potential for conflict (Homer-Dixon 1994; Homer-Dixon 1995; Baechler 1996) which manifests itself indirectly through the exacerbation of conflict-relevant social factors such as poverty and uneven income distribution. Nonetheless, the link and its relevance continue to be a source of some controversy, and particularly lack final proof as regards the modern international environmental problems (climate change, transboundary air pollution, biodiversity loss etc.) (see for instance Sprinz and Brock in this volume).[2]

As in other areas in which scientific uncertainty prevails, it is expedient and of interest in the environment and security context, too, to consider in depth the available political options before a final proof of causality has actually been delivered. It is therefore assumed here that environmental problems and problems relating to common resources are indeed *relevant* as (one) source of severe conflict. This is what is meant here by the term *environmentally-induced conflict*. The present essay asks which role attaches (or can attach) to development policy and international environmental policy in preventing environmentally-induced conflicts.

The following section examines some fundamental issues concerning the relationship between the approaches offered by security policy on the one side and environmental and development policy on the other. Subsequently, the role of development policy in preventing/resolving environmentally-induced conflicts is discussed against the backdrop of the guiding vision of sustainable development (section 3). The core of the paper is section 4, which examines the opportunities for and limits to preventing conflicts by means of international environmental policy. First the capacity of international environmental policy to perform this function is explored. This is followed by an analysis of the limits to this capacity and a discussion of selected proposals for reforming international environmental policy.

2 In view of the structure of these problems (no direct damage, unclear structures of causation and affectedness), it would appear improbable that violent conflict arises directly *about* such environmental problems (see Daase 1992; Deudney 1992).

2 Environment and development policy in relation to security policy

The demarcation made here between environment and development policy on the one side and security policy on the other should by no means be taken to mean that the juxtaposed policy domains are mutually exclusive. The demarcation is rather based on the differing weight accorded to the policy domains, both in general and in relation to environmentally-induced conflicts. While environment and development policy have traditionally been 'low politics', security policy has always been viewed as 'high politics'.

At the same time, although environment and development policy are discussed here as one 'package', the distinctions between these two policy areas shall be kept in mind. Classic development policy traditionally centers on attaining social and economic goals. In this conceptualization of development policy, environmental aspects frequently have played and continue to play a subordinate role. The environmental damage caused not least by quantity effects has potentially been and continues to be viewed as inescapable. This is why development policy has often come to be viewed from the environmental policy perspective as a vehicle for the heightened consumption of environmental goods, while from the development policy perspective environmental policy has frequently been viewed as an impediment to development. This tense relationship between development and environmental policy was first manifested at the international level in the discussions that took place upon the occasion of the United Nations Conference on the Human Environment (UNCHE) in Stockholm in 1972 (on the relationship between environmental policy and development policy in general see e.g. Sutcliffe 1995).

In contrast to the past stand-off, it has become increasingly apparent in recent years that successful environmental and development policies condition each other and that there at least is a partial overlap between their goals. Thus it is precisely a successful development policy that creates the economic, administrative, technological and other capacities that form the basis for effective environmental policies (Prittwitz 1990; Prittwitz 1993; Jänicke 1997). Due to the many forms of poverty-related environmental damage, development policy effectively becomes, in part, environmental policy in the countries of the South. Conversely, if an 'unfettered' development process is pursued without integrating environmental aspects, environmental problems frequently create development barriers. This is why, today, environmental impacts are generally an important criterion in the appraisal of development assistance projects. Overly intensive, inappropriate farming methods can for example, lead to soil erosion and thus to reduced agricultural yields. In South-East Asia and particularly in China, development limits caused by acute air and water pollution are currently becoming manifest. Effective environmental policies thus become a precondition to successful economic and social development (see e.g. Wöhlcke 1987; Sutcliffe 1995).

The altered perception of the relationship between environmental policy and development policy, in which each area is dependent upon the other in order to attain its own goals, is encapsulated in the concept of 'sustainable development', as popularized by the Brundtland Commission in 1987 and elevated to the guiding vision of global environment and development policy by the Agenda 21 "Programme of action for sustainable development" adopted at the 1992 Rio Earth Summit.

It is with this understanding that, with reference to the environment-security nexus examined here, environmental and development policies are discussed jointly in their relation to security policy and to relevant areas of foreign policy (preventive diplomacy). Figure 1 illustrates the differing weights of the juxtaposed policy domains as a function of conflict intensity. Before the open manifestation of tensions and violence, environmental and development policies have particular weight, as their instruments can be used to address environmental problems that contribute to the emergence of violent conflict. Once the violence threshold has been passed, however, sustainable environmental and development policies both become almost impossible, as then the struggle for survival becomes dominant, leading to the absence of the necessary societal framework conditions for effective environmental and development policy measures. The latter generally lose their purpose under these circumstances as they are destroyed by armed conflict, for instance when railway links, solar installations or sewage plants are destroyed in combat.

The development policy field of humanitarian disaster relief, which comes to the fore in such disaster situations, forms the exception to this rule (see Eberwein 1997). However, emergency aid can neither be termed as prevention, nor does it form a core part of classic development policy, which rather aims "to set up long-term structures for sustainable, self-supporting development" (see Schmieg in this volume). The growth of the share of emergency aid in official development assistance budgets in both absolute and relative terms (see ibid.) thus reflects not so much a reorientation of development policy but rather its decline.

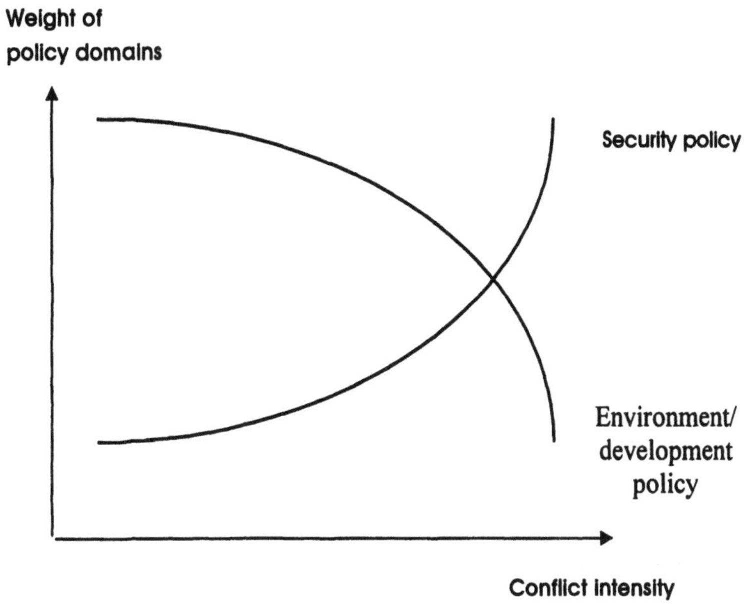

Figure 1: The weight of the environment/development policy and security policy domains as a function of conflict intensity

In an inverse relation to environment and development policy, the weight of security policy and of related areas of foreign policy rises with the intensity of environmentally-induced conflicts. Already at low environmentally-induced conflict intensities, security policy is by no means generally subordinated to environment and development policy. But it is only when tensions are manifest or even violence erupts that the whole range of security policy instruments – mediation, arbitration, peace-making, etc. – comes fully into play.

On the whole, in relation to environmentally-induced conflicts, the rather *proactive* orientation of environment and development policy can be distinguished from the rather *reactive* orientation of security policy. Security policy centers on resolving conflicts that already involve a substantial probability of violence. This is expressed by the corresponding scientific disciplines of peace and conflict research and security studies. The possible solutions examined in these disciplines generally assume the existence of a severe conflict. 'Conflict prevention' does not then mean mitigating the relevant environmental problem that contributed to causing the conflict, but preventing the deterioration of the conflict into violence (see e.g. Rohloff in this volume; Shaw 1996).

Environmental and development policy instruments, by contrast, have the greatest effect *before* an environmental problem has given rise to a severe conflict. Environmental policy options in particular refer directly to environmental problems as a specific source of conflict, and thus have a greater 'depth of effect' (on this concept, termed "Wirkungstiefe" in the German, see Prittwitz 1990: 54-58) and have a proactive or preventive thrust. The next section discusses the importance that attaches in this sense to sustainable development policy for preventing environmentally-induced conflicts.

3 Conflict prevention through sustainable development policy

As numerous as the definitions of 'sustainable development' are, as large are the differences in its conceptualization (see Pearce et al. 1989; Trzyna 1995). Nonetheless, a common core of the concept can be identified that contains a 'triad' of environment, economy and society. This is frequently depicted as a triangle of goals of sustainable development (see figure 2).

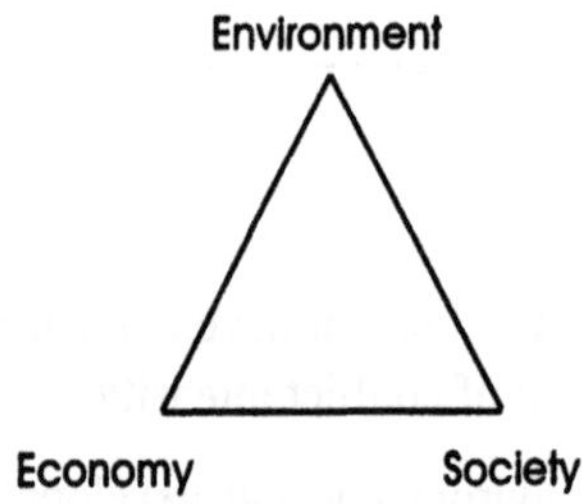

Figure 2: The triangle of goals of sustainable development

The three spheres of goals embraced by the concept of sustainability – environment, economy and society – have equal standing as a matter of principle. Ideally, this should bring about an environmentally sound integration of the three spheres. In accordance with Agenda 21, development policy would have to take environmental policy concerns into consideration in such a fashion that development policy causes of environmental problems are prevented from the outset, and such that development cooperation contributes to building capacities that can be utilized for environmental policy purposes (see Oodit and Simonis 1993: 27).

Indeed, recently environmental aspects have increasingly been integrated in the day-to-day workings of development cooperation. Today, environmental impact assessments are carried out in planned development assistance projects as a matter of routine. Moreover, a growing number of development cooperation projects is targeted directly at combating environmental problems and at improving the environmental situation in developing countries. The funding of climate protection measures and of climate-related capacity building in developing countries is no longer an exotic area of North-South cooperation.

Nonetheless, substantial room remains for the further 'greening' of development policy. Various authors have identified a trend towards a growing use of increasingly constrained development funds, particularly in financial cooperation, towards export promotion – and thus towards the pursuit of economic goals. This runs counter to the integration of environmental aspects in development policy. A further issue is that priorities in development cooperation appear to continue to depend greatly upon aspects relating to the maintenance of good relations between governments. In total, apart from emergency aid, which follows its own agenda as noted above, the concern is frequently raised that, despite all progress made, the practice of 'sustainable development assistance' continues to neglect the environmental aspect to the benefit of the economic and social aspects (e.g. Tisch and Wallace 1994).

In contrast, the security policy relevance of environmental problems would call for strengthening the environmental aspect in the triangle of goals of sustainable development. The demand raised in the relevant literature and in Agenda 21 (see above) to focus on building autonomous scientific, institutional, technological, economic and other problem-solving capacities continues to be valid. Moreover, the link between environment and security would suggest an even stronger conditionalization of development cooperation: This could increasingly be tied to the condition that minimum environmental standards are observed which facilitate conflict prevention and form the precondition for successfully mitigating security-relevant environmental hazards.[3]

Development cooperation could further contribute to preventing environmentally-induced conflicts by focusing on particularly relevant geographic and thematic areas. Here two different orientations are basically conceivable. With a *reactive* approach, economic and social development policy efforts could be increasingly undertaken with the goal of compensating for security-relevant environmental problems, in regions where these are particularly severe, by means of mitigating other societal conflict potentials (e.g. ethnic differences). Environmental problems never contribute directly and solely to the emergence of a conflict, but are always in connection with and mediated through social problems – such a compensation for the conflict potential of environmental problems might at least prevent the outbreak of violence over the short-term.

The goal of *prevention* pursued here would be better served by modifying the orientation of – both multilateral and bilateral – development cooperation with the aim of reducing and preventing the particularly security-relevant environmental problems themselves. The focus should be placed here on regions in which the emergence of environmental problems contributes to a particular degree to security hazards, and where there is a substantial potential to reduce this contribution. In

---

3 However, such a conditionalization can meet its limits: Potential recipient states can be located in priority regions in terms of environmentally-induced security hazards, but their previous environmental policy activities have not qualified them for development cooperation.

thematic terms, efforts would thus have to concentrate on mitigating environmental problems that possess a particular security hazard potential. If, for instance, it should become apparent that human-induced climate change harbors a particular danger of causing (or playing a role in causing) severe and violent conflicts, then (in addition to the urgently required curbing of greenhouse gas emissions in the industrialized countries) a purposeful political focus could be placed on low-emission energy generation and efficient energy consumption.

The elaboration of detailed guidelines for such a sustainable development policy that could develop a precautionary effect aimed at preventing security-relevant environmental problems requires precise knowledge of the potential security hazards posed directly and indirectly to certain regions by certain environmental problems. Here there is a considerable need for further research.

4 Opportunities and limits of conflict prevention through international environmental policy

International environmental policy – the transboundary control of international environmental problems – largely takes place within the framework set by international institutions. Institutions are defined here, following a social science understanding, as "recognized systems regulating the formulation and implementation of political intentions" (Prittwitz 1994: 82). The body of international institutions pertaining to the environment thus comprises both various international organizations such as the United Nations Environment Programme (UNEP) and so-called international environmental regimes (see Kilian 1987). The latter are based on international environmental agreements on certain issue areas, establish binding behavioral norms and decision-making procedures and generally continue to be developed in a lengthier ongoing negotiating process (see Gehring and Oberthür 1997).

The following section initially examines the outputs delivered by international environmental policy operating in an issue-focused manner primarily in the context of international environmental regimes. This is followed by a discussion of the limits to the capacity to provide these outputs, which also constrain the opportunities to solve security-relevant environmental problems through international environmental policy. Finally, against this background, the proposals are explored that have been put forward to further enhance the capacity of international environmental policy.

4.1 The outputs of international environmental policy

Despite all doubts, substantial progress has been made in international environmental policy over the last decades. This concerns both the agreement of wide-ranging measures and the implementation of agreed objectives for action. In many areas, environmental quality has improved as a result. Insofar as the progress made contributes to preventing environmentally-induced safety hazards, this can be viewed as the contribution of environmental policy to conflict prevention. However, it is only in the rarest of cases that the progress made to date has sufficed to solve the underlying environmental problems.

Well over 100 international environmental agreements have already been concluded. International organizations such as UNEP or the United Nations Economic Commission for Europe (UNECE) have frequently functioned as initiators of the negotiations preceding these agreements, and have provided consultative fora and supportive scientific information.

Environmental policy through environmental regimes is based on negotiations that precede the adoption of an international agreement and are generally continued afterwards in a process of ongoing refinement. These processes offer diverse opportunities to overcome the much cited 'lowest common denominator' of the participating states (see in detail Oberthür 1997; Gehring and Oberthür 1997).

Thus opportunities frequently arise in the course of these international negotiating processes to influence participating actors through the presentation of new scientific information and the formation of scientific consensus, thereby increasing the willingness to commit to vigorous environmental policy measures. An example of this is the Convention on the control of long-range transboundary air pollution in Europe: The exchange and international discussion of scientific information had a considerable impact on the development of this convention (Gehring 1997).

Moreover, far-reaching environmental protection agreements can often be achieved in international environmental policy through mutual *concessions* and *package deals* (see Scharpf 1992). Most international environmental agreements are thus compromises in which the participating governments find 'packaged solutions'. Such packaged solutions often succeed in linking the issues under negotiation in such a way that the important states see their essential interests satisfied. In the final outcome, the most reluctant actors have usually been drawn some way forward towards higher environmental standards. In order to achieve such outcomes, it has proven helpful to establish decision-making procedures that permit taking decisions by a majority and against the resistance of individual states. This has been possible in a number of environmental regimes.

Of late, progress has increasingly been made in international environmental policy through a differentiation of commitments in accordance with the differing responsibilities and capacities of the participating states. Prime examples of such a differentiation are given by the 1994 Oslo Protocol on the reduction of sulfur emissions in Europe and the 1997 Kyoto Protocol to the UN Framework

Convention on Climate Change (FCCC), which stipulate specific emissions reduction commitments for each state. Without such differentiation, the targets for action would presumably have been oriented to those parties who were the least interested or the least able to take action.

Financial compensation mechanisms are also playing a growing role, particularly in North-South relations. This is illustrated by the decision taken in 1990 to establish a Multilateral Fund for the implementation of the Montreal Protocol on Substances that Deplete the Ozone Layer, or by the international negotiations in the FCCC process and the Biodiversity Convention process. Such compensation mechanisms have in each case ensured the participation of the developing countries (Oberthür 1997; Henne 1997; Ott 1997). Compensation payments of this kind can be made between industrialized countries, too, and can contribute there to environmental protection. Thus under the Rhine River protection regime, the downstream states of Germany and the Netherlands have made earmarked payments to France as a compensation for the reduction of pollutant emissions from there (Bernauer and Moser 1997).

For several years, the intensified exertion of influence by and participation of major social groups has become an important element in international environmental policy. While states generally continue to reserve the right to take binding decisions, environmental and industry federations are playing an ever more important role. These not only frequently create a 'global public' in the various spheres, they also exert – via favorably inclined state representatives – a direct influence upon the decisions taken (e.g. Schmidt/Take 1997; for examples see the analyses in Gehring and Oberthür 1997).

Such a participation of social groups is also an important key to the successful implementation of international environmental policy, as interests crucial to implementation are thus already taken into consideration during the decision-making process. That noncompliance with internationally adopted environmental commitments seems to be rather the exception than the rule is at least in part due to this. In addition, there are a number of specific mechanisms in international environmental policy that support compliance with commitments.

A fundamental principle under international law in this context is that of *reciprocity* (see Keohane 1989: chapter 6): Fulfillment of one's own duties depends upon the partners also doing 'their' part. If states infringe upon an agreement, they risk that other governments also no longer take their commitments seriously. This reciprocity is frequently supported by specific reporting obligations, which serve to create transparency. It can be further strengthened through a joint review of such reports.

In addition, compliance with international environmental agreements is selectively supported by particular incentives, assistance and sanctions (see also Chayes and Chayes 1993). The Multilateral Fund for the Implementation of the Montreal Protocol and the Global Environment Facility (GEF) created in the early 1990s provide support in the global context to developing countries and former communist states with economies in transition to a market economy. A number of environmental regimes further provide for mechanisms aimed at capacity-building in certain countries and groups of countries.

However, the leeway to impose sanctions in order to enforce environmental measures remains strictly limited at the international level and depends upon the specific context in question. Under the Washington Convention on International Trade in Endangered Species of Wild Fauna and Flora (CITES), for instance, the contracting states have in various cases imposed import bans for protected species from states that failed to comply with the agreement (Sand 1997). There are similar examples in other issue areas.

In sum, it can be said that over the past decades the environmental situation has improved in a number of issue areas addressed by international environmental policy: The production and consumption of ozone-depleting substances has dropped considerably, as has long-range transboundary air pollution. Improvements have been brought about concerning oil pollution of the marine environment, waste disposal at sea, movements of hazardous wastes and various other internationally regulated environmental problems (see reports in Gehring and Oberthür 1997; Haas et al. 1993). Although these improvements are partially due to other factors (such as shifts in domestic policy priorities in various states), numerous studies have shown that international environmental regimes have made an important contribution (e.g. Oberthür 1997; Haas et al. 1993).

Moreover, the range of behavioral effects brought about by international environmental agreements has extended beyond the issues actually being regulated. Thus the industries affected by the ban on CFCs established by the Montreal Protocol reacted by developing substitutes and alternative technologies – although this was not explicitly required. In the final outcome, such and similar behavioral adaptations can, as was the case with the Montreal Protocol, subsequently facilitate tightening environmental regulations at the international level through actors having learned that ceasing behavior causing environmental problems is easier than it may appear at first (see on this feedback mechanism Oberthür 1997).

Nonetheless, the improvements made in the international environmental policy domain frequently fall behind what would truly be necessary. The depletion of the ozone layer has by no means been halted. Air pollution through sulfur dioxide has been reduced, but is still far above the level that can be borne by natural ecosystems. The list could be continued. As set out at the beginning of this article, the international environmental problems that continue to prevail may possibly contribute to the emergence of security hazards. Remarkable progress has been made in international environmental policy, but for a number of reasons, a sustained solution to the issues addressed has been found in the rarest cases. An important reason for this is to be sought in the structural limits to international environmental policy itself, which constrain its output. The following section examines these limits.

4.2 Limits to international environmental policy

The particular limits imposed upon the capacity of international environmental policy are mainly a result of the horizontal structure of the international system. In this system, there is no "shadow of hierarchy" (Scharpf 1991). Rather, the nation states, as basic units of international policy, are in principle sovereign and not subject to any central authority (see Milner 1991). This leads to three specific problems of international environmental policy that relate to (1) decision-making, (2) the implementation of decisions and (3) coordination (see also WBGU 1996: 70).

Decision-making in international environmental policy generally takes place through intergovernmental negotiations. Due to the traditional principle of sovereignty, which determines the structure of the system, the principle of consensus applies here, i.e. decisions must at least be passively tolerated by all parties if they are to gain binding force. Other than in nation states, dissenting minorities can scarcely be obliged to abide by majority decisions in the international system.

It has only been possible in the past to overcome the consensus principle to a very limited extent, for instance in connection with the 1987 Montreal Protocol on Substances that Deplete the Ozone Layer (see Benedick 1998; Oberthür 1997: chapter 3). But here, too, the introduction of innovative decision-making procedures required the prior consent of the states affected. In other areas, such as in international climate policy, the internationally prevailing consensus principle has at times even led to a decision-making blockade, as a number of states have reserved for themselves a veto right (see Quennet-Thielen 1996: 84; Ott 1997: 213). Although in some areas innovative acceleration mechanisms have been developed and applied, the formulation and adoption of decisions in international environmental policy do fundamentally suffer under the circumstance that the necessary negotiating processes are extremely lengthy and that in many cases major concessions need to be made to uncommitted parties in order to secure their collaboration (see Sand 1990).

A further problem that arises from the traditional understanding of the sovereignty principle is that of *implementation* (i.e. transposition and enforcement) of international environmental policies. Here, again, the international process is generally very lengthy: After the conclusion of an international environmental agreement, this must first be ratified by the nation states before it can enter into force (see Sands 1990). Subsequently, the nation states must transpose the international provisions into their national legal systems and must ensure that they are complied with (see Sands 1993). If a contracting party does not take the necessary measures or if these measures fail to secure compliance to the agreement, then, as set out above, coercive instruments and supportive measures for enforcing international commitments are only available to a very limited degree. It has albeit been variously pointed out that compliance with intrastate regulations by no means depends upon the exertion of constant coercion, either (e.g. Young 1992). Insofar, coercive instruments such as sanctions are not the *sine qua non* for attaining international environmental objectives. It can hardly be undisputed, however, that the broader scope to take coercive and supportive measures at the national level creates an improved framework within which to implement rules. Strengthening such options at the international level can therefore be expected to improve the likelihood of the objectives of international environmental policy being realized (see Downs et al. 1996).

Coordination, the third fundamental problem of international environmental policy, has only come to the fore in recent years. The rapid growth in the number of international environmental agreements over the past 20 years has led to their overlap expanding considerably. This has given rise to a growing need for coordination. There is a real danger of duplication and competency overlap. Thus the institutions and regimes that deliberate regulations of relevance to forest conservation include the International Tropical Timber Agreement (ITTA), the Food and Agricultural Organization of the United Nations (FAO), the 1992 Framework Convention on Climate Change (FCCC) and Convention on Biological Diversity (CBD) and the Intergovernmental Forum on Forests (IFF). This list alone gives an indication of the coordination problem. Moreover, the growing globalization of industry and trade is leading to growing conflicts arising between international environmental agreements and other regulatory domains, in particular that of the World Trade Organization (WTO).

5 Proposals for reforming international environmental policy

The proposals that have been put forward for addressing and overcoming the three fundamental problems of international environmental policy – which are usually joined in practice by a lack of political will to take vigorous measures – can be assigned, in a somewhat polarized form, to two categories: The fundamental restructuring of the system of institutions in international environmental policy on the one hand, and the gradual, incremental transformation of existing institutions on the other. In reality, a great number of intermediate forms are conceivable. In particular, fundamental restructuring could in fact be pursued by a sequence of small gradual steps (see WBGU 1996: 71–74). Here only the two extreme forms are distinguished, in order to outline the debate more sharply.

The goal of fundamentally *restructuring* the institutional structure of international environmental policy is above all pursued by proposals aiming at setting up a new UN institution that is responsible for environmental affairs and that has a greater weight and more competencies than the existing United Nations Environment Programme (UNEP). These proposals range from the German initiative to establish a 'world environment organization' outlined at the Special Session of the UN General Assembly (Rio+5) in June 1997 over suggestions to create a 'world environment council' at the level of the UN Economic and Social Council or even at the level of the UN Security Council, through to the proposition of an independent Global Environment Organization (GEO) with competencies in the environmental sector similar to those of the existing World Trade Organization (WTO) (see e.g. Esty 1994; Baechler et al. 1993).

All of these proposals aim at a greater centralization and channeling of international environmental policy. The objective is to give environmental policy a greater weight in international relations in general and in the United Nations system in particular. However, it has as yet remained unclear in these debates which structure, tasks and competencies could be assigned to a new global environmental institution, and how its specific relationships to the existing institutions is to be structured in detail. The clarification of these issues is crucial to any appraisal of the environmental effectiveness and political likelihood of realizing the proposals made. It is one of the tasks of future research on international environmental institutions.

Such sweeping proposals for the restructuring of international environmental policy are contrasted by proposals for a step-wise, *incremental transformation* of the existing system of institutions. These proposals include in particular suggestions aimed at introducing new procedural elements in the decision-making processes of international environmental institutions, such as expanding the scope for substantive discussion processes. A further suggestion concerns increasing the use of majority voting, for instance through 'double weighted' majority systems. Such a doubled weighted majority system, which requires an affirmative vote representing both the majority of all states and majorities of the industrialized and developing countries is already used under the Montreal Protocol (see Biermann 1996; WBGU 1996: 70). Implementational barriers can be tackled without fundamental restructuring through an improved monitoring of implementation and through the stronger introduction of measures for enforcing rules by means of sanctions and supportive measures within the existing institutional framework. The third prime problem of international environmental policy, the inadequate coordination among various policy spheres, could be overcome through intensifying, decentralized networking among existing institutions (with more intensive mutual monitoring and reporting among convention systems and organizations) and through transferring this task to the Commission on Sustainable Development (CSD) created in 1992 or to UNEP.[4]

In addition to these institutional alternatives, intensive debate has focused in recent years on a greater use of innovative, in particular market-based instruments such as ecotaxes and trading with emission certificates as a possible path to further progress in international environmental policy. The introduction of such instruments is hampered by the problems associated with decision-making and

---

4 Coordinating UN activities in the environmental sector is already one of the central tasks of UNEP. Due to its weak institutional position as an environmental programme, UNEP has not been able to do justice to this task in the past. A coordination of the various environmental agreements would moreover require an extension of UNEP's formal mandate (on the tasks and structure of UNEP see Kilian 1987).

implementation at the international level. International environmental policy has received a strong new impulse towards market-based mechanisms through the 1997 Kyoto Protocol to the Climate Convention, which provides for emissions trading among industrialized countries and for 'Joint Implementation' (JI) on both a North-North and North-South basis. Joint Implementation allows industrialized countries or individual companies to carry out CO_2 mitigation projects in Central and Eastern Europe or in developing countries, for which they receive emission credits that they can use to offset their own emissions reduction commitments. The debate on market-based instruments, which has given rise to considerable controversy for some years now suggests that their use can in principle enhance the efficiency and also the effectiveness of international environmental policies (see Simonis 1995).

The conclusion of an 'Agreement on the Law of Environmental Agreements' is an option for harmonizing advanced standards of decision-making, implementation monitoring and coordination among separate environmental institutions that has scarcely been discussed as yet. Such a step would be one level below an organizational centralization of international environmental policy, and could be understood as an extension of the Vienna Convention on the Law of Treaties of 1969. The latter provides a basic codification of the rights and duties of states as concerns international treaties. In a similar fashion, an agreement on the law of environmental agreements could stipulate basic minimum standards for the decision-making processes, implementation and coordination of environmental agreements.

Through this avenue it could for instance be agreed that, as a matter of principle, unless a treaty provides otherwise, decisions in the context of environmental agreements are adopted by a (double weighted) majority. Similarly, basic stipulations could be made concerning minimum standards of implementation monitoring (e.g. through independent advisory bodies) and for coordination among environmental institutions (such as a regular exchange of observers). However, before such a 'meta-agreement' can be concluded and applied, various issues of international law still need to be clarified. Nonetheless, a step could be taken forward by preparing corresponding guidelines that, while not binding under international law, may support the formation of a corresponding body of international custom. Examples discussed so far show that the options have not yet reached the limits of creative possibilities.

No comprehensive appraisal of the alternatives presented or of conceivable mixtures (with differing mixture ratios) can be offered here. In particular, both the political and scientific communities still need to conduct an in-depth debate on the objectives that should guide the quest for an optimum system of international environmental institutions. Possible, politically realizable paths and steps towards these objectives then need to be further discussed. It is however clear that the traditional understanding of sovereignty remains the prime barrier on the path towards a more effective structuring of international environmental policy – regardless of the objectives chosen. In order to solve the problems of international environmental policy, it would appear essential to cede to international institutions further rights that have in the past remained in the domain of state sovereignty. There is a need to clarify the question of how this could best be done – also from the aspect of a possible loss of ties to democratically legitimated national decision-making processes.

In the light of the potential of international environmental problems to generate security hazards, both the institutions and instruments of international environmental policy need to be designed as effectively as possible in order to prevent the outbreak of environmentally-induced conflicts. The security relevance of environmental problems provides a strong argument for doing so. Identifying the most effective design remains an internal question of environmental policy (analysis). Akin to the notion of sustainable development, the nexus that has emerged in the discourse on environment and security points to the necessity of examining economy, society and environment in an integrated perspective – for in conflict analysis, too, it has become clear that economic, social and environmental factors and their interrelations must be considered jointly if sustainable solutions are to be found.

It remains a matter for future research to expand the presently very limited knowledge of which environmental problems will give rise to the most serious security hazards, and where this will happen (see Carius et al. 1997). Concerning the thematic level, existing research has focused mainly on the degradation and scarcity of local and regional environmental goods (e.g. soil degradation, water scarcity) (see Baechler 1996: 61-117; Baechler in this volume). Further research efforts could be focused on determining the extent to which such resource scarcities are themselves conditioned – and will be exacerbated in the future – by more deep-seated environmental problems such as human-induced climate change. Concerning endangered regions, attention generally has concentrated on developing countries (see Baechler in this volume). This is partly because they are severely affected by environmental problems and because they generally have lower adaptation and management capacities compared to the industrialized countries. A more differentiated understanding of the regions most under threat by environmentally-induced conflict might be gained by applying the 'syndrome approach' developed by the German Advisory Council on Global Change (WBGU) to this issue (see Biermann in this volume).

6 Summary

Inasmuch as environmental problems harbor heightened security hazards, both international and national environmental policies and sustainable development policies contribute, by their very nature, to preventing conflict. In contrast to security policy, which develops its full importance when severe conflicts have already emerged and is thus more reactive in character, environmental policies and sustainable development strategies have a preventive quality – not only in terms of environmental issues, but in terms of security hazards, too.

International environmental policy, on which the present article focuses, and which operates largely in the context of international organizations such as UNEP or of international environmental regimes such as the UN Climate Convention process, has proven a remarkable capacity to improve environmental protection and thus already has a substantial potential to mitigate environmentally related security hazards. However, its effectiveness is particularly constrained by the traditional understanding of nation-state sovereignty that continues to dominate international relations. The outcome of this is that (1) on the basis of the prevailing principle of consensus, the adoption of binding decisions for the protection of the environment is hampered, as is (2) an effective implementation of decisions adopted. The system of international environmental institutions that has emerged under these conditions is (3) increasingly struggling with coordination deficits.

The available evidence provided on the security relevance of environmental problems confers particular urgency upon steps to structure international environmental policy more effectively. The answer to the question of which regions and which environmental problems should be focused upon from the 'environment and security' perspective depends upon the findings of studies that identify, in a comparative analysis, the regional exposure to environmentally-induced conflicts and the direct and indirect conflict potential of the various environmental problems.

Numerous proposals have been put forward for piece-by-piece or comprehensive reform and restructuring of the system of international environmental institutions. These have ranged from the creation of a 'world environment organization' down to the introduction of new procedural elements within specific environmental regimes. In this respect, there is a need to clarify the goals to be pursued and the steps to be taken to achieve conflict-preventing environmental policies that are as effective as possible. In any event, it appears necessary to gain a changed understanding of nation-state sovereignty that takes into account not only to the right to self-determination of each country, but also to that of the other members of the international community. The global environmental problematique calls this right to self-determination into question. Through which avenue international environmental policy can be reformed so as to strengthen international institutions while at the same time safeguarding its democratic legitimization is one of the major research tasks of environmental policy analysis.

References

Baechler, Günther et al. 1993: Umweltzerstörung: Krieg oder Kooperation? Ökologische Konflikte im internationalen System und Möglichkeiten der friedlichen Bearbeitung. Münster: agenda.

Baechler, Günther et al. 1996: Kriegsursache Umweltzerstörung: Ökologische Konflikte in der Dritten Welt und Wege ihrer friedlichen Bearbeitung. Zürich: Rüegger.

Benedick, Richard Elliot 1998: Ozone Diplomacy. New Directions in Safeguarding the Planet. 2nd edition. Cambridge: Harvard University Press.

Bernauer, Thomas and Peter Moser 1997: "Internationale Bemühungen zum Schutz des Rheins", in: Gehring, Thomas and Sebastian Oberthür (Eds.): Internationale Umweltregime. Umweltschutz durch Verhandlungen und Verträge, 147-163. Opladen: Leske+Budrich.

Biermann, Frank 1996: Financing Environmental Policies in the South: An Analysis of the Multilateral Fund and the Concept of "Full Incremental Costs". WZB Discussion Paper. No. FS II 96-406. Berlin: Wissenschaftszentrum Berlin für Sozialforschung.

Butts, Kent 1996: "National Security, the Environment and DOD", in: Environmental Change and Security Project Report, Issue 2, 22-27. Washington, DC: The Woodrow Wilson Center.

Butts, Kent 1994: "Why the Military is good for the Environment", in: Käkönen, Jyrki (Ed.): Green Security or Militarized Environment, 83-109. Aldershot: Dartmouth.

Carius, Alexander et al. 1997: "NATO CCMS Pilot Study: Environment and Security in an International Context", in: Environmental Change and Security Project Report, Issue 3, 55-65. Washington, DC: The Woodrow Wilson Center.

Chayes, Abram and Antonia Chandler Chayes 1993: "On Compliance", in: International Organization, 47/2, 175-205.

Daase, Christopher 1992: "Ökologische Sicherheit: Konzept oder Leerformel?", in: Meyer and Wellmann (Eds.): Umweltzerstörung: Kriegsfolge oder Kriegsursache, 21-52. Frankfurt am Main: Suhrkamp.

Deudney, Daniel 1992: "The Mirage of Eco-War: The Weak Relationship among Global Environmental Change, National Security and Interstate Violence", in: Rowlands, Ian H. and Malory Greene (Eds.): Global Environmental Change and International Relations, 169-191. Houndsmill: Macmillan.

Downs, George W. et al. 1996: "Is the Good News about Compliance Good News about Cooperation?" International Organization, 50/3, 379-406.

Eberwein, Wolf-Dieter 1997: Die Politik Humanitärer Hilfe: Im Spannungsfeld von Macht und Moral. WZB Discussion Paper No. P 97-301. Berlin: Wissenschaftszentrum Berlin für Sozialforschung.

Esty, D. C. 1994: Greening the GATT: Trade, Environment and the Future. Harlow Essex: Longman.

Gehring, Thomas 1997: "Das internationale Regime über weiträumige grenzüberschreitende Luftverschmutzung", in: Gehring, Thomas and Sebastian Oberthür (Eds.): Internationale Umweltregime. Umweltschutz durch Verhandlungen und Verträge, 45-62. Opladen: Leske+Budrich.

Gehring, Thomas and Sebastian Oberthür (Eds.) 1997: Internationale Umweltregime. Umweltschutz durch Verhandlungen und Verträge. Opladen: Leske+Budrich.

Haas, Peter M. et al. (Eds.) 1993: Institutions for the Earth. Sources of Effective International Environmental Protection. Cambridge: MIT Press.

Henne, Gudrun 1997: "Das Regime über die biologische Vielfalt von 1992", in: Gehring, Thomas and Sebastian Oberthür (Eds.): Internationale Umweltregime. Umweltschutz durch Verhandlungen und Verträge, 185-200. Opladen: Leske+Budrich.

Homer-Dixon, Thomas F. 1994: "Environmental Scarcities and Violent Conflict: Evidence from Cases", in: International Security, 19/1, 5-40.

Homer-Dixon, Thomas F. 1995: "On the Threshold: Environmental Changes as Causes of Acute Conflict", in: Lynn-Jones, Sean M. and Steven E. Miller (Eds.): Global Dangers, Changing Dimensions of International Securtiy, 43-83. Cambridge: MIT Press.

Jänicke, Martin 1997: "The Political System's Capacity for Environmental Policy", in: Jänicke, Martin and Helmut Weidner (Eds.): National Environmental Policies. A Comparative Study of Capacity-Building, 1-24. Berlin: Springer.

Keohane, Robert O. 1989: International Institutions and State Power. Essays in International Relations Theory. Boulder, Col.

Kilian, Michael 1987: Umweltschutz durch internationale Organisationen. Die Antwort des Völkerrechts auf die Krise der Umwelt? Berlin.

Mathews, Jessica Tuchman 1989: "Redefining Security", in: Foreign Affairs, 68/2, 162-177.

Milner, Helen 1991: "The Assumption of Anarchy in International Relations", in: Review of International Studies, 17/1, 67-85.

Myers, Norman 1989: "Environment and Security", in: Foreign Policy, 74, 23-41.

Oberthür, Sebastian 1997: Umweltschutz durch internationale Regime. Interessen, Verhandlungsprozesse, Wirkungen. Opladen: Leske+Budrich.

Oodit, Deonanan and Udo E. Simonis 1993: Poverty and Sustainable Development. WZB Discussion Paper No. FS II 93-401. Berlin: Science Center Berlin.

Ott, Hermann E. 1997: "Das internationale Regime zum Schutz des Klimas", in: Gehring, Thomas and Sebastian Oberthür (Eds.): Internationale Umweltregime. Umweltschutz durch Verhandlungen und Verträge, 201-218. Opladen: Leske+Budrich.

Pearce, David et al. 1989: A Blueprint for a Green Economy. London: Earthscan.

Prittwitz, Volker von 1990: Das Katastrophenparadox. Elemente einer Theorie der Umweltpolitik. Opladen: Leske+Budrich.

Prittwitz, Volker von 1993: "Katastrophenparadox und Handlungskapazität", in: Heritier, Adrienne (Ed.): Policy-Analyse. Kritik und Neuorientierung, 328-355. Opladen: Leske+Budrich.

Prittwitz, Volker von 1994: Politikanalyse. Opladen: Leske+Budrich.

Quennet-Thielen, Cornelia 1996: "Stand der internationalen Klimaverhandlungen nach dem Klimagipfel in Berlin", in: Brauch, Hans Günter (Ed.): Klimapolitik. Naturwissenschaftliche Grundlagen, internationale Regimebildung und Konflikte, ökonomische Analysen sowie nationale Problemerkennung und Politikumsetzung, 75-86. Berlin: Springer.

Sand, P. H. 1990: Lessons Learned in Global Environmental Governance. Washington D.C.: World Resources Institute.

Sand, Peter H. 1997: "Das Washingtoner Artenschutzabkommen (CITES) von 1973", in: Gehring, Thomas and Sebastian Oberthür (Eds.): Internationale Umweltregime. Umweltschutz durch Verhandlungen und Verträge, 165-184. Opladen: Leske+Budrich.

Sands, Philippe 1993: "Enforcing Environmental Security", in: Sands, Philippe (Ed.): Greening International Law, 50-64. London: Earthscan.

Scharpf, Fritz W. 1991: "Die Handlungsfähigkeit des Staates am Ende des zwanzigsten Jahrhunderts", in: Politische Vierteljahresschrift, 32/4, 621-634.

Scharpf, Fritz W. 1992: "Koordination durch Verhandlungssysteme: Analytische Konzepte und institutionelle Lösungen", in: Benz, Arthur et al. (Ed.): Horizontale Politikverflechtung. Zur Theorie von Verhandlungssystemen, 51-96. Frankfurt a.M.

Schmidt, Hilmar and Ingo Take 1997: More Democratic and Better? The Contribution of NGOs to the Democratization of International Politics and the Solution of Global Problems. Working Paper No. 7. World Society Research Group. Darmstadt: Technical University Darmstadt; Frankfurt: Johann Wolfgang Goethe University Frankfurt.

Shaw, Brian R. 1996: "When Are Environmental Issues Security Issues?", in: Environmental Change and Security Project Report, Issue 2, 39-44. Washington, DC: The Woodrow Wilson Center.

Simonis, Udo Ernst 1995: International handelbare Emissions-Zertifiakate. Zur Verknüpfung von Umweltschutz und Entwicklung. WZB Discussion Paper No. FS II 95-405. Berlin: Wissenschaftszentrum Berlin für Sozialforschung.

Sutcliffe, Bob 1995: "Development after Ecology", in: Bhashkar, V. and Andrew Glyn (Eds.): The North, the South and the Environment, 231-258. Tokyo: United Nations University Press.

Tisch, Sarah J. and Wallace, Michael B. (Eds.) 1994: Dilemmas of Development Assistance: The What, Why and Who of Foreign Aid. Boulder: Westview Press.

Trzyna, Thaddeus 1995: A Sustainable World. Defining and Measuring Sustainable Development. Sacramento: IUCN - The World Conservation Union.

WBGU (Wissenschaftlicher Beirat der Bundesregierung Globale Umweltveränderungen – German Advisory Council on Global Change) 1996: World in Transition: Ways Towards Global Environmental Solutions. Annual Report 1995. Berlin: Springer.

Wöhlcke, Manfred 1987: Umweltzerstörung in der Dritten Welt. Munich.

Young, Oran R. 1992: "The Effectiveness of International Institutions: Hard Cases and Critical Variables", in: Rosenau, James N. and Ernst-Otto Czempiel (Eds.): Governance without Government: Order and Change in World Politics, 160-194. Cambridge.

Development Cooperation as an Instrument of Crisis Prevention

Evita Schmieg

1 Background[1]

In recent years, social crises leading to civil war have broken out in many developing countries, frustrating decades of development efforts. Development funds are directed increasingly towards redressing the damage caused by wars, civil strife and natural disasters. Global spending on bilateral emergency aid increased between 1980 and 1995 from two percent of official development assistance (ODA) to ten percent of ODA. Since 1995, it has decreased only slightly. In essence, development funds have been withdrawn from their real task, namely to set up long-term structures for sustainable, self-supporting development.

More than 90 percent of the total 186 wars were fought between 1945 and 1994, were in economically emerging countries (Gantzel/Schlichte 1994: 10). During this time, the nature of conflict underwent considerable change. The 'classic' conflict between separate states is now no longer the most frequent conflict pattern; between 1989 and 1992, only three of 82 armed conflicts fell into this category (Amer 1994: 81). All others were internal conflicts with social, ethnic or religious backgrounds.

In some societies, conflicts and social crises clearly have to be understood as inherent to development. Events in Rwanda have shown in a particularly cruel fashion that decades of development efforts can be counteracted if basic social problems or potential conflicts are ignored. Development work designed to tackle the most important development problems thus faces the challenge of contributing to conflict prevention – as do foreign and security policy. The economic factor alone makes conflict prevention the first best solution: the costs of preventing a crisis are far lower than the costs of repairing war damage. And the humanitarian factor does not require further comment.

---

1 Large parts of this essay are based on Schmieg 1995 and BMZ aktuell 1997.

The costs of war

Human suffering: About 30 million people have died in wars since 1945; another 35.5 million people have become refugees. War, human rights violations and the systematic expulsion of communities has uprooted millions of people and destroyed their social structures.

Economic costs: The destruction of a material infrastructure, industrial production sites and the agricultural base (for example, by laying land mines) leads to a dramatic decline in agricultural production, scarcity of material goods, inflation and a serious decline in gross domestic product (GDP). In *East Timor*, agricultural production dropped to one-third during the first three years of war. Because of the conflict in *Kashmir*, the number of tourists went down from 722,000 in 1988 to only 10,000 in 1992. Foreign exchange income dropped from $200 million to nearly zero. Between 1980 and 1988, *Mozambique* suffered losses of four times its 1988 GDP; just repairing its road system would cost $600 million, or one-sixth of its GDP.

Ecological costs: Because of their irreversibility, it is almost impossible to estimate the ecological costs of war. For example, during the *Vietnam* War, 40 percent of the Vietnamese forest was irretrievably destroyed. It is nearly impossible to accurately measure the extent of damage caused by the explosion of oil wells in *Kuwait* and the bombing of oil production facilities in *Iraq* during the second Gulf War. In addition to environmental degradation caused directly by war, refugees' survival measures force them to use soil, water and forest in an ecologically damaging way (for example, war refugees clearcutting for firewood in the *Horn of Africa*).

Social effects: The social costs of war are enormous. The cost of carrying on a war is usually at the expense of public responsibilities such as education and health care. The results can be illiteracy, a lowering of job opportunities and a loss in work capabilities. Families, clans and communities are torn apart; post-war societies are characterized by a high population of orphans and uprooted and traumatized people. The emergence of black market economies and illicit trade networks (weapons, drugs, oil, diamonds, ivory, fine woods) further jeopardize inner stability. Moreover, the redistribution of land and capital due to criminal activities lead to alarming changes in social structure.

Sources: AKUF 1993; Debiel 1995: 5–15; Matthies 1994a: 141; SEF 1993: 186 f.: UNDP 1994, adapted from Wissing 1995.

The term 'ethnic' or religious conflict must be treated carefully. As a rule, the real root of such conflicts is a battle for resources or a process of nation-building. Differing languages and ethnic backgrounds or divergent religious beliefs are often used by the participants to create a target for hatred and pursue their own interests by stirring up conflict. It makes more sense to talk about ethno-political conflicts because, in general, it is only the politicization of ethnic features which gives them a key role in the conflict process (Ropers 1995: 197).

But this should not deny the existence of basic, often historically conditioned, social tensions. The dominance of one ethnic group over another or systematic repression can also be a latent cause of crisis. In other cases, partitions between conflicting parties can be in flux. Guerrilla leaders attract shifting alliances which further destabilize the social system. Often, the battle or war area is not clearly delineated geographically, either.

This is why military measures are hardly suitable for countering armed conflicts of this nature (the question of which conditions could justify military intervention will not be explored here). Joint action among development cooperation, foreign policy and security policy protagonists is called for to prevent conflicts from becoming crises or to respond adequately to crises. Different instruments in different political areas are needed during the separate phases of a conflict. In a situation that is still basically peaceful, development cooperation can appropriately address and defuse potential issues of conflict and avoid the emergence of a crisis. Once a conflict has escalated, preventive diplomacy and perhaps military intervention becomes necessary. If a conflict openly breaks out, development cooperation can be effective in providing emergency aid or, after the crisis has ended, long-term reconstruction.

2 Development cooperation approaches

2.1 Removing the sources of crisis

2.1.1 Alleviating poverty and reducing economic and social inequity

The goal of development cooperation is to improve people's living standards, in particular those of the poorer strata of society. Development efforts center on economic growth, social justice and ecological sustainability (BMZ 1996). Development cooperation also provides measures to combat poverty, to create income and safeguard the food supply help people secure their basic needs, and to reduce conflict potential. In this sense, development work helps to reduce the root causes of crisis.

An illustrative project: Development in northern Mali

The goal of this project is to promote the social and economic stabilization process in the northern region of Mali that was particularly hit hard by the Tuareg rebellion. Refugees and those who were expelled to other regions were assisted in rebuilding a life support base. Short-term emergency measures (i.e. helping people to return to their homes and providing food) were augmented by long-term development plans, especially in agriculture. Financial cooperation provided by Germany has helped reconstruct the destroyed infrastructure (water supply, schools, administration).

The project intended to strengthen the peacebuilding process by reactivating economic activities. The German effort, a part of its development cooperation program, was highly appreciated by the Mali population; peace was confirmed at a symbolic burning of weapons in the spring of 1996.

The project has been supported by 12.74 million DM in financial cooperation and 4.7 million DM in technical cooperation since 1993.

Source: German Federal Ministry for Economic Cooperation and Development (BMZ)

2.1.2 Using and preserving the natural bases of life

The use and preservation of the natural bases of life is highly conflict-laden. The availability of land, water and mineral resources combined with other factors is often the source of armed conflict. In the future, conflicts over fresh water will increase. Today up to fifty percent of the river water – or sometimes more – is used in the Middle East, North Africa and the Sahel zone. Fifty percent is a critical amount (Wöhlcke 1996: 40 ff.). A particularly serious problem arises from the fact that agricultural acreage per capita is decreasing. In industrial countries, it dropped from 0.6 to 0.5 hectares per person between 1955 and 1995; in the economically emerging countries it dropped from 0.5 to 0.2 hectares during the same time.

The Brundtland Commission (1987) and the UN Conference on Environment and Development (UNCED) (1992) both pointed out that environmental degradation is a cause of migration. The International Red Cross estimates that half a billion people are environmental refugees and that this will increase to one billion by the turn of the century (Wöhlcke 1997: 91). It is important to point out that most environmental migration is due to ecological damage caused by humans, often by the destruction of large biotopes and over-exploitation of biological resources (forest clearcutting, overgrazing, overfishing). In turn, migration is often a source of conflict in the country taking on the refugees.

Access and use of natural resources are the traditional arenas of development cooperation. Typical examples are projects to prevent erosion or provide water. In the agricultural sector, it is especially important that there be improved access for farmers to production factors such as better crop varieties with more reliable yields, plants, seeds, financial credit and marketing support.

An illustrative project: Erosion control in Rio Checua, Colombia

The goal of the project was to halt soil erosion in a designated area and enable agricultural activities that conserve the soil. Necessary investments and appropriate advice were made available. Financial investments covered the construction and sowing down of contour ditches, the planning of retention basins, the construction of stone walls for torrent control and the sowing down of erosion surfaces, slopes and starting zones of rockfall.

Investment in erosion control is only successful if farmers and other inhabitants of the threatened area are aware of the causes and effects of environmental degradation and employ sustainable and protective methods. Technical cooperation with the Colombian environmental authorities provided the intensive advice necessary.

The project succeeded in rehabilitating eroded areas and encouraging inhabitants to use protective cultivation methods. Drinking water quality also improved.

Source: BMZ: Zehnter Bericht zur Entwicklungspolitik der Bundesregierung, BT-Drucksache 13/3342, Bonn, pp. 142f.

2.1.3 Checking population growth

Population growth policy also plays an important role in preventing crises. Rapid population growth often increases competition for limited land, water and fire-wood resources. A high rate of population growth and environmental degradation can fuel each other in a vicious circle:

> If a community's resources are decreasing, more workers are needed to meet daily needs. Thus more children are brought into the world and more common property is exploited, thereby increasing pressure to enlarge the family. By the time this reciprocal escalation of fertility and environmental destruction is halted – whether through governmental measures or the dwindling advantage of having more children – sometimes millions of people can suffer deeper and deeper poverty (Dagupta 1996: 54).

Development cooperation pursues an integrated approach with population policy. Emphasis is placed on expanding family planning services and improving the population's economic and social situation, especially regarding health, education and women's position in society.

2.1.4 Encouraging regional integration

Regional integration in developing countries can significantly help reduce the crisis potential between countries. It can create not only economic and social bonds that have a stabilizing effect, but also fora for political discussions and constructive communication. The Organization of African Unity (OAU) and the Economic Community of West African States (ECOWAS), for example, have already taken steps in this direction. In the Burundi conflict, the OAU has taken on an important leading role; ECOWAS has been involved in advancing the peacebuilding process in Liberia.

Development cooperation can work to support integration at different levels:

- Direct support in formulating reciprocal commitments or treaties, especially through multilateral development cooperation. Here in particular the European Commission can provide important help based on its own experience.
- Direct support in creating administrative structures and processes enabling the implementation of peacebuilding measures.
- Support in converting regional agreements into national policy: formulating the required legislation and, in particular, providing practical assistance such as training customs officers and personnel for ministerial and administrative work.
- Support of international (including bilateral) projects, often to do with infrastructure. Areas such as education, youth exchange and cultural programs, however, may have priority as crisis prevention measures. Development work can create personal links and economic interests that transcend national borders. These are ultimately of central importance to peaceful development pathways.

2.2 Supporting societal mechanisms for resolving conflicts peacefully[2]

The previous section described how development cooperation can contribute to reducing root causes of conflict. It can also contribute to strengthening societal mechanisms that ensure conflicts are resolved peacefully.

2.2.1 Democratization, participation, civil society

A society that enables all individuals and groups to articulate their interests and that possesses the mechanisms to balance these interests, has the best conditions for peaceful coexistence. Supporting the processes of democratization and participation, and developing a civil society, should help to introduce such mechanisms and make it easier for people to use them.

The challenge of democratization cannot lie simply in transferring the Western system of democracy. The implementation of democratic systems requires access to information, citizen participation and acceptance of the system to a degree that does not exist in many cultures. In ethnically split societies especially, a multiple party system may inherently contain the risk that party platforms will coincide with ethnic differences, thus intensifying conflict (see the example of Niger discussed in

---

2 Large parts of this section are based on Schmieg 1995.

Dayak 1996: 94f.). Therefore, democracy must be seen in a historical, cultural and social context.

The development of democracy in Uganda

One example of an African model of democracy is Uganda. In the conventional sense it could be said that this model is not democratic because no opposition parties have yet been allowed. But there are elections that enable citizens to participate in political processes, starting at the lowest administrative levels. The goal is to awaken citizens' interest in their political rights and duties, for example by decentralizing administration at a regional level. Such a system can be more appropriate for avoiding conflict-laden social divisions in a society that is historically, ethnically or religiously very heterogeneous. Development cooperation donors therefore agree that it is necessary to support the political reform process in Uganda but not to insist on a multi-party system as a condition for cooperation. International observers even assume that a majority of the population would reject a multi-party system at this time.

Supporting democratization can be done in completely different ways. The most widespread form of direct assistance is in preparing and carrying out elections. But it is important to create participation structures at the local level and democratic consensus among the population.

In every society, there are groups, organizations and institutions that are particularly important for the democratization process. Any strategy must take this into account, but for long-lasting social stability it is very necessary that participation in the political process is guaranteed to all individuals and groups. Here, development cooperation has the task of supporting social groups that are marginalized or at risk. In most countries, these are women, ethnic/religious minorities, indigenous people, children, and the urban and rural poor. Often, non-governmental organizations (NGOs) support these groups (for example, in dealing with the legal system or the media, or in presenting their interests to administrative authorities). This approach is called "empowerment", which means that individuals' opportunities for effecting political and social changes are strengthened.

The importance of *the media* has been shown in an extremely negative way by the role of "Radio Milles Collines" in the Rwanda conflict. An independent and diverse media can, however, play a key role in raising awareness of important social developments and can function as a monitor. Radio and television become very significant in reaching illiterate people in rural areas. These media can contribute to defusing an impending crisis by broadcasting carefully directed information.

In addition to supporting institutions, NGOs or parties that enable broad participation in the political process, development cooperation can also encourage *dialog among different social groups*.

An illustrative project: The role of German foundations in transforming South Africa

During the apartheid era, German political foundations (Friedrich-Ebert-Stiftung, Konrad-Adenauer-Stiftung, Friedrich-Naumann-Stiftung and Hanns-Seidel-Stiftung) started social programs in South Africa with the support of the German Ministry for Economic Cooperation and Development (BMZ) in the early 1980s. They had several goals: spreading democratic values, training in democratic processes, encouraging decentralization and federalism, reforming the military, strengthening the unions and creating political parties.

Using conferences, seminars, training programs and advisory measures, the foundations addressed very different groups and organizations united in one common interest, namely to end the apartheid system and introduce democracy.

The specific instruments of these foundations and their independence from governmental organizations allowed them to influence the development of South African society under conditions which otherwise limited the options for government-financed development cooperation.

Promoting *local administration and decentralization* can also contribute significantly to participation and democratization. Setting up competent local authorities and units of self-administration is an important way in which large parts of the population can be involved in democratic processes.

The establishment of organized regional autonomy and self-administration dealing with legal, cultural and economic matters can significantly contribute to reducing potential conflicts, especially in multi-ethnic countries (see Scherrer 1995: 257 ff.).

Projects that support civil society, such as those described above, and that aim to improve social dialogue, are important approaches to conflict prevention. Support for civil society must proceed in step with the development of a political system. If support of civil society is accompanied by a growing rejection of the political system, this increasing disintegration will lead to political and social tension and instability. In this situation, weaknesses in the political system must be confronted.

A combination of governmental and NGO development cooperation is very suitable for lowering this risk. In a division of responsibility, NGOs can support civil society and governmental cooperation can support partner governments.

2.2.2 Supporting the legal system

An independent judicial system, equally accessible to the poor, in conjunction with transparency and reliability of the state as a lawmaker, constitutes a decisive precondition to the rule of law and human rights. If people trust their country's judicial system, a great deal of potential conflict does not arise. Approaches in this area include:

- Giving governmental advice on designing a constitution and laws;
- Supporting the judicial and administrative organs in applying laws: educating and training legal and administrative personnel, helping to equip courts and administrative offices, advising how to install administrative registers such as patent or land registries;
- Providing legal advice to enable individuals belonging to marginalized groups to have access to the judicial system. Included here are legal advisory and human rights centers, often for special groups such as rural women, indigenous people or inhabitants of slums.

The judicial system is a politically sensitive area that needs detailed analysis before the use of its instruments can be decided on. To make sure that undesirable power structures or old privileges are not cemented in, great care must be taken to follow a logical sequence of steps in making reforms. Giving advice on legislation may not lead to results if the partner government does not really attempt to establish a fair judicial system. Supporting the administration of justice makes no sense if the appropriate legislation is missing. Assistance in equipping courts can strengthen a system that is not constitutional if there is no guarantee that all citizens are equal before the law or if marginalized groups are not able to use courts for protection or enforcement of their rights.

Of great importance is how the governments of partner countries and administrators of justice (judges, lawyers and police) regard reform. If it is clear that reforms are not truly desired, as they are often combined with a loss of power, then it is more effective to support pressure groups that seek democracy and participation as described above.

The evaluation of projects to support the judicial system has shown that they have failed in those countries where civil society did not call for a functioning system of justice. In countries that were politically more developed, had a large NGO culture and a strong civil society, they were successful. Experts point out that clear results of judicial reform only can be expected after a very long time – 15 to 20 years (Heinz 1994).

A stable legal system can be created in multi-ethnic societies only if a balance is found between minority protection and loyalty to a national legal system (see essay by Dicke 1995). In most Latin American, African and Asian countries, however, emphasis was on nation-building rather than on institutionalizing minority protection (Ropers and Debiel 1995: 26).

2.2.3 The role of the military and police

Like a judicial system, a police system that functions properly can contribute to social stability. But in many developing countries, the military and the police are a source of social instability. They are often the very ones who violate human rights.

There is a high risk, especially after a civil war, that former members of military groups form armed bands that pose a new threat to peace. Crisis prevention therefore demands close examination of the military's role.

German development cooperation is not involved in supporting the police. However, the issues of *overarming and the position of the military in society* have become important for development cooperation. The share of state military expenditure is one of the indicators of a government's development orientation, and has been one of the criteria for the type and volume of development cooperation since 1992 (see above). On average, developing countries spend 13 percent of their national budgets ($57 billion per year) on basic needs; this is about half of their military expenditure (about $125 billion per year) (UNDP 1994:7). So there is room here for redistributing a national budget in favor of development.

It makes sense to use the issues of military expenditure and the role of the military not only as criteria for the allocation of development aid but also as areas where definite action can be taken. Instead of imposing sanctions for overarming, development cooperation now tries to support a rational security policy that contributes to national and regional stability.

An important area of support is in *demobilization.* Particularly in Africa, an effort is being made to integrate former soldiers and combatants into civilian society. This not only cuts down expenditure and reduces the military to a normal size; most importantly, it prevents former soldiers, whose profession was war, from forming bands and contributing to social instability. Cooperation projects that create employment, particularly in rural areas, or projects providing funds to start up small businesses, can ease the integration of former soldiers considerably. German cooperation projects in Uganda, Ethiopia, El Salvador, Nicaragua and Eritrea have already gained experience in this area.

Apart from the above examples, in which the significance of development cooperation lies in maintaining peace and stability and preventing new crises once an open conflict has ended, cooperation can be active with other instruments regarding the role of the military. An important area is stabilization by "civilianizing" and democratizing the military. Defense ministries and military groups must be under civil and parliamentary control. Military spending must be disclosed and also under public control. To do this, expertise must be developed in military and security issues within governments, parliaments and administrations so that the role, mission and appropriate defense spending of the armed forces can be intelligently decided upon.

Possible instruments may include education and training (seminars, informal conferences) and administrative support.

3 Crisis prevention in German development cooperation

Development cooperation is already active in many of the areas described above. Because violent conflicts are a basic problem for development in many countries, the question arises whether, for these countries, crisis prevention should not be the priority area for cooperation. The German Ministry for Economic Cooperation and Development (BMZ) is now working on methods to make cooperation more effective in preventing crises:

- Instruments of cooperation are reviewed to better recognize potential crises at an early stage.
- Crisis prevention will be integrated in existing concepts and procedures and in ongoing and future development projects. Country reports, concepts and studies should be expanded to include crisis prevention features. There should be an exchange of views on potential conflicts and possible solutions during the country talks carried out regularly by the BMZ together with representatives of development organizations and business.
- The effect of development projects on different social groups and their conflicts should be better understood. This is necessary for all institutions, including NGOs, that are active in cooperation, especially where emergency relief and aid to refugees is involved. The BMZ recently concluded an evaluation on this issue.

Indicators of crisis in a country or region need to be defined more precisely. Research on this is under way.

4 The role of NGOs

The description of possible areas of cooperation able to prevent conflict shows that NGOs play a central role, both those based in donor countries and those in the recipient countries. The professionalism of NGOs is very important, especially in crisis situations. NGOs based in donor countries have gained essential experience in promoting democratization, a civil society and participation. This experience must continue to be used. NGOs take an important role in shaping these social areas in recipient countries. It must be noted, however, that NGOs in emerging countries are not necessarily democratic or participative, as they are often dominated by members of the elite and are in some cases managed autocratically. This should not rule out cooperation, but must be taken into account.

NGOs can play a very useful role in implementing crisis prevention programs. International NGOs and institutions – such as the International Federation of Jurists, the Arab Human Rights Institute, the Interamerican Institute of Human

Rights in Costa Rica (Instituto Interamericano des Derechos Humanos), International Alert and others – can encourage debate on important issues, even in authoritarian regimes. A sub-regional or regional approach can make it easier to discuss sensitive issues such as democratization and human rights wherever national efforts aren't possible because of political oppression.

Ethno-political conflicts in particular are often deeply rooted in the society. In these situations, cooperation via NGOs has comparative advantages, as NGOs often have better access to different parts of society than state institutions do (see Ropers 1995: 217).

5 Conditions for development cooperation in support of crisis prevention

Development cooperation as a contribution to crisis prevention is particularly demanding with regard to its effectiveness and sustainability. The effectiveness of projects that support democracy, human rights, participation and systems of justice is very dependent on the political situation. If a government does not feel obliged to uphold the goals of a project, success cannot be expected for the measures taken, especially in the area of good governance and crisis prevention; support must then be limited to the activities of certain NGOs, independent media and groups adversely affected.

Measures in these areas must be supported by a coherent policy towards the partner country. The areas of cooperation should tackle the obvious deficits in a country's political and economic framework conditions; financial incentives should support the implementation of reforms. On the other hand, it must be credible to a partner country that unsatisfactory efforts on its part will threaten cooperation. In extreme cases, the donor must consider reaching mutual agreement on 'political' conditionality directed towards relaxing the problems underlying intra-state conflicts. Such a consistent and coherent strategy is possible only if domestic political developments in a partner country are observed more closely.

Regular communication on experiences must be sought with other donor countries. Common assessment of a political situation and the coordination of activities and methods are indispensable if broad effectiveness of cooperation is aimed at. It is especially necessary that national and international donors agree on evaluation standards for determining if a government is committed to reform and development.

The political arenas – development cooperation, foreign policy and security policy – must work together closely to make sure that appropriate instruments are used in a coordinated way. Active diplomacy, which can reduce conflict in peaceful ways through negotiation, mediation, or arbitration (as cited in Article 33 of the UN Charter), is especially important here. (For example, in the case of the UN, it is remarkable that the UN has relatively few resources for preventive diplomacy available. There are currently only 40 UN employees who have tasks that are directly associated with preventive diplomacy). Of course, in some situations the use of military force is a necessary option, although major efforts should concentrate on the peaceful resolution of a crisis before open conflict breaks out.

International discussion on the possibility of using development cooperation to prevent conflict has been led primarily by a working group of the OECD Development Assistance Committee (DAC). The debate demonstrates broad interest in the issue and a heightened consciousness that development cooperation can reduce conflict only if national and international actors reach agreement on the strategies and political methods (cooperation, diplomacy, security policy, economic sanctions where necessary) they use to pursue them. It can be hoped that this encouraging start to theoretical discussion will in practice lead to greater consensus among the various donors and political actors.

References

Amer, R. et al.. 1994: "Major Armed Conflict", in: *SIPRI Yearbook*. Oxford.

Ball, Nicole 1993: "Development Aid for Military Reform: A Pathway to Peace", in: *Policy Focus*, Issue 6. Washington, D.C.: Overseas Development Council.

Berthélemy, McNamara and Sen 1994: "The Disarmament Dividend: Challenges for Development Policy". OECD Development Centre. *Policy Brief*, No. 8. Paris.

BMZ – Bundesministerium für wirtschaftliche Zusammenarbeit und Entwicklung 1996: *Entwicklungspolitisches Konzept*. Bonn.

BMZ – Bundesministerium für wirtschaftliche Zusammenarbeit und Entwicklung 1997: *Entwicklungszusammenarbeit und Krisenvorbeugung*. Bonn.

Dagupta, Partha S. 1996: "Bevölkerungswachstum, Armut und Umwelt", in: *Spektrum der Wissenschaft*, Dossier 3, 50-55.

Dayak, Mano 1996: *Tuareg-Tragödie*. Unkel, Rhein.

Dicke, Klaus 1995: "Konfliktprävention durch Minderheitenschutz", in: Ropers, N. and Thomas Debiel (Eds.) 1995, 233–256.

Gantzel, Klaus-Jürgen and Schlichte 1994: *Das Kriegsgeschehen 1993*. Arbeitsgemeinschaft Kriegsursachenforschung an der Universität Hamburg. Bonn.

Goor, L.L.P. van de 1994: *Conflict and Development: The Causes of Conflict in Developing Countries*. The Hague.

Heinz, Wolfgang S. 1994: "Positive Maßnahmen zur Förderung von Demokratie und Menschenrechten als Aufgabe der Entwicklungszusammenarbeit", in: Deutsches Institut für Entwicklungspolitik. *Berichte und Gutachten*, Heft 4.

Heinz, Lingnau and Waller 1994: *Evaluation of EC Positive Measures in Favour of Human Rights and Democracy*. Berlin: Deutsches Institut für Entwicklungspolitik.

Homer-Dixon, Thomas F., Jeffrey H. Boutwell and George W. Rathjens 1996: "Politische Konflikte durch verschärfte Umweltprobleme", in: *Spektrum der Wissenschaft*, Heft 3, 112-119.

OECD, DAC 1997: *Bericht Entwicklungszusammenarbeit 1996*. Paris.

Ropers, N. 1995: "Die friedliche Bearbeitung ethno-politischer Konflikte", in: Ropers, N. and Thomas Debiel (Eds.) 1995.

Ropers, N. and Thomas Debiel (Eds.) 1995: *Friedliche Konfliktbearbeitung in der Staaten- und Gesellschaftswelt*. Bonn.

Scherrer, Christian P. 1995: "Selbstbestimmung statt Fremdherrschaft: Sezessions- und Autonomieregelungen als Wege zur konstruktiven Bearbeitung ethno-nationaler Konflikte", in: Ropers, N. and Thomas Debiel (Eds.) 1995, 257–283.

Schmieg, Evita 1995: "Entwicklungszuammenarbeit als Beitrag zur Krisenprävention?" in: *Peaceful Settlement of Conflict as a Joint Task for International Organisations*. Governments and Civil Society, 3rd International Workshop, 22–24.03.1995. Evangelische Akademie Loccum.

Tomuschat, Christian 1994: "Ein neues Modell der Friedenssicherung tut not", in: *Europa Archiv*, 24. Folge, 25. Dezember.

UNDP 1994: *Human Development Report 1994*. New York.

Wissing, Thomas 1995: *Mögliche Beiträge der Entwicklungszusammenarbeit zur Krisenprävention*. Berlin.

Wöhlcke, Manfred 1996: *Sicherheitsrisiken aus Umweltveränderungen*. SWP Arbeitspapier 2977. Ebenhausen: Stiftung Wissenschaft und Politik.

Wöhlcke, Manfred 1997: *Ökologische Sicherheit – Politisches Management von Umweltkonflikten*. SWP – S417. Ebenhausen: Stiftung Wissenschaft und Politik.

Should Security Policy Lay a New Foundation for Determining International Environmental Policy?

Sascha Müller-Kraenner

This report concerns the issue of "environment and security" as seen by a representative of environmental non-governmental organizations (NGOs). It centers on the questions of how NGOs respond to and deal with these issues, which approaches to problem-solving are preferable when ecological changes that could be relevant to security are involved, and what specific role non-governmental organizations play in identifying and implementing these approaches. It reflects the personal opinion of the author and should not be confused with any official NGO positions.

Security policy factors are increasingly drawn in to support international environmental policy.

The debate on the security effect of environmental problems and ecologically induced conflicts has been taken up in recent years by non-governmental organizations. Although the number and diversity of environmental NGOs and their work mean that they have not agreed on a single official position, they have a common approach accompanied by a common interest. Since this interest concentrates on solving the underlying ecological problem that has led to a conflict, discussion begins with the issue of resource degradation and not with what its possible effect on security could be. Security problems that arise because natural resources have been overused support the thesis that overuse is not ecologically, economically or socially sustainable.

This is the basis for understanding the concept that a link between environmental and security issues, as seen by NGOs, is a modern foundation for environmental policy, one that incorporates the precautionary principle and avoids ecological degradation. The search for a new conceptual basis is best understood in the light of environmental policy that has taken a back seat to traditional growth policy at a time when economic recession has affected many western industrial nations. Integrating environmental policy into other political areas is therefore an attempt to put the supposedly "soft" issue of environmental protection into the older, "harder" arena of politics and thus enhance its value. Linked security arguments can underline the urgent need to integrate sustainability with other political priorities and the plans of international institutes. It promotes better international cooperation on environmental issues as well. Statements made by different political actors about environmental problems that could be relevant to security are interpreted in this sense. In a speech before a special session of the UN General Assembly to analyze and evaluate the implementation of Agenda 21, made on 23 June 1997, German Chancellor Helmut Kohl linked the erosion of natural resources to the maintenance of peace:

> The crucial question is: how can we permanently safeguard the natural basis of life for a growing world population? In view of shrinking reserves of potable water, possible climate changes and the spread of deserts, this question becomes all the more urgent. We have no more time to lose. If we fail to take up this challenge, conflicts over natural resources will become even more likely (Kohl: 1997/1).

From the NGOs' perspective, this link can serve to further support international environmental protection activities.

The integration of sustainable development with foreign and security policy can prevent environmental degradation.

From an NGO point of view, the debate on environment and security seeks a solution to an environmental problem that may have induced the conflict in question. To resolve or at least prevent ecologically induced conflict, environmental NGOs discuss measures such as the consolidation of global environmental and development laws, enforced by effective institutions. It appears doubtful whether the already severely strained institutions and agreements of the process following the Earth Summit in Rio are suitable. The balance between ecological and development interests within Agenda 21 and the Climate Convention is already difficult enough to make it unseemly to attempt integrating other political interests such as security questions.

But one clear possibility for the integration of sustainable development with foreign and security policy significance, from the NGO point of view, lies within the European Union. With the anchoring of the goal of sustainable development in Article B of the EU Treaty, the task will be to achieve a new common foreign and security policy in the coming years.

Before the Amsterdam Treaty was adopted, a think tank chosen by governments and heads of state sought to define an expanded concept of security, as member states themselves sought, that would either explicitly integrate the principle of sustainable development or the ecological concerns outlined in the CFSP (Common Foreign and Security Policy) article of the treaty.

Several conferences on the issue[1] have concluded that incorporating ecological priorities in the EU's foreign relations is sensible and necessary. This necessity is seen because of the EU's geographic proximity to possible ecological crisis areas. The call for anchoring ecological interests and promoting sustainable development in foreign and security policy is grounded in the growing importance of environmental policy as an element of cooperation in the international arena.

> Next to foreign trade and security policy, environmental policy is developing appreciably as the third pillar of international cooperation. Between environment and development, environment and trade, and environment and security interests, there are intersections that are visibly gaining in importance. Therefore, the goal of sustainable development should also be anchored in CFSP. Among other things, it should be the CFSP's task to integrate environmental policy with other CFSP concerns, to properly consider the ecological roots of security questions (...) and to insist on compliance and implementation of international environmental law (DNR 1996).

One thing must be remembered in calling for the integration of sustainable development or for concrete ecological tasks for CFSP member states to fulfill, which is that common foreign and security policy is not a joint political arena but rather part of governments' cooperation. Due to different treaty agreements and institutional bases, it appears that "the integration of EC policies [such as environmental policy] in CFSP is very difficult and not appropriate (Ecologic 1996).

NGOs can contribute to avoiding conflicts by providing information and raising public awareness of the risks of possible environmental degradation

NGOs are motivated to political activity because they want to see actual environmental problems solved and not because they have broad social or comprehensive solutions to offer. Therefore, the role of the NGO is limited to drawing attention to the links between protection of the environment and other problems that affect society, such as security questions. Reference to these links can make demands to have environmental needs met more powerful, and in some cases it offers innovative solutions for other problems. One such innovative solution would be if environmental cooperation programs between different states could be used to develop security cooperation structures.

In addition to pointing out problems, environmental NGOs also have the task of taking on some responsibility for monitoring implementation of environmental

---

1 See SWP 1996; the conference of the Deutschen Naturschutzrings (DNR) (a German nature protection association) and the Evangelischen Akademie Iserlohn (Iserlohn Evangelical Academy) on "Safeguarding the natural basis of life - a cross-sectional task for foreign-, trade- and security policy of the EU" in October 1996, and the Bundestag Committee on Environment and Europe hearing on 11 November 1996 in Bonn on the role of European environmental policy within CFSP as well as its role in foreign trade and development policy.

regimes and institutions. However, NGO participation in such national and international monitoring processes quickly reaches its limits in financial and personnel terms. For this reason, NGOs must establish priorities, as they do when participating in national and international political processes. An especially important task for NGOs based in wealthy western industrial nations is to make sure that NGOs based in eastern European states undergoing transformation and nations of the South are included equally in international negotiation and monitoring processes.

References

Deutscher Naturschutzring (DNR) 1996: *Stellungnahme des Deutschen Naturschutzring (DNR) zur gemeinsamen Anhörung des Europa- und Umweltausschusses am 11. November 1996 zum Thema "EU-Umweltpolitik vor und nach Maastricht II"*. Bonn: Deutscher Bundestag.

Ecologic 1996: *EU-Umweltpolitik vor und nach Maastricht II. Stellungnahme zur gemeinsamen Anhörung des Europa- und des Umweltausschusss des Deutschen Bundestages.* Bonn: Deutscher Bundestag.

Kohl, Helmut 1977: *Rede von Dr. Helmut Kohl vor der Sondersitzung der Vereinten Nationen.* 23 June 1997, New York.

Environment and Security in the Context of the Debate on UN Reform

Irene Freudenschuß-Reichl

1 The UN system's capabilities and limits

What role do the United Nations today play at the point where environment and security intersect? Can reform of the United Nations help strengthen its role in this issue? What need is there for institutional action to effectively ensure that environmental problems do not become a threat to security? This discussion paper aims to go into these questions in more detail. Its perspective is that of a practician who, as a representative of her country, has for years been involved in efforts to reform the UN in diverse UN bodies.

1.1 The UN is an international, not a supranational organization

Discussions on United Nations reform are apt to overlook the fact that the United Nations is an international organization in the classical mold, not a supranational organization. Its Member States accordingly bear full responsibility for effectively arriving at and implementing decisions.

The fact that arriving at decisions is often arduous, and implementing them often slow and less than adequate, is a direct consequence of the nature of the United Nations as an international organization. The UN cannot be anything other than a faithful reflection of the actual differences in interests of its Member States.

The United Nations is thus as a rule very successful as a global platform for discussion in developing new ideas and ideas which have been well enough prepared. But the UN is often excruciatingly slow at implementing actual measures which affect the interests of groups of states.

1.2 The UN is a small but complex organization weak in resources

The discussion on reform which has been going on for years has seen the media bestow on the United Nations the dubious reputation of being a huge, rampantly growing bureaucracy. In reality the UN has a small staff in every sense. In 1990 a total of 3,265 posts for officials were approved for the UN Secretariat in New York, Geneva and Vienna, and for the five Regional Commissions and other central units (Childers and Urquhart 1994: 27). This means the UN makes do with fewer personnel than the Winnipeg city administration in Canada or the advertising agency of Saatchi & Saatchi.

The varying significance Member States ascribe to the United Nations' different fields of activity is very clearly reflected in its staff appointments. The International Court of Justice in The Hague, for example, has to make do with 15 judges and four posts for officials. The International Atomic Energy Agency, on the other hand, has been assigned 684 professional posts by the nuclear nations.

The United Nations budget is also relatively small. The regular budget including that for peace-keeping operations is approximately US $ 4 billion per year, making it smaller than the costs of fire services and police in New York City.

Also significant for the status the community of states ascribes to the United Nations' different fields of activity is its expenditure seen in relation to the world's population.

Via the United Nations the community of states annually spends approximately $ 0.14 per inhabitant of the earth on development, as much as $ 0.25 per capita on peace-keeping measures, and a full $ 0.49 on humanitarian aid. This means more and more of the UN's financial resources are going into short-term disaster aid and are thus lacking for long-term socio-economic and political development which might help prevent civil war or conflicts between states (Kamal 1996; Jenks and Neufing 1996).

With little willingness to exercise preventive solidarity through long-term aid at present existing, disasters very much 'help' mobilize financial resources in the short-term (Childers and Urquhart 1994: 144).

1.3 The UN's twofold objective

The United Nations has a twofold goal – maintaining peace and international security on the one hand and international cooperation for the common good on the other. The United Nations Charter formulates the organization's main aims in Article 1:

> The purposes of the United Nations are:
>
> - To maintain international peace and security, and to that end: to take effective collective measures for the prevention and removal of threats to the peace, and for the suppression of acts of aggression or other breaches of the peace, and to bring about by peaceful means, and in conformity with the principles of justice and international law, adjustment or settlement of international disputes or situations which might lead to a breach of the peace;
> - To develop friendly relations among nations based on respect for the principle of equal rights and self-determination of peoples, and to take other appropriate measures to strengthen universal peace;
> - To achieve international cooperation in solving international problems of an economic, social, cultural, or humanitarian character, and in promoting and encouraging respect for human rights and for fundamental freedoms for all without distinction as to race, sex, language, or religion; and
> - To be a center for harmonizing the actions of nations in the attainment of these common ends.

Depending on the goals to be pursued, the UN Charter ascribes the main responsibility for realizing them to one or other of its organs:

- Articles 33 and 34 declare the responsibility of the Security Council and General Assembly in disputes, while Articles 11 and 12 establish the precedence of the Security Council over the General Assembly; further to which, in the event of threats, the provisions of Chapter VII of the Charter gain ground, whereby the Security Council may impose sanctions;
- Cooperation in economic and social matters – the UN Charter makes no explicit mention of environmental matters – is further dealt with in Article 55; the Economic and Social Council (ECOSOC), which is under the overall authority of the General Assembly, has the main responsibility for achieving goals in this area.

With the 'one country, one vote' system, developing countries are in the majority when votes are taken in ECOSOC and the General Assembly; but, as is well known, nothing can become accepted if it is against the will of one of the Security Council's permanent members.

The uneven situation concerning these institutions is significant when it comes to questions of 'environment and security'. At least where the terms are concerned, the 'environment and security' nexus draws environmental issues into the jurisdictional orbit of the Security Council. This means that the option of sanctions as set forth in Chapter VII of the Charter is theoretically conceivable. This alone may be sufficient to produce fierce resistance on the part of the Group of 77, who systematically oppose everything they see as restricting their sovereignty in economic and social matters.

1.4 The UN at odds with Bretton Woods institutions and the ,World Trade Organization

Despite the clear commitments made by the UN Charter the OECD countries systematically withhold macroeconomic clout from the United Nations. This is to all intents and purposes in the hands of Bretton Woods institutions (the World Bank and International Monetary Fund) and the WTO.

The notion of sustainable development requires, however, that the three essential strands of economy, ecology and social security be integrated in accordance with agreement reached by all countries. But, with macroeconomic clout withheld, it is in fact impossible for the United Nations to integrate the three elements of sustainable development. German Chancellor Kohl's proposal in June 1997 to the UN General Assembly Special Session reviewing the commitments made at the Earth Summit in Rio, that the term 'sustainable development' be incorporated into the United Nations Charter, would in practice require enormous efforts if macroeconomic mandates were to be given in the way the Charter would foresee.

2 The ongoing discussion on reform

Plans at reforming the United Nations have for years enjoyed great popularity among career diplomats, specialists in international law and political scientists. There have been a whole host of efforts, which may be divided into three categories:

- Processes between countries;
- Reform of the Secretariat; and
- Proposals on a theoretical level.

2.1 Processes between countries

No claim to completeness is made regarding processes between countries:

2.1.1 Reform of the Security Council

Reform of the Security Council should on the one hand improve its methods of work. At the same time it should extend its membership in such a way as to take better account of actual, current political circumstances, without impairing the Council's ability to act.

There have been a host of proposals as to how this expansion should be made, and how it might be geographically balanced. No agreement has been reached, however, partly because a fierce competitive struggle for Council membership for the African, Asian and Latin American regions has been unleashed among the possible candidates.

The President of the 51st General Assembly, Razali Ismail, tried to overcome the impasse in this situation in the spring of 1997. In his two-phase proposal the General Assembly would first decide on the number of new members (five new permanent and four non-permanent seats). In the second phase, a few months later, decisions would be made on the occupants of the seats. Whether membership would rotate between different countries or be permanently assigned to one of them was a matter to be left to the regions themselves. Razali's proposal was however shot down at, if not before, the meeting of non-aligned countries' Foreign Ministers in New Delhi in April 1997.

It is not expected that the 52nd General Assembly beginning in September 1997 will be able to achieve big breakthroughs in the matter of Security Council reform.

2.1.2 Financial situation

Since the 49th General Assembly a high-level working group co-chaired by Austria has concentrated its efforts on securing reliable US budget contributions. Up to and including 1996, the USA owed about $1.3 billion in contributions (Zumach 1997: 6). The USA makes payment of these sums, only part of which it acknowledges it owes, dependent on its compulsory contribution being reduced from 25 per cent to a maximum 20 per cent, with a maximum of 25 per cent for peacekeeping operations. The other member countries see this as blackmail and reject it. No progress is in sight in the short term.

2.1.3 Strengthening the UN System

Discussions in another high-level Working Group on Strengthening the UN were conducted with little vigor after the new UN Secretary-General took up office and announced a willingness to propose reforms, as it seemed sensible to wait for the

Secretary-General's package of reforms. These proposals were presented in the middle of July, and are due to be discussed at the 52nd General Assembly.

2.1.4 Agenda for Development and follow-up to Resolution 48/162 – ECOSOC subsidiary structure

Small advances in economic and social areas have been achieved in the 'Follow-up to resolution 48/162 – ECOSOC subsidiary structure', with progress made in terms of improved coordination, a more economical use of resources, and the UN's work becoming more oriented on results.

Work on the Agenda for Development was concluded in the spring of 1997. It was concerned less with actual reforms than with giving an account of multilateral development cooperation in the light of the consensus which had emerged at international conferences since 1992.[1] These conferences have substantially reinforced the principle of sustainable development, emphasizing that development cooperation must above all serve to develop human potential – by satisfying people's basic needs for food, health and education.

2.1.5 Agenda for Peace

Work on the Agenda for Peace was divided into four subgroups, two of which (coordination and sanctions) have already finished their work. In the 'Preventive Diplomacy' working group the idea of the same name has continued to be rejected by China and a number of G77 countries, who fear violation of the principle of non-intervention in their internal affairs were the idea put into practice. The Preventive Diplomacy working group was disbanded in September 1997 without achieving any notable results.

2.1.6 Reform of particular programmes or special organizations (UNIDO, UNEP)

In addition to discussions on reforms at the General Assembly level, efforts at reform are also being made in the UN programmes' administrative councils and in special organizations, with very varying degrees of progress (e.g. with matters well advanced in UNIDO, and just beginning in UNEP).

---

1 I include in these international conferences the Conference on Environment and Development (UNCED, Rio de Janeiro, 1992), the World Conference on Human Rights (Vienna, 1993), the International Population Conference (Cairo, 1994), the World Social Summit (Copenhagen, 1995), the Fourth World Conference on Women (Beijing, 1995), the Conference on Human Settlements (Habitat II, Istanbul, 1996) and the World Food Summit (Rome, 1996).

2.2 Reform of the Secretariat

In the ideas for reform he put forward in July 1997, the UN Secretary-General links the goals of improved management of the UN to putting the organization on a more secure financial footing and intensifying its work in the development sphere.

The Secretary-General proposes a budget of the same amount as the previous year and a reduction of 1,000 posts in the Secretariat.

Administration costs are to be cut by 33%. Consumption of paper for UN reports is to be reduced by 30%. These savings should be to the benefit of development cooperation. To put an end to the situation in which the UN has been for years, namely of being on the verge of insolvency, the Secretary-General has made new proposals on financing. These include setting up a Revolving Credit Fund. In his speech to the General Assembly on 16 July Kofi Annan said:

> Let me be clear: assuring the Organization's financial viability is not only an essential part of reform, it is a condition for the very success of reform.

The UN's activities in development should be consolidated in a United Nations Development Group, and locally all UN organizations should operate from the common base of a United Nations House.

The UN Secretary-General's package of reforms is extremely far-reaching. He summarized it thus himself in his speech to the General Assembly:

> The programme of reform you have in your hands will affect virtually every department and every activity of the United Nations. It contains proposals for increasing the speed with which we can deploy peace-keeping and other field operations.
>
> It focuses on improving our capacity for peace-building, advancing the disarmament agenda, and strengthening the environmental dimension of United Nations activities. It proposes ways to combat the scourge of "uncivil society" – criminals, drug pushers and terrorists.
>
> It reorients our public information activities so that the world's peoples better understand our goals, our role and our range of activities.
>
> It calls for simplified administrative procedures and for a thorough overhaul of human resources policies and practices.
>
> It advocates major restructuring in several areas, including economic and social affairs, human rights and humanitarian affairs. The advancement of human rights needs to be integrated into all principal United Nations activities and programmes. We need to deal more effectively with complex humanitarian emergencies. Accordingly a new Emergency Relief Coordination Office will be established to replace the Department of Humanitarian Affairs.
>
> The natural complement to these proposals would be certain changes of a more fundamental nature, which can be undertaken only by Member States.

How the Member States deal with these proposals for reform during the 52nd General Assembly will show how much genuine will for reform can be mustered among countries' representatives. The UN Secretary-General has at all events put the ball back in the Member States' court.

2.3 Academic proposals

All kinds of academic proposals have been made. These include revising the UN Charter, creating major new bodies (such as an Economic Security Council), the issuing of new mandates to existing organs, international taxation, and reorganization of multilateral negotiations (Bertrand 1997: 88-111; Susskind 1994).

It is difficult to judge how seriously these proposals are taken by countries' representatives. The studies on reform published in the run-up to the UN's 50th anniversary celebrations – such as those by the Commission on Global Governance or the Ford Foundation – have certainly never been systematically discussed in the General Assembly.

One proposal which could also be of interest in the 'environment and security' issue, and which has been discussed in academic circles for years now, has found its way into the Secretary-General's ideas for reform. It envisages that the Trusteeship Council, practically redundant since the disappearance of almost all colonial areas, should now devote itself to the 'Global Commons'. The UN Secretary-General writes:

> 84. Although the United Nations was established primarily to serve Member States, it also expresses the highest aspirations of men, women, and children around the world. Indeed, the Charter begins by declaring the determination of 'we the peoples of the United Nations' to achieve a peaceful and just world order. Relations between the United Nations and agencies of civil society are growing in salience in every major sector of the United Nations agenda. The global commons are the policy domain in which this intermingling of sectors and institutions is most advanced.
>
> 85. Member States appear to have decided to retain the Trusteeship Council. The Secretary-General proposes, therefore, that it be reconstituted as the forum through which Member states exercise their collective trusteeship for the integrity of the global environment and common areas such as the oceans, atmosphere, and outer space. At the same time, it should serve to link the United Nations and civil society in addressing these areas of global concern, which require the active contribution of public, private, and voluntary sectors.

3 Areas of tension

Despite all the UN reform called for in the last few years, very little has been achieved in constructively and creatively overcoming the actual interacting tensions which have become clearly manifest during the efforts at reform.

I would like to use the following examples by way of illustration. Tension lies in the interaction between:

- Reform and finances;
- Governments and other actors;
- Sovereignty and global responsibility; and
- Specialization and integration.

3.1 Reform versus finances

The very country which has for years been pressing the United Nations to become less bureaucratic and more efficiently organized – the USA – has over years piled up debts in unpaid contributions; these have resulted in financial problems which have from time to time paralyzed the organization and must have had a disastrous impact on the motivation of United Nations staff.

Despite all the lip-service paid to maintaining the contrary, the target of 0.7% of GDP for official development aid (ODA), which has been set repeatedly since the 1960s, has never been attained. The OECD countries are today further from this goal than they were at the Conference on Environment and Development in 1992, when the target was reaffirmed at head of state and government level by over 100 countries.

> Developed countries reaffirm their commitments to reach the accepted United Nations target of 0.7 per cent of GNP for ODA and, to the extent that they have not yet achieved that target, agree to augment their aid programmes in order to reach that target as soon as possible and to ensure prompt and effective implementation of Agenda 21. (Agenda 21, Paragraph 33.13).

Funds flowing into official development aid are infinitely small in comparison to direct foreign investment, with as a rule little or no value placed on the lofty aims of sustainable development in multilateral investment agreements.

In many respects the United Nations is dependent on voluntary contributions. This means fluctuations and unpredictability in its funds, and renders goal-oriented management quite impossible; a whole array of donor countries nonetheless adamantly refuse to switch to assessed contributions (or a mix of voluntary and assessed contributions).

All in all there is, despite all the reformers' rhetoric, no sign of any real willingness to achieve greater reliability and fair distribution of the burdens where financing the United Nations' activities is concerned.

3.2 Governments versus other actors

From the theoretical standpoint it has long been recognized that enacting sustainable development is way beyond the capacities of Governments. Governments rely, rather, on the participation of other actors – in science and research, the private sector, media, civic society, and so forth. Solutions can only be arrived at in close alignment with all the other actors. The United Nations nonetheless adheres firmly to the principle that states alone are the subject of rights and duties. Conditions for 'observers' from the spheres of NGOs, the private sector, media, science and academia, etc., have of course improved enormously. But nothing has altered in their basic status as observers. In many major areas the United Nations continues to produce resolutions which no one outside the circle of delegates in the Second Committee[2] knows, and which make no impact on the outside world.

There are of course counter-examples. The decisions taken at conferences of parties to the Montreal Protocol (on the protection of the stratospheric ozone layer), for example, are systematically prepared by reports made by scientific advisory councils.

The question of the transfer of technology, namely whether it is possible for developing countries to avoid the mistakes the industrialized countries made during industrialization – a key issue in sustainable development – is also insoluble unless procedures which are strictly intergovernmental are transcended. Governments in modern states have little modern technology of their own at their disposal, and so cannot play an important role in technology transfer. They can however create the necessary conditions under which the private sector can transfer technologies to developing countries. This and the close interplay between the economy and environmental policy is making a major contribution to the (relative) success of the Montreal Protocol.

3.3 Sovereignty versus global responsibility

The idea of sovereignty has decreased in force in the last few years. The obligations to make reports as required by the diverse environmental conventions; the annual reports to the Commission on Sustainable Development (CSD); the June 1997 recommendation of the UN General Assembly Special Session for the general review of the commitments made at Rio; the introduction of regional peer reviews following the lines of OECD countries' reports; bilateral agreements on

---

2 The United Nations General Assembly works in both a plenum and a number of Committees. The Second Committee is concerned above all with macroeconomic questions, development policy, environment and disaster aid; it is notorious for long procedural debates and a host of resolutions which are negotiated down to the last comma and have very little effect in the 'real world'.

sustainable development, etc. – require more and more areas to be organized so as to serve the good of the international community, and thus not be solely subject to the control of the nation states involved. This expectation has not yet been met, however, nor has it been linked to incisive sanctions – as is normal, for example, in the field of disarmament.

Much more far-reaching progress in communally incorporating areas of interest into international relations will probably not be achieved until account is taken of developing countries' demands for international relations to take a more democratic form and international solidarity to become more effective (being manifested in transfer of finances and technology).

3.4 Specialization versus integration

The international system reflects the national divisions of responsibilities among a number of specialist ministers. Rapid scientific and technological development calls for increasing specialization. And yet the catchword today, more than ever, is 'integration'. Different areas have to be connected to one another. The environment can only be protected effectively if ecological criteria are systematically integrated into all political areas.

The re-organization of the CSD's program of work ought to help effect this 'mainstreaming'. But the CSD can only be successful when Member States pose the same challenge to their national decision-making structures.

4 Tasks for the future

I am convinced that the most important task in the next fifty years is to implement sustainable development. I see sustainable development as the best prescription for peace and security. Where there is no opportunity for sustainable development there is, in the medium term, no possibility of peace either.

To come closer to fulfilling this task, we must succeed in:

- Effecting a transition from environmental protection as it is traditionally understood to sustainable management of natural resources (such as water, energy, fish stocks, fertile soil, etc.);
- Arranging an equitable balance between interests so that certain sections of the population, or particular countries, do not have to pay disproportionately for the transition to sustainable economic management;
- Making institutional arrangements for effecting this equitable balance of interests and
- Integrating ecological criteria in the prevention of conflicts.

The United Nations have an indispensable role in international efforts at sustainable development. This should be taken into account in all efforts at reform.

It is not necessary to negotiate new documents for which there is consensus in order to lend the United Nations more weight in achieving sustainable development. It would be quite sufficient to implement the programs of action which have already been adopted by the Conference on Environment and Development (UNCED, Rio de Janeiro, 1992), the World Conference on Human Rights (Vienna, 1993), the International Population Conference (Cairo, 1994), the World Social Summit (Copenhagen, 1995), the World Conference on Women (Beijing, 1995), the Conference on Human Settlements (Habitat II, Istanbul, 1996) and the World Food Summit (Rome, 1996).

It would of course also be necessary for the financial commitments to development cooperation to be discharged at long last.

The United Nations must also be allowed to act in the macroeconomic fields, and Member States must speak with one tongue in UN forums and those of the WTO or Bretton Woods institutions.

Trade liberalization should not be conducted as an aim in itself; it must be put at the service of sustainable development. This would mean that external costs (including environmental costs) would have to be internalized and social and environmental standards adhered to.

To avoid ecological disasters and potential conflicts, the United Nations should:

- Develop early-warning systems which help identify threatening trends in time; and
- Couple these early-warning systems to effective mechanisms for reacting, so that the international community really does act swiftly when crises threaten.

5 Final remarks

A key point in answering the question why putting sustainable development into practice is so laborious – both within countries and internationally – may lie in the fact that sustainable development requires thinking and planning for the medium-term, while in our democratically legitimized processes, politicians and the electorate prefer political decisions which are made for the short term and success which is immediately attainable.

It is probably only possible for political decisions to be made for the medium term when voters have a well-developed ecological awareness. Such a change in awareness can probably only be arrived at when it is got across to voters that they and subsequent generations are directly threatened by ecological disasters. Apart from reinforcing efforts at creating awareness in general, it is necessary to make the potential impacts of global environmental issues, the dimension of the environment and security included, clear to voters.

I steadfastly retain the hope that well-informed, mature citizens will be prepared to support medium-term strategies for implementing sustainable development –

including measures strengthening institutions – as these strategies are developed by responsible politicians.

In my view this would be the best option for avoiding conflicts, including those in which environmental resources are involved.

References

Bertrand, Maurice 1995: "Future", in: Bertrand and Gerbet (Eds.): *The United Nations Fifty Years on.*

Childers, Erskine and Brian Urquhart 1994: *Renewing the United Nations System*, 27.

Jenks, Bruce and Thomas Neufing 1996: "The United Nations and Development Co-operation", in: *Law and State,* No. 53/54, 159–163.

Kamal, Ahmad 1996: "The United Nations and the Crisis of Development", in: *Internationale Politik und Gesellschaft*, Heft 2, 167–172.

Susskind, Lawrence 1994: *Environmental Diplomacy.* New York: Oxford University Press.

Zumach, Andreas 1997: "Reformdebatte in der UNO stockt", in: *Die Presse*, 19 Sept. 1997, 6.

Annex A
Research Institutes in German-Speaking Regions that Work on the Environment and Security Nexus

Ecologic
Gesellschaft für Internationale
und Europäische Umweltforschung
(Centre for International and European Environmental Research)
Friedrichstraße 165
D-10117 Berlin

Tel.: + 49 (30) 2265 1135
Fax: + 49 (30) 2265 1136
e-mail: office@ecologic.de

Directors: Alexander Carius; R. Andreas Kraemer
Contact person: Alexander Carius

Forschungsstelle für
Sicherheitspolitik und Konfliktanalyse
der ETH Zürich
(Center for Security Studies and Conflict Research at the
Swiss Federal Institute of Technology)
Scheichzerstr. 20
CH-8092 Zürich

Tel.: + 41 (1) 6324025
Fax: + 41 (1) 6321941
e-mail postmaster@sipo.reok.ethz.ch

Director: Prof. Dr. Kurt R. Spillmann

Heidelberger Institut
für internationale Konfliktforschung e.V.
Universität Heidelberg
(Heidelberg Institute for International Conflict Research
at Heidelberg University)
Marstallstr. 6
D-69117 Heidelberg

Tel.: + 49 (6221) 54 3198
Fax: + 49 (6221) 54 2896
e-mail: jh1@ix.urz.uni-heidelberg

Director: Christoph Rohloff
Contact persons: Michael Windfuhr + Christoph Rohloff

Hessische Stiftung Friedens- und Konfliktforschung
(Peace Research Institute Frankfurt)
Leimenrode 29
D-60322 Frankfurt

Tel.: + 49 (69) 95 91 04 - 0
Fax: + 49 (69) 55 84 81
e-mail: hsfk@em.uni-frankfurt.de

Director: Dr. Harald Müller
Contact person: Prof. Dr. Lothar Brock

Institut für Friedensforschung und Sicherheitspolitik
an der Universität Hamburg (IFSH)
(Institute for Peace Research and Security Policy at the
University of Hamburg)
Falkenstein 1
D-22587 Hamburg

Tel.: + 49 (40) 866 077 - 0
Fax: + 49 (40) 866 36 15
e-mail: ifsh@rrz.uni-hamburg.de

Director: Dr. Dieter S. Lutz
Contact person: Dr. Hans-Joachim Gießmann

Interdisziplinäre Arbeitsgruppe
Naturwissenschaft, Technik und Sicherheit
(IANUS)
(Research Group Science, Technology and Security
at the Darmstadt University)
Schloßgartenstr. 9
D-64289 Darmstadt

Tel.: + 49 (6151)-16-4368
Fax: + 49 (6151)-16-6039
e-mail: scheffran@hrzpub.th-darmstadt.de

Director (Spokesman): Prof. Werner Krabs
Contact person: Dr. Jürgen Scheffran

Potsdam-Institut für Klimafolgenforschung e.V.
(Potsdam Institute for Climate Impact Research)
Postfach 60 12 03
D-14412 Potsdam

Tel.: + 49 (331) 288-2500
Fax: + 49 (331) 288-2600
e-mail: dsprinz@pik-potsdam.de

Director: Prof. Dr. H.-J. Schellnhuber
Contact person: D. Sprinz, Ph.D + Dr. Fritz Reußwig

Schweizerische Friedensstiftung Bern
(Swiss Peace Foundation)
Gerechtigkeitsgasse 12
Postfach
CH-3000 Bern 8

Tel.: + 41 (31) 310 27 27
Fax: + 41 (31) 310 27 28
e-mail: swisspeace@dial.eunet.ch

Director: Dr. Günther Baechler

Annex B
Internet Sites on Environment and Security

The following listing of relevant Internet sites is intended to offer a first point of access to the burgeoning literature and documentation, with valuable links to further sources. The list makes no claim to being exhaustive.

Center for Security Studies and Conflict Research
http://www.fsk.ethz.ch

This home page gives an overview of the "Environment and Conflicts Project" (ENCOP), offering a complete list of project reports. It further offers links to other projects and sources of information on the environment and security nexus.

International Relations and Security Network/Center for Security Studies
http://www.isn.ethz.ch

The International Security Network site leads to numerous web pages on security policy issues and on peace and conflict research. It also offers information on environmentally conditioned conflicts. The site has a search engine, with information sorted by topic, region, institution and event.

North Atlantic Treaty Organization (NATO),
Committee on the Challenges of Modern Society (CCMS)
http://www.nato.int/ccms

The page provides information on the pilot studies of the CCMS and events and publications relating to its topics. It further contains, in the context of the NATO Environmental Clearinghouse System (ECHS), environmental data, reports and studies. It serves as a coordination platform for the various CCMS pilot studies and the participating states.

North Atlantic Treaty Organization (NATO),
Scientific and Environmental Affairs
http://www.nato.int/science

The NATO scientific and environmental affairs program web site offers newsletters, press releases, meetings and information on current activities. It highlights in particular the work of the NATO Science Committee and environmental projects of the NATO Committee on the Challenges of Modern Society (CCMS).

United States Department of State,
Bureau of Oceans and International Environmental and Scientific Affairs
http://www.state.gov/www/global/oes

This site informs about the US State Department's foreign policy development and implementation in global environment, science and technology issues. It also features the State Department's April 1997 Environmental Diplomacy Report.

Demographic, Environmental and Security Issues Project (DESIP)
http://www.igc.apc.org/desip

This database lists ongoing conflicts and identifies their environment and population aspects. It presents to the user the linkages between environmental scarcities and political conflicts.

International Human Dimensions Programme (IHDP)/
Global Environmental Change and Human Security
http://www.uni-bonn.de/ihdp/

This home page presents IHDP activities, offering access to reports produced by IDHP and other basic research organizations, an on-line bibliography and links to the global change research community.

Stockholm International Peace Research Institute (SIPRI)
http://www.sipri.se/

This site lists projects, conferences and publications, and offers links to web pages related to environment and security.

University of Toronto/Peace and Conflict Studies
http://utl1.library.utoronto.ca/WWW/pcs/pcs.htm

This page contains information on the programs and objectives of the Peace and Conflict Studies Program of the University of Toronto. It further offers links to the program's projects on "Environment, Population and Security" and "Environmental Scarcity, State Capacity and Civil Violence".

University of Toronto/Environmental Security Library & Database
http://www.library.utoronto.ca /pcs/database/libintro.htm

This page gives access to a database and library on environment and security containing a wide range of information on topics relating to environmental degradation and violent conflict in developing countries.

Woodrow Wilson International Center for Scholars
http://ecsp.si.edu

The site gives access to the "ECSP Report" and to the "China Environment Series". It further contains various discussion forums on environment, population and security issues.

Annex C
Journals that Publish Articles on Environment and Security Regularly or Occasionally

- *Environmental Change and Security Report Project*
  Woodrow Wilson Center. Washington,DC.
  e-mail: ecsp@erols.com

- *Journal of Environment and Security*
  The International Institute for Environmental Strategies and Security.
  Quebec City.
  e-mail: es.gerpe@fss.ulaval.ca

- *Journal of Environment and Development*
  The Graduate School of International Relations and Pacific Studies. University of California at San Diego.
  e-mail: envdev@ucsd.edu

- *International Security*
  Center for Science and International Affairs and the JFK School of Government.
  Harvard University. Cambridge.
  e-mail: journals-orders@mit.edu

- *Journal of Peace Research*
  International Peace Research Institute. Oslo.
  e-mail: info@prio.no

- *Security Dialogue (formerly: Bulletin of Peace Proposals)*
  International Peace Research Institute. Oslo.
  e-mail: info@prio.no

The Authors

Günther Baechler, Dr., is director of the Swiss Peace Foundation in Berne. Together with Professor Spillmann, he was co-director of the international "Environment and Conflicts Project" (ENCOP), which deals, by means of a broadly-based, systematic study, with the emergence and management of ecologically induced conflicts. He has published numerous works in the field of peace and conflict research.

Frank Biermann, Dr. phil. LL.M., is Research Fellow with the Global Environmental Assessment Project, Belfer Center for Science and International Affairs, John F. Kennedy School of Government, Harvard University. He pursued research and/or taught at the University of Aberdeen, Free University of Berlin, Jawaharlal Nehru University (New Delhi), Stanford University (Berlin Program) and in the Secretariat of the German Advisory Council on Global Change. He has published three books and numerous articles on international environmental policy.

Lothar Brock, Dipl.-Pol., Dr. phil., professor of political science specializing in international relations at the Johann Wolfgang Goethe University in Frankfurt and director of research at the Peace Research Institute Frankfurt (Hessische Stiftung Friedens- und Konfliktforschung). Together with Klaus Dieter Wolf (Darmstadt Technical University), he runs the inter-university research group on the global community (Forschungsgruppe Weltgesellschaft). He is particularly interested in the implications of new developments in relations between states and societies for theory formation in the study of international relations.

Alexander Carius, Dipl.-Pol., is the director of Ecologic, Centre for International and European Environmental Research. He works on questions of European and international environmental policy, particularly on the links between environmental policy and development, foreign and security policy. He acts as a consultant and adviser to various ministries and international organizations.

Wolf-Dieter Eberwein, Dr. habil., is the head of the department of international politics at the Science Centre Berlin (WZB). He specializes in international conflict and security. At present he is working on this question in the context of his project on humanitarian aid policy.

Irene Freudenschuß-Reichl has been a member of the Austrian diplomatic service since 1982. She works mainly on multilateral issues and has extensive experience in the areas of economics, environment, social affairs and emergency relief. Until spring 1998 she was assigned to the Austrian federal ministry for environment, youth and family affairs, where she ran the international department. Since then she has been the Austrian ambassador to the United Nations in Vienna.

Kerstin Imbusch, has a degree in public administration and is research fellow and project coordinator at the Institute for East.-European Studies at the Free University of Berlin. She works on European and international politics. The topic of her Ph.D is the Eastern Enlargement of the European Union.

Sascha Müller-Kraenner has a degree in biology and public law and is the director of international affairs for the environmental umbrella organization, the German League for Nature and Environment (Deutscher Naturschutzring, DNR), and head of its Berlin office. His specialist field is EU environmental policy and the Rio follow-up process.

Kurt M. Lietzmann, Ass. jur., is the head of the department for Central and Eastern Europe and the Newly Independent States at the German Ministry for the Environment, Nature Conservation and Nuclear Safety. Since 1996 he is the director of the NATO/CCMS Pilot Study on "Environment and Security in an International Context".

Sebastian Oberthür, Dr. phil., is a senior fellow at Ecologic, Centre for International and European Environmental Research. He specializes in institutional questions relating to international environmental and climate policy and the environmental foreign policy of the European Union. He has published numerous works on international environmental policy and climate and ozone protection.

Volker R. Quante, Lieutenant Colonel (G.S), has a degree in information science and until September 1997, worked as senior researcher in the field of security policy and military strategy at the German Army office of studies and exercises in Waldbröl. Under the authority of the German Ministry of Defense, he has dealt with the following fields: the foreign and security policy of the countries of North America and Western Europe, the strategy of the North Atlantic Alliance, tactical nuclear weapons and arms control, and environment and security in an international context. At present he works as German liasion officer at the Polish National Defense Academy in Warsaw.

Christoph Rohloff has an M.A. and a B.A. from the Sophia University of Tokyo (1991) and a masters degree from the University of Heidelberg (1995). Since 1997 he has been the head of the Heidelberg Institute for International Conflict Research (HIIK) at Heidelberg University's institute of political science. He is writing a doctoral thesis on international organisations and foreign policy formulation. Since 1999 he has also been research fellow at the Institute for Development and Peace at the University of Duisburg. He has published numerous works in the field of peace and conflict research.

Dr. Jürgen Scheffran has a background in mathematics and physics and works as a scientific assistant in the Research Group Science, Technology and Security (IA-

NUS) at Darmstadt University, Germany. His specialist areas are arms and disarmament, the ambivalence of research and technology and the modeling of conflicts in the area of environmental and security policy. He has written a series of books, has co-organized several conferences and is the publisher and editor of two periodicals on peace studies.

Evita Schmieg has a social sciences degree in economics and is assistant ministerial counselor (Regierungsdirektorin) in the planning department of the German Ministry for Economic Cooperation and Development (BMZ). Her work covers questions relating to the contribution of development cooperation to crisis prevention.

Detlef F. Sprinz has a Ph.D. in political science and a degree in economics. He works as a senior fellow at the Potsdam Institute for Climate Impact Research (PIK) and teaches international environmental policy at Potsdam University's faculty of political science. In his research he specializes in the fields of international environmental policy, environment and security, international climate policy, the effectiveness of international organizations and social science methodology.

Michael Windfuhr has an M.A. in political science and is currently a university assistant at Heidelberg University's institute of political science. He is concerned with general questions relating to the development of international relations, specializing in international economic policy, human rights and general trends in international law, international environmental policy and peace and conflict research. He also works for the Heidelberg Institute for International Conflict Research (HIIK) as a scientific adviser.

Manfred Woehlcke, Dr. habil., is a research fellow at the Stiftung Wissenschaft und Politik (Research Institute for International Affairs) in Ebenhausen, Germany. His specialist fields are Latin American politics, international development issues and political implications of global hazards, including ecological risks and demographic change.

Dr. Bernd Wulffen is a top-ranking diplomat (Vortragender Legationsrat 1. Klasse) and, since 1993, has been the head of the German foreign office division for international environmental and resource policy.

Index

MIX
Papier aus verantwortungsvollen Quellen
Paper from responsible sources
FSC® C105338

If you have any concerns about our products, you can contact us on
ProductSafety@springernature.com

In case Publisher is established outside the EU, the EU authorized representative is:
Springer Nature Customer Service Center GmbH
Europaplatz 3, 69115 Heidelberg, Germany

Printed by Libri Plureos GmbH
in Hamburg, Germany